# VARIANCE-CONSTRAINED MULTI-OBJECTIVE STOCHASTIC CONTROL AND FILTERING

# Wiley Series in Dynamics and Control of Electromechanical Systems

# VARIANCE-CONSTRAINED MULTI-OBJECTIVE STOCHASTIC CONTROL AND FILTERING

**Lifeng Ma**

*Nanjing University of Science and Technology, China*

**Zidong Wang**

*Brunel University, United Kingdom*

**Yuming Bo**

*Nanjing University of Science and Technology, China*

This edition first published 2015
© 2015 John Wiley & Sons Ltd

*Registered office*
John Wiley & Sons Ltd, The Atrium, Southern Gate, Chichester, West Sussex, PO19 8SQ, United
Kingdom

For details of our global editorial offices, for customer services and for information about how to apply
for permission to reuse the copyright material in this book please see our website at www.wiley.com.

*Library of Congress Cataloging-in-Publication Data applied for.*

A catalogue record for this book is available from the British Library.

ISBN: 9781118929490

Typeset in 11/13pt TimesLTStd by Laserwords Private Limited, Chennai, India
Printed and bound in Singapore by Markono Print Media Pte Ltd

1    2015

# Contents

# Preface

Nonlinearity and stochasticity are arguably two of the main resources in reality that have resulted in considerable system complexity. Therefore, recently, control and filtering of nonlinear stochastic systems have been an active branch within the general research area of nonlinear control problems. In engineering practice, it is always desirable to design systems capable of simultaneously guaranteeing various performance requirements to meet the ever-increasing practical demands toward the simultaneous satisfaction of performances such as stability, robustness, precision, and reliability, among which the system covariance plays a vital role in system analysis and synthesis due to the fact that several design objectives, such as stability, time-domain and frequency-domain performance specifications, robustness, and reliability, can be directly related to steady-state covariance of the closed-loop systems.

In this book, we discuss the multi-objective control and filtering problems for a class of nonlinear stochastic systems with variance constraints. The stochastic nonlinearities taken into consideration are quite general and could cover several classes of well-studied nonlinear stochastic systems. The content of this book is divided mainly into two parts. In the first part, we focus on the variance-constrained control and filtering problems for time-invariant nonlinear stochastic systems subject to different kinds of complex situations, including measurements missing, actuator failures, output degradation, etc. Some sufficient conditions are derived for the existence of the desired controllers and filters in terms of the linear matrix inequalities (LMIs). The control and filtering problems with multiple performance specifications are considered in the second part for time-varying nonlinear stochastic systems. In this part, several design techniques including recursive linear matrix inequalities (RLMIs), game theory, and gradient method have been employed to develop the desired controllers and filters capable of simultaneously achieving multiple pre-specified performance requirements.

The compendious frame and description of the book are given as follows: Chapter 1 introduces the recent advances on variance-constrained multi-objective control and filtering problems for nonlinear stochastic systems and the outline of the book. Chapter 2 is concerned with the $H_\infty$ control problem for a class of nonlinear stochastic systems with variance constraints. Chapter 3 deals with the mixed $H_2/H_\infty$ filtering problem for a type of time-invariant nonlinear stochastic systems.

In Chapter 4, the variance-constrained filtering problem is solved in the case of missing measurements. Chapter 5 discusses the controller design problem with variance constraints when the actuator is confronted with possible failures. The sliding mode control problem is investigated in Chapter 6 for a class of nonlinear discrete-time stochastic systems with $H_2$ specification. In Chapter 7, the dissipativity performance is taken into consideration with variance performance and the desired control scheme is given. For a special type of nonlinear stochastic system, namely, systems with multiplicative noises, Chapter 8 deals with the robust controller design problem with simultaneous consideration of variance constraints and $H_\infty$ requirement. For time-varying nonlinear stochastic systems, Chapters 9 and 10 investigate the $H_\infty$ control and filtering problems over a finite horizon, respectively. Chapters 11 and 12 discuss the mixed $H_2/H_\infty$ control problems, taking the randomly occurring nonlinearities (RONs) and Markovian jump parameters into consideration, respectively. Chapters 13 and 14 give the solutions to the multi-objective control problems for time-varying nonlinear stochastic systems in the presence of sensor and actuator failures, respectively. Chapter 15 gives the conclusions and some possible future research topics.

This book is a research monograph whose intended audience is graduate and post-graduate students as well as researchers.

# Series Preface

Electromechanical Systems permeate the engineering and technology fields in aerospace, automotive, mechanical, biomedical, civil/structural, electrical, environmental, and industrial systems. The Wiley Book Series on dynamics and control of electromechanical systems covers a broad range of engineering and technology in these fields. As demand increases for innovation in these areas, feedback control of these systems is becoming essential for increased productivity, precision operation, load mitigation, and safe operation. Furthermore, new applications in these areas require a reevaluation of existing control methodologies to meet evolving technological requirements. An example involves distributed control of energy systems. The basics of distributed control systems are well documented in several textbooks, but the nuances of its use for future applications in the evolving area of energy system applications, such as wind turbines and wind farm operations, solar energy systems, smart grids, and energy generation, storage and distribution, require an amelioration of existing distributed control theory to specific energy system needs. The book series serves two main purposes: (1) a delineation and explication of theoretical advancements in electromechanical system dynamics and control and (2) a presentation of application driven technologies in evolving electromechanical systems.

This book series embraces the full spectrum of dynamics and control of electromechanical systems from theoretical foundations to real world applications. The level of the presentation should be accessible to senior undergraduate and first-year graduate students, and should prove especially well suited as a self-study guide for practicing professionals in the fields of mechanical, aerospace, automotive, biomedical, and civil/structural engineering. The aim is to provide an interdisciplinary series ranging from high-level undergraduate/graduate texts, explanation and dissemination of science and technology and good practice, through to important research that is immediately relevant to industrial development and practical applications. It is hoped that this new and unique perspective will be of perennial interest to students, scholars, and employees in these engineering disciplines. Suggestions for new topics and authors for the series are always welcome.

This book, *Variance-Constrained Multi-Objective Stochastic Control and Filtering*, has the objective of providing a theoretical foundation as well as practical insights on the topic at hand. It is broken down into two essential parts: (1) variance-constrained

control and filtering problems for time-invariant nonlinear stochastic systems and
(2) designing controllers and filters capable of simultaneously achieving multiple
pre-specified performance requirements. The book is accessible to readers who
have a basic understanding of stochastic processes, control, and filtering theory. It
provides detailed derivations from first principles to allow the reader to thoroughly
understand the particular topic. It also provides several illustrative examples to bridge
the gap between theory and practice. This book is a welcome addition to the Wiley
Electromechanical Systems Series because no other book is focused on the topic
of stochastic control and filtering with a specific emphasis on variance-constrained
multi-objective systems.

Mark J. Balas, John L. Crassidis, and Florian Holzapfel

# Acknowledgements

The authors would like to express their deep appreciation to those who have been directly involved in various aspects of the research leading to this book. Special thanks go to Professor James Lam from the University of Hong Kong and Professor Xiaohui Liu from Brunel University of the United Kingdom for their valuable suggestions, constructive comments, and support. We also extend our thanks to many colleagues who have offered support and encouragement throughout this research effort. In particular, we would like to acknowledge the contributions from Derui Ding, Hongli Dong, Xiao He, Jun Hu, Liang Hu, Xiu Kan, Zhenna Li, Jinling Liang, Qinyuan Liu, Yang Liu, Yurong Liu, Bo Shen, Guoliang Wei, Nianyin Zeng, Sunjie Zhang, and Lei Zou. Finally, the authors are especially grateful to their families for their encouragement and never-ending support when it was most required.

The writing of this book was supported in part by the National Natural Science Foundation of China under Grants 61304010, 61273156, 61134009, 61004067, and 61104125, the Natural Science Foundation of Jiangsu Province under Grant BK20130766, the Postdoctoral Science Foundation of China under Grant 2014M551598, the International Postdoctoral Exchange Fellowship Program from China Postdoctoral Council, the Engineering and Physical Sciences Research Council (EPSRC) of the UK, the Royal Society of the UK, and the Alexander von Humboldt Foundation of Germany. The support of these organizations is gratefully acknowledged.

# List of Abbreviations

| | |
|---:|:---|
| $\mathbb{R}^n$ | The $n$-dimensional Euclidean space. |
| $\mathbb{I}^+$ | The set of non-negative integers. |
| $\mathbb{R}^{n \times m}$ | The set of all $n \times m$ real matrices. |
| $\lvert \cdot \rvert$ | The Euclidean norm in $\mathbb{R}^n$. |
| $L_2[0, \infty)$ | The space of square-integrable vector functions over $[0, \infty)$. |
| $\rho(A)$ | The spectral radius of matrix $A$. |
| $\lambda(A)$ | The eigenvalue of matrix $A$. |
| $\mathrm{tr}(A)$ | The trace of matrix $A$. |
| $\otimes$ | The Kronecker product of matrices. |
| $\mathrm{st}(A)$ | The stack that forms a vector out of the columns of matrix $A$. |
| $\lVert a \rVert_A^2$ | Equal to $a^{\mathrm{T}} A a$. |
| $\mathrm{Prob}\{\,\cdot\,\}$ | The occurrence probability of the event "$\cdot$". |
| $\mathbb{E}\{x\}$ | The expectation of stochastic variable $x$. |
| $\mathbb{E}\{x \vert y\}$ | The expectation of $x$ conditional on $y$, $x$, and $y$ are both stochastic variables. |
| $I$ | The identity matrix of compatible dimension. |
| $X > Y$ | The $X - Y$ is positive definite, where $X$ and $Y$ are symmetric matrices. |
| $X \geq Y$ | The $X - Y$ is positive semi-definite, where $X$ and $Y$ are symmetric matrices. |
| $M^{\mathrm{T}}$ | The transpose matrix of $M$. |
| $\mathrm{diag}\{M_1, \ldots, M_n\}$ | The block diagonal matrix with diagonal blocks being the matrices $M_1, \ldots, M_n$. |
| $*$ | The ellipsis for terms induced by symmetry, in symmetric block matrices. |

# List of Figures

# 1

# Introduction

It is widely recognized that in almost all engineering applications, nonlinearities are inevitable and could not be eliminated thoroughly. Hence, the nonlinear systems have gained more and more research attention, and many results have been published. On the other hand, due to the wide appearance of stochastic phenomena in almost every aspect of our daily lives, stochastic systems that have found successful applications in many branches of science and engineering practice have stirred quite a lot of research interest during the past few decades. Therefore, control and filtering problems for nonlinear stochastic systems have been studied extensively in order to meet an ever-increasing demand toward systems with both nonlinearities and stochasticity.

In many engineering control/filtering problems, the performance requirements are naturally expressed by the upper bounds on the steady-state covariance, which is usually applied to scale the control/estimation precision, one of the most important performance indices of stochastic design problems. As a result, a large number of control and filtering methodologies have been developed to seek a convenient way to solve the variance-constrained design problems, among which the linear quadratic Gaussian(LQG) control and Kalman filtering are two representative minimum variance design algorithms.

On the other hand, in addition to the variance constraints, real-world engineering practice also desires the simultaneous satisfaction of many other frequently seen performance requirements, including stability, robustness, reliability, energy constraints, to name but a few key ones. This gives rise to the so-called multi-objective design problems, in which multiple cost functions or performance requirements are simultaneously considered with constraints being imposed on the system. An example of multi-objective control design would be to minimize the system steady-state variance indicating the performance of control precision, subject to a pre-specified external disturbance attenuation level evaluating system robustness. Obviously, multi-objective design methods have the ability to provide more flexibility in dealing with the trade-offs and constraints in a much more explicit manner on the pre-specified performance

*Variance-Constrained Multi-Objective Stochastic Control and Filtering*, First Edition.
Lifeng Ma, Zidong Wang and Yuming Bo.
© 2015 John Wiley & Sons, Ltd. Published 2015 by John Wiley & Sons, Ltd.

requirements than those conventional optimization methodologies like the LQG control scheme or $H_\infty$ design technique, which do not seem to have the ability of handling multiple performance specifications.

When coping with the multi-objective design problem with variance constraints for stochastic systems, the well-known covariance control theory provides us with a useful tool for system analysis and synthesis. For linear stochastic systems, it has been shown that multi-objective control/filtering problems can be formulated using linear matrix inequalities (LMIs), due to their ability to include desirable performance objectives such as variance constraints, $H_2$ performance, $H_\infty$ performance, and pole placement as convex constraints. However, as nonlinear stochastic systems are concerned, the relevant progress so far has been very slow due primarily to the difficulties in dealing with the variance-related problems resulting from the complexity of the nonlinear dynamics. A key issue for the nonlinear covariance control study is the existence of the covariance of nonlinear stochastic systems and its mathematical expression, which is extremely difficult to investigate because of the complex coupling of nonlinearities and stochasticity. Therefore, it is not surprising that the multi-objective control and filtering problems for nonlinear stochastic systems with variance constraints have not been adequately investigated despite the clear engineering insights and good application prospect.

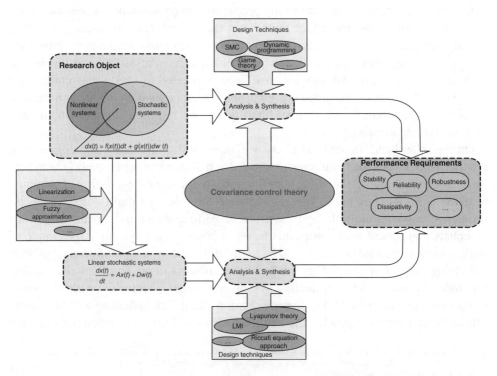

**Figure 1.1**  Architecture of surveyed contents.

In this chapter, we focus mainly on the multi-objective control and filtering problems for nonlinear systems with variance constraints and aim to give a survey on some recent advances in this area. We shall give a comprehensive discussion from three aspects, i.e., design objects (nonlinear stochastic system), design requirements (multiple performance specifications including variance constraints), and several design techniques. Then, as a special case of the addressed problem, mixed $H_2/H_\infty$ design problems have been discussed in great detail with some recent advances. Subsequently, the outline of this book is given. The contents that are reviewed in this chapter and the architecture are shown in Figure. 1.1.

## 1.1 Analysis and Synthesis of Nonlinear Stochastic Systems

For several decades, nonlinear stochastic systems have been attracting increasing attention in the system and control community due to their extensive applications in a variety of areas ranging from communication and transportation to manufacturing, building automation, computing, automotive, and chemical industries, to mention just a few key areas. In this section, the analysis and synthesis problems for nonlinear systems and stochastic systems are recalled respectively, and some recent advances in these areas are also given.

### 1.1.1 Nonlinear Systems

It is well recognized that in almost all engineering applications, nonlinearities are inevitable and could not be eliminated thoroughly. Hence, nonlinear systems have gained more and more research attention, and many results have been reported; see, for example, Refs [1–3]. When analyzing and designing nonlinear dynamical systems, there are a wide range of nonlinear analysis tools, among which the most common and widely used is linearization because of the powerful tools we know for linear systems. It should be pointed out that, however, there are two basic limitations of linearization [4]. (1) As is well known, linearization is an approximation in the neighborhood of certain operating points. Thus, the resulting linearized system can only show the local behavior of the nonlinear system in the vicinity of those points. Neither non-local behavior of the original nonlinear system far away from those operating points nor global behavior throughout the entire state space can be correctly revealed after linearization. (2) The dynamics of a nonlinear system are much richer than that of a linear system. There are essentially nonlinear phenomena, like finite escape time, multiple isolated equilibria, subharmonic, harmonic or almost periodic oscillations, to name just a few key ones that can take place only in the presence of nonlinearity; hence, they cannot be described by linear models [5–8]. Therefore, as a compromise, during the past few decades, there has been tremendous interest in studying nonlinear systems, with nonlinearities being taken as the exogenous disturbance input to a linear system, since it could better illustrate the dynamics of the original nonlinear system than the linearized one with less sacrifice of the convenience on the application of existing mathematical tools. The nonlinearities emerging in such systems may arise from the linearization process of an originally highly nonlinear plant or may be an external nonlinear input, which would drastically degrade the system performance or even cause instability; see, for example, Refs [9–11].

On the other hand, in real-world applications, one of the most inevitable and physically important features of some sensors and actuators is that they are always corrupted by different kinds of nonlinearities, either from within the device themselves or from the external disturbances. Such nonlinearities generally result from equipment limitations as well as the harsh environments such as uncontrollable elements (e.g., variations in flow rates, temperature, etc.) and aggressive conditions (e.g., corrosion, erosion, and fouling, etc.) [12]. Since the sensor/actuator nonlinearity cannot be simply ignored and often leads to poor performance of the controlled system, a great deal of effort in investigating the analysis and synthesis problems has been devoted by many researchers to the study of various systems with sensor/actuator nonlinearities; see Refs [13–18].

Recently, the systems with randomly occurring nonlinearities (RONs) have started to stir quite a lot of research interest as it reveals an appealing fact that, instead of occurring in a deterministic way, a large amount of nonlinearities in real-world systems would probably take place in a random way. Some of the representative publications can be discussed as follows. The problem of randomly occurring nonlinearities was raised in Ref. [19], where an iterative filtering algorithm has been proposed for the stochastic nonlinear system in the presence of both RONs and output quantization effects. The filter parameters can be obtained by resorting to solving certain recursive linear matrix inequalities. The obtained results have quickly been extended to the case of multiple randomly occurring nonlinearities [20]. Such a breakthrough on how to deal with nonlinear systems with RONs has been well recognized and quickly followed by other researchers in the area. Using similar techniques, the filtering as well as control problems have been solved for a wide range of systems containing nonlinearities that are occurring randomly, like Markovian jump systems [21, 22], sliding mode control systems [23], discrete-time complex networks [24], sensor networks [25], time-delay systems [26], and other types of nonlinear systems [27–29], which therefore has proven that the method developed in Ref. [19] is quite general and is applicable to the analysis and synthesis of many different kinds of nonlinear systems.

It should be emphasized that, for nonlinearities, there are many different constraint conditions for certain aims, such as Lipschitz conditions, among which the kind of stochastic nonlinearities described by statistical means has drawn particular research focus since it covers several well-studied nonlinearities in stochastic systems; see Refs [27, 30–33] and the references therein. Several techniques for analysis and synthesis of this type of nonlinear system have been exploited, including the linear matrix inequality approach [30], the Riccati equation method [31], the recursive matrix inequality approach [32], gradient method [33], sliding mode control scheme [34], and the game theory approach [27].

## 1.1.2 Stochastic Systems

As is well known, in the past few decades there have been extensive studies and applications of stochastic systems because the stochastic phenomenon is inevitable and cannot be avoided in real-world systems. When modeling such kinds of systems, the way of neglecting the stochastic disturbances, which is a conventional technique in traditional control theory for deterministic systems, is no longer suitable. Having realized the necessity of introducing more realistic models, nowadays a great number of real-world systems such as physical systems, financial systems, ecological systems, as well as social systems are more suitable to be modeled by stochastic systems. Therefore, the stochastic control problem, which deals with dynamical systems described by difference or differential equations, and subject to disturbances characterized as stochastic processes, has drawn much research attention; see Ref. [35] and the references therein. It is worth mentioning that a kind of stochastic system represented as a deterministic system adding a stochastic disturbance characterized as white noise has gained special research interest and found extensive applications in engineering based on the fact that it is possible to generate stochastic processes with covariance functions belonging to a large class simply by sending white noise through a linear system; hence a large class of problems can be reduced to the analysis of linear systems with white noise inputs; see Refs [36–40] for examples.

Parallel to the control problems, the filtering and prediction theory for stochastic systems, which aims to extract a signal from observations of signal and disturbances, has been well studied and found to be widely applied in many engineering fields. It also plays a very important role in the solution of the stochastic optimal control problem. Research on the filtering problem originated in Ref. [41], where the well-known Wiener–Kolmogorov filter has been proposed. However, the Wiener–Kolmogorov filtering theory has not been widely applied mainly because it requires the solution of an integral equation (the Wiener–Hopf equation), which is not easy to solve either analytically or numerically. In Refs [42, 43], Kalman and Bucy gave a significant contribution to the filtering problem by giving the celebrated Kalman–Bucy filter, which could solve the filtering problem recursively. The Kalman–Bucy filter (also known as the $H_2$ filter) has been extensively adopted and widely used in many branches of stochastic control theory, due to the fast development of digital computers recently; see Refs [44–47] and the references therein.

## 1.2 Multi-Objective Control and Filtering with Variance Constraints

In this section, we first review the covariance control theory, which provides us with a powerful tool in variance-constrained design problems with multiple requirements specified by engineering practice. Then, we discuss several important performance specifications including robustness, reliability, and dissipativity. Two common techniques for solving the addressed problems for nonlinear stochastic systems are introduced. The mixed $H_2/H_\infty$ design problem is reviewed in great detail as a special case of the multi-objective control/filtering problem with variance constraints.

### 1.2.1   Covariance Control Theory

As we have stated in the previous section, engineering control problems always require upper bounds on the steady-state covariances [39, 48, 49]. However, many control design techniques used in both theoretical analysis and engineering practice, such as LQG and $H_\infty$ design, do not seem to give a direct solution to this kind of design problem since they lack a convenient avenue for imposing design objectives stated in terms of upper bounds on the variance values. For example, the LQG controllers minimize a linear quadratic performance index without guaranteeing the variance constraints with respect to individual system states. The covariance control theory [50] developed in late 1980s has provided a more direct methodology for achieving the individual variance constraints than the LQG control theory. The covariance control theory aims to solve the variance-constrained control problems while satisfying other performance indices [38, 45, 50, 51]. It has been shown that the covariance control approach is capable of solving multi-objective design problems, which has found applications in dealing with transient responses, round-off errors in digital control, residence time/probability in aiming control problems, and stability and robustness in the presence of parameter perturbations [51]. Such an advantage is based on the fact that several control design objectives, such as stability, time-domain and frequency-domain performance specifications, robustness and pole location, can be directly related to steady-state covariances of the closed-loop systems. Therefore, covariance control theory serves as a practical method for multi-objective control design as well as a foundation for linear system theory.

On the other hand, it is always the case in real-world applications such as the tracking of a maneuvering target that the filtering precision is characterized by the error variance of estimation [51, 52]. Considering its clear engineering insights, in the past few years the filtering problem with error variance constraints has received much interest and a great many research findings have been reported in the literature [42, 43, 53, 54]. The celebrated Kalman filtering approach is a typical method that aims to obtain state information based on the minimization of the variance of the estimation error [42, 43]. Nevertheless, the strict request of a highly accurate model seriously impedes the application of Kalman filtering as in many cases only an approximate model of the system is available. It therefore has brought about remarkable research interest to the robust filtering method, which aims to minimize the error variance of estimation against the system uncertainties or external unknown disturbances [55, 56]. Despite certain merits and successful applications, as in the case of the LQG control problem, the traditional minimum variance filtering techniques cannot directly impose the designing objectives stated in terms of upper bounds on the error variance values, by which we mean that those techniques try to minimize the filtering error variance in a mean-square sense rather than to constrain it within a pre-specified bound, which is obviously better able to meet the requirements of practical engineering. Motivated by the covariance control theory, in Ref. [57] the authors have proposed a more direct design procedure for achieving the individual variance constraint in filtering problems. Due to its design flexibility, the covariance control theory is capable of solving the error variance-constrained filtering problem while guaranteeing other multiple designing objectives [58]. Therefore, it always serves as one of the most powerful tools in dealing with multi-objective filtering as well as control problems [59].

It should be pointed out that most available literature regarding covariance control theory has been concerned with *linear time invariant* stochastic systems using the linear matrix inequality (LMI) approach. Moreover, when it comes to the variance-constrained controller/filter design problems for much more complicated systems such as *time-varying systems, nonlinear systems, Markovian Jump systems*, etc., unfortunately, the relevant results have been very few due primarily to the difficulties in dealing with the existence problem of the steady-state covariances and their mathematical expressions for those above-mentioned systems with complex dynamics. With the hope of resolving such difficulties, in recent years, special effort has been devoted to studying variance-constrained multi-objective design problems for systems of complex dynamics, and several methodologies for analysis and synthesis have been developed. For example, in Ref. [45], a Riccati equation method has been proposed to solve the filtering problem for linear time-varying stochastic systems with pre-specified error variance bounds. In Refs [60–62], by means of the technique of sliding mode control (SMC), the robust controller design problem has been solved for linear parameter perturbed systems, since SMC has strong robustness to matched disturbances or parameter perturbations. We shall return to this SMC problem later and more details will be discussed in the following section.

When it comes to nonlinear stochastic systems, limited work has been done in the covariance-constrained analysis and design problems, just as we have anticipated. A multi-objective filter has been designed in Ref. [63] for systems with Lipschitz nonlinearity, but the variance bounds cannot be pre-specified. Strictly speaking, such an algorithm cannot be referred to as variance-constrained filtering in view of the lack of capability for directly imposing specified constraints on variance. An LMI approach has been proposed in Ref. [30] to cope with robust filtering problems for a class of stochastic systems with nonlinearities characterized by statistical means, attaining an assignable $H_2$ performance index. In Ref. [59], for a special class of nonlinear stochastic systems, namely, systems with multiplicative noises (also called bilinear systems or systems with state/control dependent noises), a state feedback controller has been put forward in a unified LMI framework in order to ensure that the multiple objectives, including stability, $H_\infty$ specification, and variance constraints, are simultaneously satisfied. This paper is always regarded as the origination of covariance control theory for nonlinear systems, for within the established theoretical framework quite a lot of performance requirements can be taken into consideration simultaneously. Furthermore, with the developed techniques, the obtained elegant results could be easily extended to a wide range of nonlinear stochastic systems; see, for example, Refs [27, 33, 64–66].

We shall return to such types of nonlinear stochastic systems later to present more details of recent progress in Section 1.2.3.

## 1.2.2 Multiple Performance Requirements

In the following, several performance indices originating from engineering practice and frequently applied in multi-objective design problems are introduced.

### 1.2.2.1 Robustness

In real-world engineering practice, various reasons, such as variations of the operating point, aging of devices, identification errors, etc., would lead to the parameter uncertainties that result in the perturbations of the elements of a system matrix when modeling the system in a state-space form. Such a perturbation in system parameters cannot be avoided and would cause degradation (sometimes even instability) to the system. Therefore, in the past decade, considerable attention has been devoted to different issues for linear or nonlinear uncertain systems, and a great number of papers have been published; see Refs [2, 46, 67–70] for some recent results.

On another research frontier of robust control, the $H_\infty$ design method, which is used to design controller/filter with guaranteed performances with respect to the external disturbances as well as internal perturbations, has received an appealing research interest during the past decades; see Refs [71–74] for instance. Since Zames' original work [71], significant advances have been made in the research area of $H_\infty$ control and filtering. The standard $H_\infty$ control problem has been completely solved by Doyle *et al.* for linear systems by deriving simple state-space formulas for all controllers [72]. For nonlinear systems, the $H_\infty$ performance evaluation can be conducted through analyzing the $L_2$ gain of the relationship from the external disturbance to the system output, which is a necessary step when deciding whether further controller design is needed. In the past years, the nonlinear $H_\infty$ control problem has also received considerable research attention, and many results have been available in the literature [73–77]. On the other hand, the $H_\infty$ filtering problem has also gained considerable research interest along with the development of $H_\infty$ control theory; see Refs [75, 78–82]. It is well known that the existence of a solution to the $H_\infty$ filtering problem is in fact associated with the solvability of an appropriate algebraic Riccati equality (for linear cases) or a so-called Hamilton–Jacobi equation (for nonlinear ones). So far, there have been several approaches for providing solutions to nonlinear $H_\infty$ filtering problems, few of which, however, is capable of handling multiple performance requirements in an $H_\infty$ optimization framework.

It is worth mentioning that, in contrast to the $H_\infty$ design framework within which multiple requirements can hardly be under simultaneous consideration, the covariance control theory has provided a convenient avenue for the robustness specifications to be perfectly integrated into the multi-objective design procedure; see Refs [59, 76] for example. For nonlinear stochastic systems, control and filtering problems have been solved with the occurrence of parameter uncertainties and stochastic nonlinearities while guaranteeing the $H_\infty$ and variance specifications; see Refs [33, 64, 65, 76] for some recent publications.

### 1.2.2.2 Reliability

In practical control systems, especially networked control systems (NCSs), due to a variety of reasons, including erosion caused by severe circumstances, abrupt changes of working conditions, intense external disturbances, and the internal physical equipment constraints and aging, the process of signal sampling and transmission has always been confronted with different kinds of failures, such as measurements missing, signal quantization, sensor and actuator saturations, and so on. Such a phenomenon is always referred to as incomplete information, which would drastically degrade the system performance. In recent years, as requirements increase toward the reliability of engineering systems, the reliable control problem, which aims to stabilize the systems accurately and precisely in spite of incomplete information caused by possible failures, has therefore attracted considerable attention. In Refs [83, 84], binary switching sequences and Markovian jump parameters have been introduced to model the missing measurements phenomena. A more general model called the multiple missing measurements model has been proposed in Ref. [85] by employing a diagonal matrix to characterize the different missing probabilities for individual sensors. The incomplete information caused by sensor and actuator saturations has also received considerable research attention and some results have been reported in the literature[18, 86, 87], where the saturation has been modeled as so-called sector-bound nonlinearities. As far as signal quantization is mentioned, in Ref. [88] a sector-bound scheme has been proposed to handle the logarithmic quantization effects in feedback control systems, and such an elegant scheme has then been extensively employed later on; see, for example, Refs [89, 90] and the references therein.

It should be pointed out that for nonlinear stochastic systems, the relevant results of reliable control/filtering with variance constraints are relatively fewer, and some representative results can be summarized as follows. By means of the linear matrix inequality approach, a reliable controller has been designed for a nonlinear stochastic system in Ref. [64] against actuator faults with variance constraints. In the case of sensor failures, the gradient method and the LMI method have been applied respectively in Refs [33] and [65] to design multi-objective filters respectively satisfying multiple requirements including variance specifications simultaneously. However, despite its clear physical insight and importance in engineering applications, the control problem for nonlinear stochastic systems with incomplete information has not yet been studied sufficiently.

### 1.2.2.3 Dissipativity

In recent years, the theory of dissipative systems, which plays an important role in system and control areas, has been attracting a great deal of research interest and many results have been reported so far; see Refs [91–96]. Originating in Ref. [95], the dissipative theory serves as a powerful tool in characterizing important system behaviors such as stability and passivity, and has close connections with bounded real lemma, passitivity lemma and circle criterion. It is worth mentioning that, due to its simplicity in analysis and convenience in simulation, the LMI method has gained particular attention in dissipative control problems. For example, in Refs [94, 96], an LMI method was used to design the state feedback controller, ensuring both asymptotic stability and strictly quadratic dissipativity. For singular systems, Ref. [91] has established a unified LMI framework to satisfy admissibility and dissipativity of the system simultaneously. In Ref. [93], the dissipative control problem has been solved for time-delay systems.

Although the dissipativity theory provides us with a useful tool for the analysis of systems with multiple performance criteria, unfortunately, when it comes to nonlinear stochastic systems, not much literature has been concerned with the multi-objective design problem for nonlinear stochastic systems, except for Ref. [97], where a multi-objective control law has been proposed to simultaneously meet the stability, variance constraints, and dissipativity of a closed-loop system. So far, the variance-constrained design problem with dissipativity taken into consideration has not yet been studied adequately and still remains challenging.

## 1.2.3 Design Techniques for Nonlinear Stochastic Systems with Variance Constraints

The complexity of nonlinear system dynamics challenges us to come up with systematic design procedures to meet control objectives and design specifications. It is clear that we cannot expect one particular procedure to apply to all nonlinear systems; therefore, quite a lot of tools have been developed to deal with control and filtering problems for nonlinear stochastic systems, including the Takagi–Sugeno (T-S) fuzzy model approximation approach, linearization, gain scheduling, sliding mode control, backstepping, to name but a few of the key ones. In the sequel, we will investigate two nonlinear design tools that can be easily combined with the covariance control theory for the purpose of providing a theoretical framework within which the variance-constrained control and filtering problems can be solved systematically for nonlinear stochastic systems.

### 1.2.3.1 Takagi–Sugeno Fuzzy Model

The T-S fuzzy model approach occupies an important place in the study of nonlinear systems for its excellent capability in nonlinear system descriptions. Such a model allows one to perfectly approximate a nonlinear system by a set of local linear sub systems with certain fuzzy rules, thereby carrying out the analysis and synthesis work within the linear system framework. Therefore, the T-S fuzzy model is extensively applied in both theoretical research and engineering practice of nonlinear systems; see Refs [98–101] for some of the latest publications. However, despite its engineering significance, not much literature has taken the system state variance into consideration, mainly due to the technical difficulties in dealing with the variance-related problems. Some tentative work can be summarized as follows. In Ref. [102], a minimum variance control algorithm as well as direct adaptive control scheme has been applied in a stochastic T-S fuzzy ARMAX model to track the desired reference signal. However, as we mentioned above, such a minimum variance control algorithm lacks the ability of directly imposing design requirements on the variances of an individual state component. Therefore, in order to cope with this problem, in Ref. [103] a fuzzy controller has been designed to stabilize a nonlinear continuous-time system, while simultaneously minimizing the control input energy and satisfying variance constraints placed on the system state. The result has then been extended in Ref. [104] to the output variance constraints case. Recently, such a T-S fuzzy model based variance-constrained algorithm has found successful application in nonlinear synchronous generator systems; see Ref. [105] for more details.

### 1.2.3.2 Sliding Mode Control

In the past few decades, the sliding mode control (also known as the variable structure control) problem, originated in Ref. [106], has been extensively studied and widely applied, because of its advantage of strong robustness against model uncertainties, parameter variations, and external disturbances. In the sliding mode control, trajectories are forced to reach a sliding manifold in finite time and to stay on the manifold for all future time. It is worth mentioning that in the existing literature about the sliding mode control problem for nonlinear systems, the nonlinearities and uncertainties taken into consideration are mainly under matching conditions, that is, when nonlinear and uncertain terms enter the state equation at the same point as the control input, the motion on the sliding manifold is independent of those matched terms; see Refs [107, 108] for examples. Under such an assumption, the covariance-constrained control problems have been solved in Refs [60–62] for a type of continuous stochastic system with matching condition nonlinearities.

Along with the development of continuous-time sliding mode control theory, in recent years, as most control strategies are implemented in a discrete-time setting (e.g., networked control systems), the sliding mode control problem for discrete-time systems has gained considerable research interest and a large amount of literature has appeared on this topic. For example, in Refs [109, 110] the integral type SMC schemes have been proposed for sample-data systems and a class of nonlinear discrete-time systems respectively. Adaptive laws were applied in Refs [111, 112] to synthesize sliding mode controllers for discrete-time systems with stochastic as well as deterministic disturbances. In Ref. [113], a simple methodology for designing sliding mode controllers was proposed for a class of linear multi-input discrete-time systems with matching perturbations. Using the dead-beat control technique, Ref. [114] presented a discrete variable structure control method with a finite-time step to reach the switching surface. In cases when the system states were not available, the discrete-time SMC problems were solved in Refs [115, 116] via output feedback. It is worth mentioning that in Ref. [117], the discrete-time sliding mode reaching condition was first revised and then a new reaching law was developed, which has proven to be a convenient way to handle robust control problems; see Refs [118, 119] for some of the latest publications. Recently, for discrete-time systems that are not only confronted with nonlinearities but also corrupted by more complicated situations like propagation time-delays, randomly occurring parameter uncertainties, and multiple data packet dropouts, the SMC strategies have been designed in Refs [23, 77, 80] to solve the robust control problems and have shown good performances against all the mentioned negative factors. Currently, the sliding mode control problems for discrete-time systems still remain a hotspot in systems and control science; however, when it comes to variance-constrained problems, the related work is much fewer. As preliminary work, Ref. [34] has proposed an SMC algorithm guaranteeing the required $H_2$ specification for discrete-time stochastic systems in the presence of both matched and unmatched nonlinearities. In this paper, although only the $H_2$ performance is handled, it is worth mentioning that with the proposed method, other performance indices can be considered simultaneously within the established unified framework by similar design techniques.

### 1.2.4   A Special Case of Multi-Objective Design: Mixed $H_2/H_\infty$ Control/Filtering

As a special case of multi-objective control problem, mixed $H_2/H_\infty$ control/filtering has gained a great deal of research interest for several decades. So far, there have been several approaches to tackle the mixed $H_2/H_\infty$ control/filtering problem. For linear deterministic systems, the mixed $H_2/H_\infty$ control problems have been extensively studied. For example, an algebraic approach has been presented in Ref. [120] and a time-domain Nash game approach has been proposed in Refs. [37, 121] to solve the addressed mixed $H_2/H_\infty$ control/filtering problems respectively. Moreover, some efficient numerical methods for mixed $H_2/H_\infty$ control problems have been developed based on a convex optimization approach in Refs [40, 122–124], among which the linear matrix inequality approach has been employed widely to design both linear state feedback and output feedback controllers subject to $H_2/H_\infty$ criterion, due to its effectiveness in numerical optimization. It is noted that the mixed $H_2/H_\infty$ control theories have already been applied to various engineering fields [47, 125, 126].

Parallel to the mixed $H_2/H_\infty$ control problem, the mixed $H_2/H_\infty$ filtering problem has also been well studied and several approaches have been proposed to tackle the problem. For example, Bernstein and Haddad [120] transformed the mixed $H_2/H_\infty$ filtering problem into an auxiliary minimization problem. Then, by using the Lagrange multiplier technique, they gave the solutions in terms of an upper bound on the $H_2$

filtering error. In Refs [127] and [128], a time-domain game approach was proposed to solve the mixed $H_2/H_\infty$ filtering problem through a set of coupled Riccati equations. Recently, the LMI method has been widely employed to solve the multi-objective mixed $H_2$ and $H_\infty$ filtering problems; see Refs [58, 129] for examples.

As far as nonlinear systems are concerned, the mixed $H_2/H_\infty$ control problem as well as the filtering problem have gained some research interest; see, for example, Refs [130–132]. For nonlinear deterministic systems, the mixed $H_2/H_\infty$ control problem has been solved with the solutions characterized in terms of the cross-coupled Hamilton–Jacobi-Issacs (HJI) partial differential equations. Since it is difficult to solve the cross-coupled HJI partial differential equations either analytically or numerically, in Ref. [131] the authors have used the Takagi and Sugeno (T-S) fuzzy linear model to approximate the nonlinear system, and solutions to the mixed $H_2/H_\infty$ fuzzy output feedback control problem has been obtained via an LMI approach. For nonlinear stochastic systems, unfortunately, the mixed $H_2/H_\infty$ control and filtering problem has not received a full investigation and few results have been reported. In Ref. [36], for a special type of nonlinear stochastic system, which is known as a bilinear system (also called systems with state-dependent noise or systems with multiplicative noise), a stochastic mixed $H_2/H_\infty$ control problem has been solved and sufficient conditions have been provided in terms of the existence of the solutions of cross-coupled Ricatti equations. Very recently, an LMI approach has been proposed in Ref. [132] to solve the mixed $H_2/H_\infty$ control problem for a class of nonlinear stochastic system that includes several well-studied types of nonlinear systems. For the stochastic systems with much more complicated nonlinearities, by means of the game theory approach, the mixed $H_2/H_\infty$ control problem has been solved for systems with RONs in Ref. [27] and Markovian jump parameters in Ref. [66] respectively. Nevertheless, to the best of authors' knowledge, the mixed $H_2/H_\infty$ control and filtering problems for general nonlinear systems have not yet received enough investigation and still remain challenging topics.

## 1.3 Outline

The outline of this book is given as follows. It it worth mentioning that Chapter 2 to Chapter 7 mainly focus on the variance-constrained control and filtering problems for nonlinear time-invariant systems, while in Chapter 9 to Chapter 12, the research objects are nonlinear time-varying systems. The framework of this book is shown in Figure 1.2.

- In Chapter 1, the research background is firstly introduced, which mainly involves the nonlinear stochastic systems, covariance control, multiple performance requirements, several design techniques including the T-S fuzzy model and sliding mode control, and a special case of variance-constrained multi-objective design problem–mixed $H_2/H_\infty$ control/filtering problem. Then the outline of the book is listed.

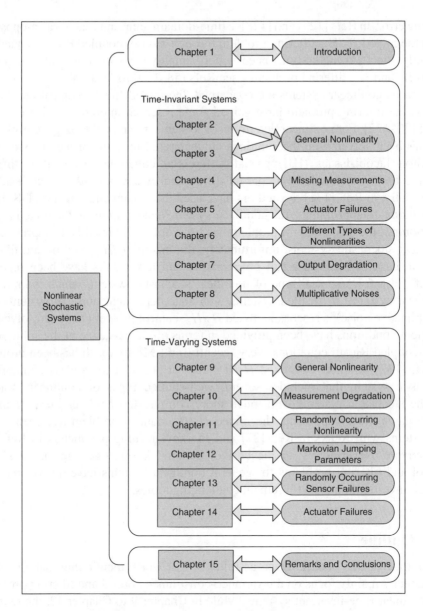

**Figure 1.2**    The framework of this book

- In Chapter 2, a robust $H_\infty$ variance-constrained controller has been designed for a class of uncertain discrete-time nonlinear stochastic systems. A unified framework has been established to solve the controller design problem, which requires the simultaneous satisfaction of exponential stability, $H_\infty$ performance index, and individual variance constraints. Two types of optimization problems have been proposed either optimizing $H_\infty$ performance or the system state variances.

- Chapter 3 investigates the mixed $H_2/H_\infty$ filtering problem for systems with deterministic uncertainties and stochastic nonlinearities, where the nonlinearities are characterized by statistical means. It is the objective to design a filter such that, for all admissible stochastic nonlinearities and deterministic uncertainties, the overall filtering process is exponentially mean-square quadratically stable, the $H_2$ filtering performance is achieved, and the prescribed disturbance attenuation level is guaranteed in an $H_\infty$ sense.

- Chapter 4 is concerned with the robust filtering problem for a class of nonlinear stochastic systems with missing measurements and parameter uncertainties. The missing measurements are described by a binary switching sequence satisfying a conditional probability distribution and the nonlinearities are expressed by statistical means. We aim to seek a sufficient condition for the exponential mean-square stability of the filtering error system, which is first derived and then an upper bound of the state estimation error variance is obtained, in order to obtain the explicit expression of the desired filters parameterized.

- In Chapter 5, we deal with a robust fault-tolerant controller design problem with variance constraints for a class of nonlinear stochastic systems with both norm-bounded parameter uncertainties and possible actuator failures. The nonlinearity taken into consideration can be expressed by statistic means, which can cover several types of important nonlinearities, and the actuator failure model adopted here is more practical than the conventional outage case.

- Chapter 6 is concerned with a robust SMC design for uncertain nonlinear discrete-time stochastic systems and the $H_2$ performance constraint has been studied. The nonlinearities considered in this chapter contain both matched and unmatched forms. We also have proposed an improved form of discrete switching function with which several typical classes of stochastic nonlinearities can be dealt with via the SMC method.

- In Chapter 7, a dissipative control problem has been solved for a class of nonlinear stochastic systems while guaranteeing simultaneously exponential mean-square stability, variance constraints, system dissipativity, and reliability. An algorithm has been proposed to convert the original nonconvex feasibility problem into an optimal minimization problem, which is much easier to solve by standard numerical software.

- In Chapter 8, the robust variance-constrained $H_\infty$ control problem is considered for uncertain stochastic systems with multiplicative noises. The norm-bounded parametric uncertainties enter into both the system and output matrices. The purpose of the problem is to design a state feedback controller such that, for all admissible parameter uncertainties, (1) the closed-loop system is exponentially mean-square quadratically stable; (2) the individual steady-state variance satisfies given upper bound constraints; and (3) the prescribed noise attenuation level is guaranteed in an $H_\infty$ sense with respect to the additive noise disturbances. A general framework is established to solve the addressed multi-objective problem by using a linear matrix inequality (LMI) approach, where the required stability, the $H_\infty$ characterization, and variance constraints are all easily enforced. Within such a framework, two additional optimization problems are formulated: one is to optimize the $H_\infty$ performance and the other is to minimize the weighted sum of the system state variances. A numerical example is provided to illustrate the effectiveness of the proposed design algorithm.

- In Chapter 9, we are concerned with the robust $H_\infty$ control problem for a class of uncertain nonlinear discrete time-varying stochastic systems with the covariance constraint. All the system parameters are time-varying and the uncertainty enters into the state matrix. An output feedback controller has been designed by means of recursive linear matrix inequalities (RLMIs), satisfying the $H_\infty$ disturbance rejection attenuation level in the finite horizon and an individual upper bound of state covariance at each time point.

- When it comes to the finite-horizon multi-objective filtering for time-varying nonlinear stochastic systems, Chapter 10 has proposed a technique that could handle $H_\infty$ performance and variance constraint at the same time. It is worth mentioning that the design algorithm developed in this chapter is forward in time, which is different from those in most of the existing literature, where the $H_\infty$ problem can be solved only backward in time and thus can be combined with the variance design and is suitable for online design.

- In Chapter 11, the mixed $H_2/H_\infty$ controller design problem has been dealt with for a class of nonlinear stochastic systems with randomly occurring nonlinearities that are characterized by two Bernoulli distributed white sequences with known probabilities. For the multi-objective controller design problem, the sufficient condition of the solvability of the mixed $H_2/H_\infty$ control problem has been established by means of the solvability of four coupled matrix-valued equations. A recursive algorithm has been developed to obtain the value of the feedback controller step by step at every sampling instant.

- Chapter 12 is concerned with the mixed $H_2/H_\infty$ control problem over a finite horizon for a class of nonlinear Markovian jump systems with both stochastic nonlinearities and probabilistic sensor failures. The failure probability for each sensor is governed by an individual random variable satisfying a certain probability distribution over a given interval for each mode. The purpose of the addressed problem is to design a controller such that the closed-loop system achieves the expected $H_2$ performance requirement with a guaranteed $H_\infty$ disturbance attenuation level. The solvability of the addressed control problem is expressed as the feasibility of certain coupled matrix equations. The controller gain at each time instant $k$ can be obtained by solving the corresponding set of matrix equations. A numerical example is given to illustrate the effectiveness and applicability of the proposed algorithm.

- In Chapter 13, the robust variance-constrained $H_\infty$ control problem is concerned for a class of nonlinear systems with possible sensor failures occurring in a random way, called randomly occurring sensor failures (ROSFs). The occurrence of the nonlinearities under consideration is governed by a Bernoulli distributed variable. The purpose of the addressed problem is to design an output-feedback controller such that, for certain systems with ROSFs, (1) the closed-loop systems meets the desired $H_\infty$ performance in the finite horizon and (2) the state covariance is under an individual upper bound at each time point. A sufficient condition for the existence of the addressed controller is given and a numerical computing algorithm is developed to meet the aforementioned requirements by means of recursive linear matrix inequalities (RLMIs). An illustrative simulation example is provided to show the applicability of the proposed algorithm.

- Chapter 14 deals with the fault-tolerant control problem for a class of nonlinear stochastic time-varying systems with actuator failures. Both $H_2$ and $H_\infty$ performance requirements are taken into consideration. The proposed actuator failure model is quite general and covers several frequently seen actuator failure phenomena as special cases. The stochastic nonlinearities are quite general and could stand for several nonlinear systems. It is the purpose of this chapter to find equilibrium strategies of a two-player Nash game, and meanwhile both $H_2$ and $H_\infty$ performances are achieved via a proposed state feedback control scheme, which is characterized by the solution to a set of coupled matrix equations. The feedback gains can be solved recursively backward in $k$. A numerical computing algorithm is presented and then a simulation example is given to illustrate the effectiveness and applicability of the proposed algorithm.
- In Chapter 15, the conclusions and some potential topics for future work are given.

# 2

# Robust $H_\infty$ Control with Variance Constraints

In stochastic control problems, the performance requirements of engineering systems are usually expressed as upper bounds on steady-state variances. Unfortunately, current control design techniques, such as LQG and $H_\infty$ design, do not seem to give a direct solution to this kind of design problem, since they lack a convenient avenue for imposing design objectives stated in terms of upper bounds on the variance values. For example, the LQG controllers minimize a linear quadratic performance index that lacks guarantee variance constraints with respect to individual system states.

The covariance control theory developed in late 1980s has provided a more direct methodology for achieving the individual variance constraints than the LQG control theory. Covariance control theory is capable of dealing with variance-constrained control problems and, at the same time, considering other multiple performance objectives due to its design flexibility. It has been shown that the covariance control approach is capable of solving multi-objective design problems, which has found applications in dealing with transient responses, round-off errors in digital control, residence time/probability in aiming control problems, and stability robustness in the presence of parameter perturbations. Such an advantage is based on the fact that several control design objectives, such as stability, time-domain and frequency-domain performance specifications, robustness, and pole location, can be directly related to steady-state covariance of the closed-loop systems. Therefore, the idea of covariance control theory has been widely applied in solving multi-objective control problems as well as filtering problems.

In recent years, there has been an increasing research interest in robust controller and filter design problems with variance constraints, and several approaches have been developed. Most of the literature, however, has been concerned with *linear* stochastic systems using the linear matrix inequality (LMI) approach. As far as *nonlinear* stochastic systems are concerned, unfortunately, the relevant results have been very few, mainly due to the complexity in dealing with the existence and expression of the steady-state variance for nonlinear stochastic systems. Up to now, the robust $H_\infty$ variance-constrained control problem for uncertain stochastic systems with nonlinearity has not yet been thoroughly investigated.

Motivated by the above discussions, this chapter is concerned with the controller design problem for a class of discrete-time uncertain nonlinear stochastic systems with variance constraints. The nonlinearities described by statistical means are quite general, which include several well-studied classes of nonlinearities as special cases. The parameter uncertainties are assumed to be norm-bounded and enter into both the system and output matrices. A linear matrix inequality (LMI) approach is proposed to solve the addressed multi-objective controller design problem where the exponential stability, $H_\infty$ performance requirement, and individual variance constraints are achieved simultaneously for all admissible parameter uncertainties. Within such an LMI framework, two additional optimization problems are also discussed, which demonstrate the flexibility of the proposed design technique.

The remainder of this chapter is organized as follows: In Section 2.1, the multi-objective control problem for the nonlinear stochastic system is formulated. In Section 2.2, the stability, $H_\infty$ performance, and steady-state variance are firstly analyzed separately and then synthesized in a unified LMI framework. In Section 2.3, an LMI algorithm is developed for the controller design problem and two types of optimization problems are considered. In Section 2.4, an illustrative example is presented to show the effectiveness of the proposed algorithm. In Section 2.5, summary remarks are provided.

## 2.1 Problem Formulation

Consider the following system:

$$x(k + 1) = (A + H_1 FE)x(k) + B_1 u(k) + f(x(k)) + D_1 \omega(k),$$

$$y(k) = (C + H_2 FE)x(k) + B_2 u(k) + D_2 \omega(k), \tag{2.1}$$

where $x(k) \in \mathbb{R}^n$ is the state, $u(k) \in \mathbb{R}^r$ is the control input, $\omega(k) \in \mathbb{R}^m$ is a zero mean Gaussian white noise sequence with covariance $R$, and $A, B_1, B_2, D_1, D_2, H_1, H_2,$ and $E$ are known real matrices with appropriate dimensions.

The real matrix $F \in \mathbb{R}^{i \times j}$, which could be time-varying, represents the deterministic norm-bounded parameter uncertainty and satisfies

$$FF^T \leq I. \tag{2.2}$$

The parameter uncertainty $F$ is said to be admissible if it satisfies condition (2.2).

The function $f(x(k))$ is a stochastic nonlinear function of the system states and is assumed to have the following first moment for all $x(k)$:

$$\mathbb{E}\left\{f(x(k))|x(k)\right\} = 0, \tag{2.3}$$

$$\mathbb{E}\left\{f(x(k))f^{\mathrm{T}}(x(k))\big|\, x(k)\right\} = \sum_{i=1}^{q} \pi_i \pi_i^{\mathrm{T}}\left(x^{\mathrm{T}}(k)\Omega_i x(k)\right), \tag{2.4}$$

where $\pi_i$ $(i = 1, \dots, q)$ is a known column vector and, $\Omega_i$ $(i = 1, \dots, q)$ is a known positive definite matrix with appropriate dimensions.

**Remark 2.1** *The nonlinearity description in (2.3) and (2.4) covers several classes of well-studied nonlinear systems such as:*

- *Linear systems with state-dependent multiplicative noises $D_1 x(k)\xi(k)$, where $\xi(k)$ is a zero mean, uncorrelated noise sequence.*
- *Nonlinear systems with random vectors that depend on the state norm, i.e., $\|x(k)\|D_1\xi(k)$, where $\xi(k)$ is a zero mean, uncorrelated noise sequence.*
- *Nonlinear systems with a random sequence dependent on the sign of a nonlinear function of the states, i.e., $sign[\phi(x(k))](D_1 x(k)\xi(k))$, where $\xi(k)$ is a zero mean, uncorrelated noise sequence.*
- *Other models that have been discussed in Refs [132, 133].*

*One can see that some of the most important uncertain nonlinear stochastic models can be special cases of the system given in (2.3) and (2.4). For example, the stochastic linear systems with state-dependent noise (also called stochastic bilinear systems) can be treated within this framework.*

Now, let us consider the following state feedback controller for system (2.1):

$$u(k) = Kx(k), \tag{2.5}$$

where $K$ is the feedback gain to be designed. Applying (2.5) to the system (2.1), we obtain the following closed-loop system:

$$\begin{aligned} x(k+1) &= A_K x(k) + f(x(k)) + D_1\omega(k), \\ y(k) &= C_K x(k) + D_2\omega(k), \end{aligned} \tag{2.6}$$

where

$$A_K \triangleq A + H_1 FE + B_1 K,$$

$$C_K \triangleq C + H_2 FE + B_2 K.$$

Before giving our design goal, we introduce the following stability concept for the system (2.6).

**Definition 2.1.1** *The system* (2.6) *is said to be exponentially mean-square stable if, with* $\omega(k) = 0$, *there exist constants* $\zeta \geq 1$ *and* $\tau \in (0, 1)$ *such that*

$$\mathbb{E}\left\{\|x(k)\|^2\right\} \leq \zeta \tau^k \mathbb{E}\left\{\|x_0\|^2\right\}, \quad \forall x_0 \in \mathbb{R}^n, \quad k \in \mathbb{I}^+, \tag{2.7}$$

*for all admissible uncertainties.*

This chapter aims to design a state feedback controller of form (2.5), such that for all admissible parameter uncertainties, the following three requirements are simultaneously satisfied:

(*R*1)  The system (2.6) is exponentially mean-square stable.
(*R*2)  For a given constant $\gamma > 0$, with $\omega(k) \neq 0$, the controlled output $y(k)$ satisfies

$$\sum_{k=0}^{\infty} \mathbb{E}\left\{\|y(k)\|^2\right\} \leq \gamma^2 \sum_{k=0}^{\infty} \mathbb{E}\left\{\|\omega(k)\|^2\right\}, \tag{2.8}$$

under the zero initial condition $x_0 = 0$.
(*R*3)  The individual steady-state variances satisfy the following constraints:

$$\text{Var}\{x_i(k)\} \triangleq \lim_{k \to \infty} \mathbb{E}\left\{x_i(k)x_i^{\mathrm{T}}(k)\right\} \leq \sigma_i^2, \tag{2.9}$$

where $x(k) = \begin{bmatrix} x_1(k)\, x_2(k) \cdots x_n(k) \end{bmatrix}^{\mathrm{T}}$ and $\sigma_i^2 > 0 \quad (i = 1, 2, \ldots, n)$ are given pre-specified constants.

## 2.2  Stability, $H_\infty$ Performance, and Variance Analysis

Before the analysis, several useful lemmas are needed.

**Lemma 2.2.1** *Given any real matrices* $\mathcal{N}_1$, $\mathcal{N}_2$, $\mathcal{N}_3$ *with appropriate dimensions, such that* $0 < \mathcal{N}_3 = \mathcal{N}_3^{\mathrm{T}}$, *the following inequality holds:*

$$\mathcal{N}_1^{\mathrm{T}}\mathcal{N}_2 + \mathcal{N}_2^{\mathrm{T}}\mathcal{N}_1 \leq \mathcal{N}_1^{\mathrm{T}}\mathcal{N}_3\mathcal{N}_1 + \mathcal{N}_2^{\mathrm{T}}\mathcal{N}_3^{-1}\mathcal{N}_2. \tag{2.10}$$

**Lemma 2.2.2** *(Schur Complement Lemma) Given constant matrices* $\mathcal{S}_1$, $\mathcal{S}_2$, $\mathcal{S}_3$ *where* $\mathcal{S}_1 = \mathcal{S}_1^{\mathrm{T}}$ *and* $0 < \mathcal{S}_2 = \mathcal{S}_2^{\mathrm{T}}$, *then* $\mathcal{S}_1 + \mathcal{S}_3^{\mathrm{T}}\mathcal{S}_2^{-1}\mathcal{S}_3 < 0$ *if and only if*

$$\begin{bmatrix} \mathcal{S}_1 & \mathcal{S}_3^{\mathrm{T}} \\ \mathcal{S}_3 & -\mathcal{S}_2 \end{bmatrix} < 0, \tag{2.11}$$

*or, equivalently,*

$$\begin{bmatrix} -\mathcal{S}_2 & \mathcal{S}_3 \\ \mathcal{S}_3^{\mathrm{T}} & \mathcal{S}_1 \end{bmatrix} < 0. \tag{2.12}$$

**Lemma 2.2.3** *[132] Given the state feedback gain matrix K, the following statements are equivalent.*

*1.*
$$\rho(\Psi) < 1, \tag{2.13}$$

*where*
$$\Psi = A_K \otimes A_K + \sum_{i=1}^{q} \mathrm{st}(\pi_i \pi_i^{\mathrm{T}}) \mathrm{st}^{\mathrm{T}}(\Omega_i)$$

*or*
$$\Psi = A_K^{\mathrm{T}} \otimes A_K^{\mathrm{T}} + \sum_{i=1}^{q} \mathrm{st}(\Omega_i) \mathrm{st}^{\mathrm{T}}(\pi_i \pi_i^{\mathrm{T}}).$$

*2. There exists $P > 0$ such that*
$$A_K^{\mathrm{T}} P A_K - P + \sum_{i=1}^{q} \Omega_i \mathrm{tr}[\pi_i \pi_i^{\mathrm{T}} P] < 0. \tag{2.14}$$

*3. There exists $Q > 0$ such that*
$$A_K Q A_K^{\mathrm{T}} - Q + \sum_{i=1}^{q} \pi_i \pi_i^{\mathrm{T}} \mathrm{tr}[Q \Omega_i] < 0. \tag{2.15}$$

*4. The system (2.6) is exponentially mean-square stable.*

**Lemma 2.2.4** *Consider the system*
$$\xi(k+1) = N \xi(k) + f(\xi(k)), \tag{2.16}$$

*where*
$$\mathbb{E}\{f(\xi(k)) | \xi(k)\} = 0$$

*and*
$$\mathbb{E}\{f(\xi(k)) f^{\mathrm{T}}(\xi(k)) | x(k)\} = \sum_{i=1}^{q} \pi_i \pi_i^{\mathrm{T}} \xi^{\mathrm{T}}(k) \Xi_i \xi(k),$$

*where $\Xi_i$ $(i = 1, 2, \ldots, q)$ are known positive definite matrices with appropriate dimensions. If the system (2.16) is exponentially mean-square stable and there exists a symmetric matrix X satisfying*
$$N X N^{\mathrm{T}} - X + \sum_{i=1}^{q} \pi_i \pi_i^{\mathrm{T}} \mathrm{tr}[X \Xi_i] < 0, \tag{2.17}$$

*then $X \geq 0$.*

*Proof.* The proof can be carried out along a similar line to that of Lemma 2 in Ref. [59], and is therefore omitted here.                                                                                    □

### 2.2.1  Stability and $H_\infty$ Performance Analysis

**Theorem 2.2.5** *Given a scalar $\gamma > 0$ and a feedback gain matrix $K$. The system (2.6) is exponentially mean-square stable and the $H_\infty$ norm constraint (2.8) is achieved for all $\omega(k) \neq 0$ if there exits a positive definite matrix $P$ satisfying*

$$\begin{bmatrix} A_K^\mathrm{T} P A_K - P + \sum_{i=1}^{q} \Omega_i \mathrm{tr}[\pi_i \pi_i^\mathrm{T} P] + C_K^\mathrm{T} C_K & A_K^\mathrm{T} P D_1 + C_K^\mathrm{T} D_2 \\ D_1^\mathrm{T} P A_K + D_2^\mathrm{T} C_K & D_1^\mathrm{T} P D_1 + D_2^\mathrm{T} D_2 - \gamma^2 I \end{bmatrix} < 0, \qquad (2.18)$$

*for all admissible parameter uncertainties.*

*Proof.* Firstly, from (2.18),

$$A_K^\mathrm{T} P A_K - P + \sum_{i=1}^{q} \Omega_i \mathrm{tr}\left[\pi_i \pi_i^\mathrm{T} P\right] < -C_K^\mathrm{T} C_K \leq 0. \qquad (2.19)$$

Hence, it follows directly from Lemma 2.2.3 that the system (2.6) is exponentially mean-square stable.

Define a Lyapunov functional $V(x(k)) = x^\mathrm{T}(k) P x(k)$, where $P > 0$ is the solution to (2.18). Then, for any nonzero $\omega(k)$, it follows from (2.18) that

$$\begin{aligned}
& \mathbb{E}\left\{ V(x(k+1))| x(k) \right\} - V(x(k)) + \mathbb{E}\left\{ y^\mathrm{T}(k)y(k) \right\} - \gamma^2 \mathbb{E}\left\{ \omega^\mathrm{T}(k)\omega(k) \right\} \\
&= \mathbb{E}\left\{ \left( x^\mathrm{T}(k)A_K^\mathrm{T} + f^\mathrm{T} + \omega^\mathrm{T}(k)D_1^\mathrm{T} \right) P \left( A_K + f + D_1\omega(k) \right)| x(k) \right\} \\
&\quad - x^\mathrm{T}(k)Px(k) + (x^\mathrm{T}(k)C_K^\mathrm{T} + \omega^\mathrm{T}(k)D_2^\mathrm{T})(C_K x(k) + D_2\omega(k)) - \gamma^2 \omega^\mathrm{T}(k)\omega(k) \\
&= x^\mathrm{T}(k) \left( A_K^\mathrm{T} P A_K - P + \sum_{i=1}^{q} \Omega_i \mathrm{tr}(\pi_i \pi_i^\mathrm{T} P) + C_K^\mathrm{T} C_K \right) x(k) \\
&\quad + x^\mathrm{T}(k) \left( A_K^\mathrm{T} P D_1 + C_K^\mathrm{T} D_2 \right) \omega(k) + \omega^\mathrm{T}(k) \left( D_1^\mathrm{T} P A_K + D_2^\mathrm{T} C_K \right) x(k) \\
&\quad + \omega^\mathrm{T}(k) \left( D_1^\mathrm{T} P D_1 + D_2^\mathrm{T} D_2 - \gamma^2 I \right) \omega(k) \\
&= \begin{bmatrix} x(k) \\ \omega(k) \end{bmatrix}^\mathrm{T} \Lambda \begin{bmatrix} x(k) \\ \omega(k) \end{bmatrix} \\
&< 0, \hspace{9cm} (2.20)
\end{aligned}$$

where

$$\Lambda = \begin{bmatrix} A_K^\mathrm{T} P A_K - P + \sum_{i=1}^{q} \Omega_i \mathrm{tr}[\pi_i \pi_i^\mathrm{T} P] + C_K^\mathrm{T} C_K & A_K^\mathrm{T} P D_1 + C_K^\mathrm{T} D_2 \\ D_1^\mathrm{T} P A_K + D_2^\mathrm{T} C_K & D_1^\mathrm{T} P D_1 + D_2^\mathrm{T} D_2 - \gamma^2 I \end{bmatrix}.$$

Summing (2.20) on both sides with respect to $k$ from 0 to $\infty$, we obtain

$$\sum_{k=0}^{\infty} (\mathbb{E}\{V(x(k+1))|x(k)\} - V(x(k))$$

$$+ \mathbb{E}\{y^T(k)y(k)\} - \gamma^2\mathbb{E}\{\omega^T(k)\omega(k)\}) < 0 \qquad (2.21)$$

or

$$\sum_{k=0}^{\infty} \mathbb{E}\{\|y(k)\|^2\} < \gamma^2 \sum_{k=0}^{\infty} \mathbb{E}\{\|\omega(k)\|^2\} + V(x_0) - V(x_\infty). \qquad (2.22)$$

Since $x_0 = 0$ and the system (2.6) is exponentially mean-square stable, it is easy to see that

$$\sum_{k=0}^{\infty} \mathbb{E}\{\|y(k)\|^2\} < \gamma^2 \sum_{k=0}^{\infty} \mathbb{E}\{\|\omega(k)\|^2\}, \qquad (2.23)$$

which completes the proof.                                                         □

### 2.2.2 Variance Analysis

Define the state covariance by

$$Q(k) \triangleq \mathbb{E}\{x(k)x^T(k)\}$$
$$= \mathbb{E}\left\{[x_1(k)\ x_2(k)\ \cdots\ x_n(k)][x_1(k)\ x_2(k)\ \cdots\ x_n(k)]^T\right\}. \qquad (2.24)$$

Then, the Lyapunov-type equation that governs the evolution of the state covariance matrix $Q(k)$ can be derived as follows:

$$Q(k+1) = A_K Q(k)A_K^T + \sum_{i=1}^{q} \pi_i \pi_i^T \text{tr}[Q(k)\Omega_i] + D_1 R D_1^T. \qquad (2.25)$$

Rewriting (2.25) in the form of stack matrix results in

$$\text{st}(Q(k+1)) = \Psi\, \text{st}(Q(k)) + \text{st}(D_1 R D_1^T), \qquad (2.26)$$

where

$$\Psi = A_K \otimes A_K + \sum_{i=1}^{q} \text{st}(\pi_i \pi_i^T)\text{st}^T(\Omega_i).$$

If the system (2.6) is exponentially mean-square stable, it follows from Lemma 2.2.3 that $\rho(\Psi) < 1$, which means that $Q(k)$ in (2.26) converges to a constant matrix $\hat{Q}$ when $k \to \infty$, that is

$$\hat{Q} = \lim_{k\to\infty} Q(k), \qquad (2.27)$$

and, in the steady state, (2.25) can be rewritten as

$$A_K \hat{Q} A_K^T - \hat{Q} + \sum_{i=1}^{q} \pi_i \pi_i^T \text{tr}[\hat{Q}\Omega_i] + D_1 R D_1^T = 0. \tag{2.28}$$

**Theorem 2.2.6** *If there exits a positive definite matrix Q satisfying*

$$A_K Q A_K^T - Q + \sum_{i=1}^{q} \pi_i \pi_i^T \text{tr}[Q\Omega_i] + D_1 R D_1^T < 0, \tag{2.29}$$

*for all admissible parameter uncertainties, then the system (2.6) is exponentially mean-square stable and $\hat{Q} \leq Q$.*

*Proof.* From Lemma 2.2.3, the system (2.6) is exponentially mean-square stable. Therefore, the system steady-state state covariance defined by (2.27) exists and satisfies (2.28).

Subtracting (2.28) from (2.29) gives

$$A_K (Q - \hat{Q}) A_K^T - (Q - \hat{Q}) + \sum_{i=1}^{q} \pi_i \pi_i^T \text{tr}[(Q - \hat{Q})\Omega_i] < 0, \tag{2.30}$$

which indicates from Lemma 2.2.4 that $Q - \hat{Q} \geq 0$. The proof ends.      □

Up to now, the stability, $H_\infty$ norm performance, and variance of the system (2.6) have been considered separately. Before giving our main results, we first restate Theorem 2.2.5 and Theorem 2.2.6 in a unified LMI framework as follows.

**Theorem 2.2.7** *Given a scalar $\gamma > 0$ and a feedback gain matrix $K$. Then the requirements (R1) to (R3) are satisfied simultaneously for all admissible uncertainties if there exist a positive definite matrix $Q > 0$ and positive scalars $\alpha_i > 0$, $\beta_i > 0$ ($i = 1, \ldots, q$) such that the following set of linear matrix inequalities (LMIs) are feasible:*

$$\begin{bmatrix} -\alpha_i & \alpha_i \pi_i^T \\ \alpha_i \pi_i & -Q \end{bmatrix} < 0, \tag{2.31}$$

$$\begin{bmatrix} -\beta_i & 0 & \mu\beta_i & 0 \\ 0 & -Q & 0 & Q\theta_i \\ \mu\beta_i & 0 & -\mu & 0 \\ 0 & \theta_i^T Q & 0 & -\mu I \end{bmatrix} < 0, \tag{2.32}$$

$$\begin{bmatrix} -Q & 0 & QA_K^T & Q\theta_1 & \cdots & Q\theta_q & QC_K^T \\ 0 & -\gamma^2 I & D_1^T & 0 & \cdots & 0 & D_2^T \\ A_K Q & D_1 & -Q & 0 & \cdots & 0 & 0 \\ \theta_1^T Q & 0 & 0 & -\alpha_1 I & \cdots & 0 & 0 \\ \cdots & \cdots & \cdots & \cdots & \cdots & \cdots & \cdots \\ \theta_q^T Q & 0 & 0 & 0 & \cdots & -\alpha_q I & 0 \\ C_K Q & D_2 & 0 & 0 & \cdots & 0 & -I \end{bmatrix} < 0, \tag{2.33}$$

$$\begin{bmatrix} -Q & A_K Q & \pi_1 & \cdots & \pi_q & D_1 \\ QA_K^T & -Q & 0 & \cdots & 0 & 0 \\ \pi_1^T & 0 & -\beta_1 I & \cdots & 0 & 0 \\ \cdots & \cdots & \cdots & \cdots & \cdots & \cdots \\ \pi_q^T & 0 & 0 & \cdots & -\beta_q I & 0 \\ D_1^T & 0 & 0 & \cdots & 0 & -R^{-1} \end{bmatrix} < 0, \tag{2.34}$$

$$\begin{bmatrix} 1 & 0 & 0 & \cdots & 0 \end{bmatrix} Q \begin{bmatrix} 1 & 0 & 0 & \cdots & 0 \end{bmatrix}^T \le \sigma_1^2,$$

$$\begin{bmatrix} 0 & 1 & 0 & \cdots & 0 \end{bmatrix} Q \begin{bmatrix} 0 & 1 & 0 & \cdots & 0 \end{bmatrix}^T \le \sigma_2^2,$$

$$\vdots$$

$$\begin{bmatrix} 0 & 0 & \cdots & 0 & 1 \end{bmatrix} Q \begin{bmatrix} 0 & 0 & \cdots & 0 & 1 \end{bmatrix}^T \le \sigma_n^2, \tag{2.35}$$

*where $\mu > 0$ is any known positive scalar and*

$$\theta_i \theta_i^T = \Omega_i. \tag{2.36}$$

Before giving the proof of Theorem 2.2.7, it is worth paying some attention to (2.36). We know that a symmetric matrix $\Gamma \in \mathbb{R}^{n \times n}$ is positive definite if and only if there exist $\delta_j \in \mathbb{R}^n$ ($j = 1, 2, \ldots, n$) such that $\Gamma = \sum_{j=1}^{n} \delta_j \delta_j^T$ and rank$[\delta_1, \delta_2, \ldots, \delta_n] = n$. The reason why we here use (2.36) to decompose $\Omega_i$ is to avoid unnecessarily complicated notations and this does not lose any generality.

*Proof.* First, let us prove that (2.31) together with (2.33) guarantees the exponential stability of the system (2.6) and, meanwhile, ensures that the $H_\infty$ norm constraint is satisfied.

Define new variables $\alpha_i > 0$ ($i = 1, 2, \ldots, q$) satisfying

$$\alpha_i < (\text{tr}[\pi_i \pi_i^T P])^{-1}, \quad i = 1, 2, \ldots, q, \tag{2.37}$$

where $P = Q^{-1}$ is a positive definite matrix.

Using the property of matrix trace and the Schur Complement Lemma, we obtain

$$\text{tr}[\pi_i \pi_i^T P] = \pi_i^T P \pi_i < \alpha_i^{-1} \iff \begin{bmatrix} -\alpha_i & \alpha_i \pi_i^T \\ \alpha_i \pi_i & -P^{-1} \end{bmatrix} < 0. \tag{2.38}$$

Moreover, we have

$$\begin{bmatrix} A_K^T P A_K - P + \sum_{i=1}^{q} \Omega_i \text{tr}[\pi_i \pi_i^T P] + C_K^T C_K & A_K^T P D_1 + C_K^T D_2 \\ D_1^T P A_K + D_2^T C_K & D_1^T P D_1 + D_2^T D_2 - \gamma^2 I \end{bmatrix}$$

$$< \begin{bmatrix} A_K^T P A_K - P + \sum_{i=1}^{q} \theta_i \theta_i^T \alpha_i^{-1} + C_K^T C_K & A_K^T P D_1 + C_K^T D_2 \\ D_1^T P A_K + D_2^T C_K & D_1^T P D_1 + D_2^T D_2 - \gamma^2 I \end{bmatrix} \triangleq \Omega_a. \tag{2.39}$$

Now, we intend to prove that (2.33) is equivalent to $\Omega_a < 0$. Using Lemma 2.2.2, we find that $\Omega_a < 0$ is equivalent to

$$
\begin{bmatrix}
-P & 0 & A_K^T & \theta_1 & \cdots & \theta_q & C_K^T \\
0 & -\gamma^2 I & D_1^T & 0 & \cdots & 0 & D_2^T \\
A_K & D_1 & -P^{-1} & 0 & \cdots & 0 & 0 \\
\theta_1^T & 0 & 0 & -\alpha_1 I & \cdots & 0 & 0 \\
\cdots & \cdots & \cdots & \cdots & \cdots & \cdots & \cdots \\
\theta_q^T & 0 & 0 & 0 & \cdots & -\alpha_q I & 0 \\
C_K & D_2 & 0 & 0 & \cdots & 0 & -I
\end{bmatrix} < 0. \tag{2.40}
$$

Since $P^{-1} = Q$, then pre- and post-multiplying (2.40) by $\mathrm{diag}\{Q, I, I, I, \ldots, I, I\}$, we arrive at (2.33). Thus, with (2.39) and from Theorem 2.2.5, the system (2.6) is exponentially mean-square stable and achieves the $H_\infty$ norm constraint.

In the next stage, we shall prove that the exponential stability and variance performance index (2.9) can be guaranteed simultaneously by (2.32), (2.34), and (2.35) with any known scalar $\mu > 0$.

Firstly, similarly as in (2.37), define another series of new variables $\beta_i > 0$ ($i = 1, 2, \ldots, q$) satisfying

$$
\beta_i < \mathrm{tr}[\theta_i \theta_i^T Q]^{-1}, \quad i = 1, 2, \ldots, q. \tag{2.41}
$$

Hence, by the Schur Complement Lemma we have

$$
\mathrm{tr}(\theta_i \theta_i^T Q) = \theta_i^T Q \theta_i < \beta_i^{-1} \iff \begin{bmatrix} -\beta_i & \beta_i \theta_i^T \\ \beta_i \theta_i & -Q^{-1} \end{bmatrix} < 0. \tag{2.42}
$$

Performing the congruence transformation to (2.42) by $\mathrm{diag}\{I, Q\}$, we can get

$$
\begin{bmatrix} -\beta_i & \beta_i \theta_i^T Q \\ \beta_i Q \theta_i & -Q \end{bmatrix} < 0. \tag{2.43}
$$

It follows from Lemma 2.2.1 that, for any given positive scalar $\mu > 0$, the following inequality holds:

$$
\begin{aligned}
& \begin{bmatrix} -\beta_i & \beta_i \theta_i^T Q \\ \beta_i Q \theta_i & -Q \end{bmatrix} \\
&= \begin{bmatrix} -\beta_i & 0 \\ 0 & -Q \end{bmatrix} + \begin{bmatrix} 0 & \beta_i \theta_i^T Q \\ \beta_i Q \theta_i & 0 \end{bmatrix} \\
&= \begin{bmatrix} -\beta_i & 0 \\ 0 & -Q \end{bmatrix} + \begin{bmatrix} \beta_i \\ 0 \end{bmatrix} [0 \ \theta_i^T Q] + \left( \begin{bmatrix} \beta_i \\ 0 \end{bmatrix} [0 \ \theta_i^T Q] \right)^T \\
&\leq \begin{bmatrix} -\beta_i & 0 \\ 0 & -Q \end{bmatrix} + \mu \begin{bmatrix} \beta_i \\ 0 \end{bmatrix} \begin{bmatrix} \beta_i \\ 0 \end{bmatrix}^T + \mu^{-1} \begin{bmatrix} 0 \\ Q\theta_i \end{bmatrix} \begin{bmatrix} 0 \\ Q\theta_i \end{bmatrix}^T \\
&= \begin{bmatrix} -\beta_i + \mu \beta_i \beta_i & 0 \\ 0 & -Q + \mu^{-1} Q\theta_i \theta_i^T Q \end{bmatrix}.
\end{aligned} \tag{2.44}
$$

By the Schur Complement Lemma, (2.32) is equivalent to

$$\begin{bmatrix} -\beta_i + \mu\beta_i\beta_i & 0 \\ 0 & -Q + \mu^{-1}Q\theta_i\theta_i^T Q \end{bmatrix} < 0 \tag{2.45}$$

and thus (2.41) is implied by (2.32).

Secondly, inequality (2.41) also indicates that

$$A_K QA_K^T - Q + \sum_{i=1}^{q} \pi_i\pi_i^T \text{tr}[Q\Omega_i] + D_1 RD_1^T$$

$$< A_K QA_K^T - Q + \sum_{i=1}^{q} \pi_i\pi_i^T \beta_i^{-1} + D_1 RD_1^T. \tag{2.46}$$

Therefore, by the Schur Complement Lemma again, the following relationships are true:

$$(2.34) \iff A_K QA_K^T - Q + \sum_{i=1}^{q} \pi_i\pi_i^T \beta_i^{-1} + D_1 RD_1^T < 0$$

$$\Rightarrow A_K QA_K^T - Q + \sum_{i=1}^{q} \pi_i\pi_i^T \text{tr}[Q\Omega_i] + D_1 RD_1^T < 0. \tag{2.47}$$

It then follows from Theorem 2.2.6 that the system (2.6) is exponentially mean-square stable and the steady-state state covariance defined in (2.27) exists and satisfies $\hat{Q} \leq Q$. Moreover, together with (2.35), the variance performance index (2.9) is met. The proof is complete. □

**Remark 2.2** *In Theorem 2.2.7, a sufficient condition, which is essential for designing the controllers in next section, has been proposed. The possible conservatism caused probably by introducing (2.37) and (2.41) can be reduced by making the values of* $\text{tr}[\pi_i\pi_i^T P]$ *and* $\text{tr}[\theta_i\theta_i^T Q]$ *respectively as close as possible to the values* $\alpha_i^{-1}$ *and* $\beta_i^{-1}$ *when solving the LMIs. On the other hand, the positive scalar* $\mu$ *can be determined by using the optimization approach proposed in Ref. [134] in order to reduce the possible conservatism resulting from inequality (2.44).*

We are now ready to derive the sufficient conditions for the robust controller designing problem.

## 2.3   Robust Controller Design

In this section, a sufficient condition for solvability of the robust $H_\infty$ variance-constrained control problem for a class of nonlinear stochastic system is proposed. To start with, we introduce the well-known $S$-procedure.

**Lemma 2.3.1** *($S$-procedure) Let* $L = L^T$, *$H$, and $E$ be real matrices of appropriate dimensions, and $F$ satisfy (2.3). Then*

$$L + HFE + E^T F^T H^T < 0 \tag{2.48}$$

*if and only if there exists a positive scalar $\varepsilon$ such that*

$$L + \varepsilon HH^{\mathrm{T}} + \varepsilon^{-1} E^{\mathrm{T}} E < 0 \tag{2.49}$$

*or, equivalently,*

$$\begin{bmatrix} L & \varepsilon H & E^{\mathrm{T}} \\ \varepsilon H^{\mathrm{T}} & -\varepsilon I & 0 \\ E & 0 & -\varepsilon I \end{bmatrix} < 0. \tag{2.50}$$

The following theorem provides an LMI approach to the multi-objective state feedback controller design problem for the nonlinear stochastic system.

**Theorem 2.3.2** *Given $\gamma > 0$, $\sigma_j^2 > 0$ ($j = 1, 2, \ldots, n$) and any positive scalar $\mu > 0$. If there exist a positive definite matrix $Q > 0$, a real matrix $M$, positive scalars $\varepsilon_1$, $\varepsilon_2$, $\alpha_i$ and $\beta_i$ ($i = 1, 2, \ldots, q$) such that the following set of LMIs*

$$\begin{bmatrix} -\alpha_i & \alpha_i \pi_i^{\mathrm{T}} \\ \alpha_i \pi_i & -Q \end{bmatrix} < 0, \tag{2.51}$$

$$\begin{bmatrix} -\beta_i & 0 & \mu\beta_i & 0 \\ 0 & -Q & 0 & Q\theta_i \\ \mu\beta_i & 0 & -\mu & 0 \\ 0 & \theta_i^{\mathrm{T}} Q & 0 & -\mu I \end{bmatrix} < 0, \tag{2.52}$$

$$\begin{bmatrix} -Q & 0 & QA^{\mathrm{T}} + M^{\mathrm{T}} B_1^{\mathrm{T}} & \Upsilon_{14} & QC^{\mathrm{T}} + M^{\mathrm{T}} B_2^{\mathrm{T}} & 0 & QE^{\mathrm{T}} \\ * & -\gamma^2 I & D_1^{\mathrm{T}} & 0 & D_2^{\mathrm{T}} & 0 & 0 \\ * & * & -Q & 0 & 0 & \varepsilon_1 H_1 & 0 \\ * & * & * & \Upsilon_{44} & 0 & 0 & 0 \\ * & * & * & * & -I & \varepsilon_1 H_2 & 0 \\ * & * & * & * & * & -\varepsilon_1 I & 0 \\ * & * & * & * & * & * & -\varepsilon_1 I \end{bmatrix} < 0, \tag{2.53}$$

$$\begin{bmatrix} -Q & AQ + B_1 M & \pi_1 & \cdots & \pi_q & D_1 & \varepsilon_2 H_1 & 0 \\ * & -Q & 0 & \cdots & 0 & 0 & 0 & QE^{\mathrm{T}} \\ * & * & -\beta_1 I & \cdots & 0 & 0 & 0 & 0 \\ * & * & * & \cdots & & \cdots & \cdots & \cdots \\ * & * & * & * & -\beta_q I & 0 & 0 & 0 \\ * & * & * & * & * & -R^{-1} & 0 & 0 \\ * & * & * & * & * & * & -\varepsilon_2 I & 0 \\ * & * & * & * & * & * & * & -\varepsilon_2 I \end{bmatrix} < 0, \tag{2.54}$$

$$\begin{bmatrix} 1 & 0 & 0 & \cdots & 0 \end{bmatrix} Q \begin{bmatrix} 1 & 0 & 0 & \cdots & 0 \end{bmatrix}^{\mathrm{T}} \le \sigma_1^2,$$

$$\begin{bmatrix} 0 & 1 & 0 & \cdots & 0 \end{bmatrix} Q \begin{bmatrix} 0 & 1 & 0 & \cdots & 0 \end{bmatrix}^{\mathrm{T}} \le \sigma_2^2,$$

$$\vdots$$

$$\begin{bmatrix} 0 & 0 & \cdots & 0 & 1 \end{bmatrix} Q \begin{bmatrix} 0 & 0 & \cdots & 0 & 1 \end{bmatrix}^{\mathrm{T}} \le \sigma_n^2, \tag{2.55}$$

*where*

$$\Upsilon_{14} = \begin{bmatrix} Q\theta_1 & Q\theta_2 & \cdots & Q\theta_q \end{bmatrix},$$

$$\Upsilon_{44} = \mathrm{diag}\{-\alpha_1 I, -\alpha_2 I, \cdots, -\alpha_q I\}$$

*is feasible, then there exists a state feedback controller of the form* (2.5) *such that the three requirements* (R1) *to* (R3) *are satisfied for all admissible parameter uncertainties. Moreover, the desired controller* (2.5) *can be determined by*

$$K = MQ^{-1}. \tag{2.56}$$

*Proof.* In view of Theorem 2.3.1, we just need to prove that (2.33) is true if and only if (2.53) is true, while (2.34) holds if and only if (2.54) holds.

To eliminate the uncertainty $F$ contained in (2.33) and (2.34), rewrite (2.33) as follows:

$$\begin{bmatrix} -Q & 0 & Q(A+B_1K)^{\mathrm{T}} & \Upsilon_{14} & Q(C+B_2K)^{\mathrm{T}} \\ * & -\gamma^2 I & D_1^{\mathrm{T}} & 0 & D_2^{\mathrm{T}} \\ * & * & -Q & 0 & 0 \\ * & * & * & \Upsilon_{44} & 0 \\ * & * & * & * & -I \end{bmatrix}$$

$$+ \begin{bmatrix} 0 \\ 0 \\ H_1 \\ 0 \\ \vdots \\ 0 \\ H_2 \end{bmatrix} F \begin{bmatrix} QE^{\mathrm{T}} \\ 0 \\ 0 \\ 0 \\ \vdots \\ 0 \\ 0 \end{bmatrix}^{\mathrm{T}} + \begin{bmatrix} QE^{\mathrm{T}} \\ 0 \\ 0 \\ 0 \\ \vdots \\ 0 \\ 0 \end{bmatrix} F \begin{bmatrix} 0 \\ 0 \\ H_1 \\ 0 \\ \vdots \\ 0 \\ H_2 \end{bmatrix}^{\mathrm{T}} < 0. \tag{2.57}$$

Applying Lemma 2.3.1 to (2.57) and noticing $FF^{\mathrm{T}} \le I$, we know that (2.33) holds if and only if there exits a positive scalar $\varepsilon_1 > 0$ such that the following matrix inequality holds:

$$
\begin{bmatrix}
-Q & 0 & Q(A+B_1K)^{\mathrm{T}} & \Upsilon_{14} & Q(C+B_2K)^{\mathrm{T}} \\
* & -\gamma^2 I & D_1^{\mathrm{T}} & 0 & D_2^{\mathrm{T}} \\
* & * & -Q & 0 & 0 \\
* & * & * & \Upsilon_{44} & 0 \\
* & * & * & * & -I
\end{bmatrix}
$$

$$
+\varepsilon_1
\begin{bmatrix} 0 \\ 0 \\ H_1 \\ 0 \\ \vdots \\ 0 \\ H_2 \end{bmatrix}
\begin{bmatrix} 0 \\ 0 \\ H_1 \\ 0 \\ \vdots \\ 0 \\ H_2 \end{bmatrix}^{\mathrm{T}}
+\varepsilon_1^{-1}
\begin{bmatrix} QE^{\mathrm{T}} \\ 0 \\ 0 \\ 0 \\ \vdots \\ 0 \\ 0 \end{bmatrix}
\begin{bmatrix} QE^{\mathrm{T}} \\ 0 \\ 0 \\ 0 \\ \vdots \\ 0 \\ 0 \end{bmatrix}^{\mathrm{T}}
< 0. \tag{2.58}
$$

After some straightforward calculations and using the Schur Complement Lemma, we could find out that the matrix inequality (2.58) is equivalent to the following linear matrix inequality:

$$
\begin{bmatrix}
-Q & 0 & Q(A+B_1K)^{\mathrm{T}} & \Upsilon_{14} & Q(C+B_2K)^{\mathrm{T}} & 0 & QE^{\mathrm{T}} \\
* & -\gamma^2 I & D_1^{\mathrm{T}} & 0 & D_2^{\mathrm{T}} & 0 & 0 \\
* & * & -Q & 0 & 0 & \varepsilon_1 H_1 & 0 \\
* & * & * & \Upsilon_{44} & 0 & 0 & 0 \\
* & * & * & * & -I & \varepsilon_1 H_2 & 0 \\
* & * & * & * & * & -\varepsilon_1 I & 0 \\
* & * & * & * & * & * & -\varepsilon_1 I
\end{bmatrix}
< 0. \tag{2.59}
$$

Letting

$$
M = KQ, \tag{2.60}
$$

then it is easy to see that (2.59) is identical to (2.53).

Similarly, we rewrite (2.34) as follows:

$$
\begin{bmatrix}
-Q & AQ+B_1M & \pi_1 & \cdots & \pi_q & D_1 \\
(AQ+B_1M)^{\mathrm{T}} & -Q & 0 & \cdots & 0 & 0 \\
\pi_1^{\mathrm{T}} & 0 & -\beta_1 I & \cdots & 0 & 0 \\
\cdots & \cdots & \cdots & \cdots & \cdots & \cdots \\
\pi_q^{\mathrm{T}} & 0 & 0 & \cdots & -\beta_q I & 0 \\
D_1^{\mathrm{T}} & 0 & 0 & \cdots & 0 & -R^{-1}
\end{bmatrix}
$$

$$
+
\begin{bmatrix} H_1 \\ 0 \\ 0 \\ \vdots \\ 0 \\ 0 \end{bmatrix}
F
\begin{bmatrix} 0 \\ E^{\mathrm{T}} \\ 0 \\ \vdots \\ 0 \\ 0 \end{bmatrix}^{\mathrm{T}}
+
\begin{bmatrix} 0 \\ E^{\mathrm{T}} \\ 0 \\ \vdots \\ 0 \\ 0 \end{bmatrix}
F
\begin{bmatrix} H_1 \\ 0 \\ 0 \\ \vdots \\ 0 \\ 0 \end{bmatrix}^{\mathrm{T}}
< 0. \tag{2.61}
$$

Applying Lemma 2.3.1, we can see the matrix inequality (2.61) holds if and only if there exists a positive scalar $\varepsilon_2$ such that

$$
\begin{bmatrix}
-Q & AQ+B_1M & \pi_1 & \cdots & \pi_q & D_1 \\
(AQ+B_1M)^T & -Q & 0 & \cdots & 0 & 0 \\
\pi_1^T & 0 & -\beta_1 I & \cdots & 0 & 0 \\
\cdots & \cdots & \cdots & \cdots & \cdots & \cdots \\
\pi_q^T & 0 & 0 & \cdots & -\beta_q I & 0 \\
D_1^T & 0 & 0 & \cdots & 0 & -R^{-1}
\end{bmatrix}
$$

$$
+\varepsilon_2
\begin{bmatrix} H_1 \\ 0 \\ 0 \\ \vdots \\ 0 \\ 0 \end{bmatrix}
\begin{bmatrix} H_1 \\ 0 \\ 0 \\ \vdots \\ 0 \\ 0 \end{bmatrix}^T
+\varepsilon_2^{-1}
\begin{bmatrix} 0 \\ E^T \\ 0 \\ \vdots \\ 0 \\ 0 \end{bmatrix}
\begin{bmatrix} 0 \\ E^T \\ 0 \\ \vdots \\ 0 \\ 0 \end{bmatrix}^T
< 0,
\tag{2.62}
$$

which, after some tedious but straightforward calculations, is equivalent to the LMI (2.54). This completes the proof. $\qquad\square$

To this end, we now discuss the following two kinds of optimal controller design problems that would be encountered in engineering:

*OPT*1: The optimal $H_\infty$ controller design problem with given bounded variances:

$$
\min_{\{Q>0, M, \alpha_1, \alpha_2, \dots, \alpha_q, \beta_1, \beta_2, \dots, \beta_q, \varepsilon_1>0, \varepsilon_2>0\}} \gamma
$$

subject to (2.51) to (2.55)

for given $\sigma_1^2, \sigma_2^2, \dots, \sigma_n^2$.      (2.63)

*OPT*2: The optimal variance-constrained controller design problem with given noise attenuation level:

$$
\min_{\{Q>0, M, \alpha_1, \alpha_2, \dots, \alpha_q, \beta_1, \beta_2, \dots, \beta_q, \varepsilon_1>0, \varepsilon_2>0\}} \sum_{j=1}^{n} \delta_j \sigma_j^2
$$

subject to (2.51) to (2.55)

for given $\gamma$,      (2.64)

where $\delta_j$ $(j=1, 2, \dots, n)$ are given weighting coefficients for variances corresponding to their importance in real engineering systems that satisfy $\sum_{j=1}^{n} \delta_j = 1$.

## 2.4   Numerical Example

In this section, we present an illustrative example to demonstrate the effectiveness of the proposed algorithms.

Consider the following discrete uncertain system with stochastic nonlinearities:

$$
x(k+1) = \left( \begin{bmatrix} 0.8 & -0.5 & 0 \\ -0.2 & 0.8 & 0.21 \\ 0.1 & -0.38 & -0.55 \end{bmatrix} + \begin{bmatrix} 0.5 \\ 0.6 \\ 0 \end{bmatrix} F(k) \begin{bmatrix} 0.8 & 0 & 0 \end{bmatrix} \right) x(k)
$$

$$
+ f(x(k)) + \begin{bmatrix} 0.5 \\ 0 \\ 0.1 \end{bmatrix} u(k) + \begin{bmatrix} 0.3 \\ 0 \\ 0.2 \end{bmatrix} \omega(k),
$$

$$
y(k) = \begin{bmatrix} 1 & -0.6 & 2 \end{bmatrix} x(k) + 0.5u(k) + 0.6\omega(k),
$$

where $F(k) = \sin(0.6k)$ is a perturbation matrix satisfying $F(k)F^{\mathrm{T}}(k) \leq I$ and $\omega(k)$ is a zero mean Gaussian white noise process with unity covariance. As mentioned in Remark 2.1, we now consider the stochastic nonlinearity $f(x(k))$ in the following three cases.

**Case 1:** $f(x(k))$ is a nonlinear function with multiplicative noise of the following form:

$$
f(x(k)) = a_f \sum_{i=1}^{n} \alpha_{fi} x_i(k) \xi_i(k), \quad i = 1, 2, \dots, n,
$$

where $a_f$ is a known column vector, $\alpha_{fi}$ ($i = 1, 2, \dots, n$) are known coefficients, $x_i(k)$ is the $i$th member of $x(k)$, and $\xi_i(k)$ are zero mean, uncorrelated Gaussian white noise processes with unity covariances. In this case, we assume

$$
f(x(k)) = \begin{bmatrix} 0.2 \\ 0.3 \\ 0.5 \end{bmatrix} (0.3x_1(k)\xi_1(k) + 0.4x_2(k)\xi_2(k) + 0.5x_3(k)\xi_3(k)).
$$

**Case 2:** $f(x(k))$ is a nonlinearity with the sign of a function

$$
f(x(k)) = b_f \sum_{i=1}^{n} \beta_{fi} \operatorname{sign}[x_i(k)] x_i(k) \xi_i(k), \quad i = 1, 2, \dots, n.
$$

In this case, we assume

$$
f(x(k)) = \begin{bmatrix} 0.2 \\ 0.3 \\ 0.5 \end{bmatrix} (0.3 \operatorname{sign}[x_1(k)] x_1(k) \xi_1(k)
$$

$$
+ 0.4 \operatorname{sign}[x_2(k)] x_2(k) \xi_2(k) + 0.5 \operatorname{sign}[x_3(k)] x_3(k) \xi_3(k)).
$$

**Case 3:** $f(x(k))$ is a nonlinear function with the following form:

$$f(x(k)) = c_f \sum_{i=1}^{n} \rho_{fi} x_i(k)(\sin(x_i(k))\xi_i(k) + \cos(x_i(k))\eta_i(k)),$$

where $\eta_i(k)(i = 1, 2, \ldots, n)$ are zero mean Gaussian white noise processes with unity covariances, which are mutually uncorrelated and also uncorrelated with $\xi_i(k)$. In this case, we assume

$$f(x(k)) = \begin{bmatrix} 0.2 \\ 0.3 \\ 0.5 \end{bmatrix} \Big( 0.3x_1(k)\,(\sin(x_1(k))\xi_1(k) + \cos(x_1(k))\eta_1(k))$$

$$+ 0.4x_2(k)\,(\sin(x_2(k))\xi_2(k) + \cos(x_2(k))\eta_2(k))$$

$$+ 0.5x_3(k)\,\big(\sin(x_3(k))\xi_3(k) + \cos(x_3(k))\eta_3(k)\big) \Big).$$

We can easily check that all the above three classes of stochastic nonlinearities satisfy the following equality:

$$\mathbb{E}\{f(x(k))f^{\mathrm{T}}(x(k))|x(k)\}$$

$$= \begin{bmatrix} 0.04 & 0.06 & 0.10 \\ 0.06 & 0.09 & 0.15 \\ 0.10 & 0.15 & 0.25 \end{bmatrix} x^{\mathrm{T}}(k) \begin{bmatrix} 0.09 & 0 & 0 \\ 0 & 0.16 & 0 \\ 0 & 0 & 0.25 \end{bmatrix} x(k).$$

Now, we will examine the following three controller design problems.

**Problem 1.** $\gamma^2 = 1.2$, $\sigma_1^2 = 0.5$, $\sigma_2^2 = 0.8$, $\sigma_3^2 = 0.8$.

This case is concerned with the addressed robust $H_\infty$ variance-constrained control problem and hence can be tackled by using Theorem 2.3.2 with $n = 3$ and $q = 3$. By employing the Matlab LMI Toolbox, the solution is given by

$$Q = \begin{bmatrix} 0.46 & 0.2064 & -0.0365 \\ 0.2064 & 0.7654 & -0.1154 \\ -0.0365 & -0.1154 & 0.1778 \end{bmatrix},$$

$$M = \begin{bmatrix} -0.4373 & -0.6526 & -0.1441 \end{bmatrix},$$

$$K = \begin{bmatrix} -1.5213 & 1.2122 & -0.3357 \end{bmatrix},$$

$$\alpha_1 = 8.9878, \quad \alpha_2 = 8.4487, \quad \alpha_3 = 33.7209,$$

$$\beta_1 = \beta_2 = \beta_3 = 0.8431, \quad \varepsilon_1 = 0.5548, \quad \varepsilon_2 = 0.3920.$$

The simulation results are shown in Figures 2.1 to 2.13.

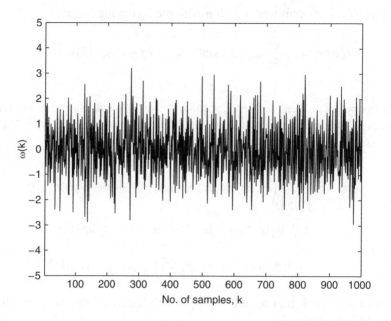

**Figure 2.1** The white sequence $\omega(k)$

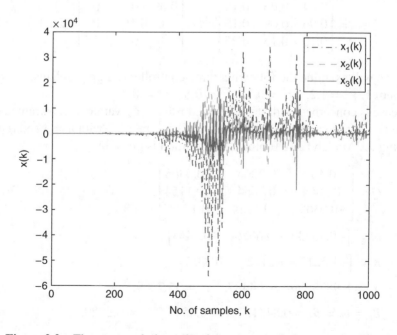

**Figure 2.2** The state evolution $x(k)$ of the uncontrolled system for Case 1

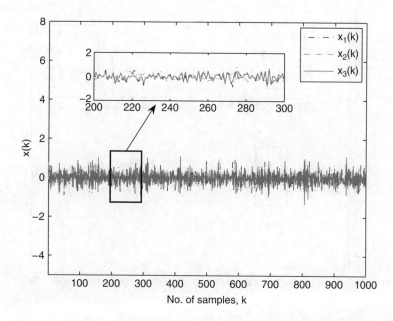

**Figure 2.3** The state evolution $x(k)$ of the controlled system for Case 1

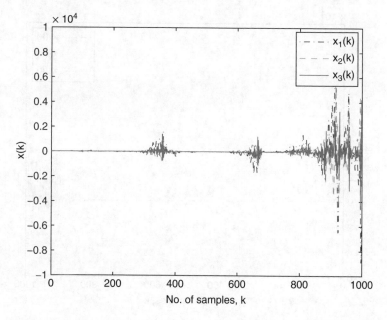

**Figure 2.4** The state evolution $x(k)$ of the uncontrolled system for Case 2

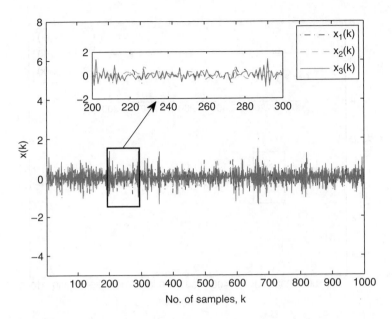

**Figure 2.5**   The state evolution $x(k)$ of the controlled system for Case 2

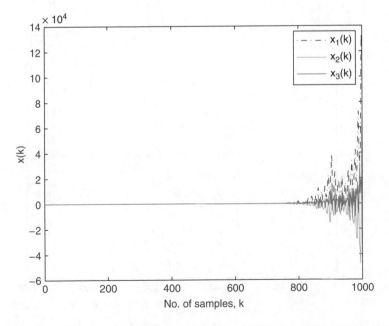

**Figure 2.6**   The state evolution $x(k)$ of the uncontrolled system for Case 3

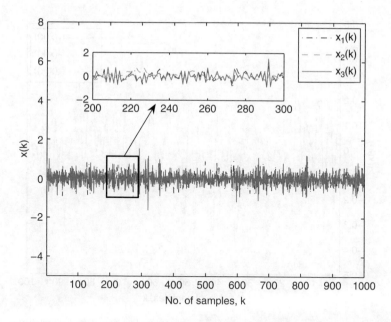

**Figure 2.7**  The state evolution $x(k)$ of the controlled system for Case 3

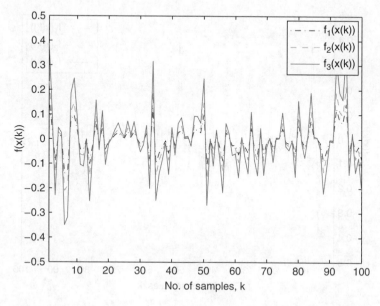

**Figure 2.8**  The stochastic nonlinearity $f(x(k))$ for Case 1, $k \in [0, 100]$

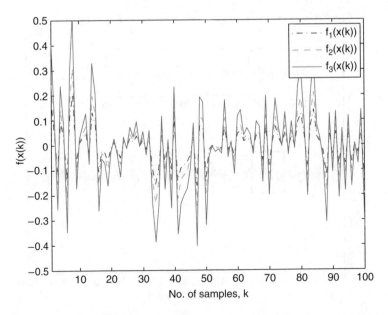

**Figure 2.9**   The stochastic nonlinearity $f(x(k))$ for Case 2, $k \in [0, 100]$

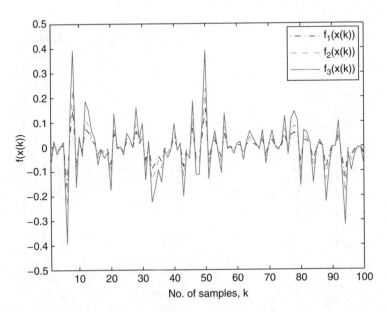

**Figure 2.10**   The stochastic nonlinearity $f(x(k))$ for Case 3, $k \in [0, 100]$

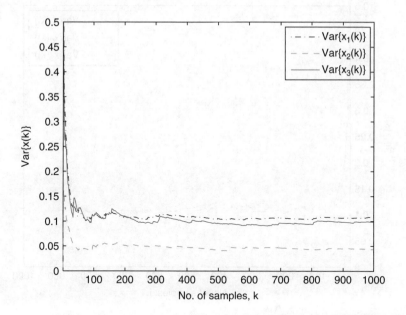

**Figure 2.11**    The state variance evolution for Case 1

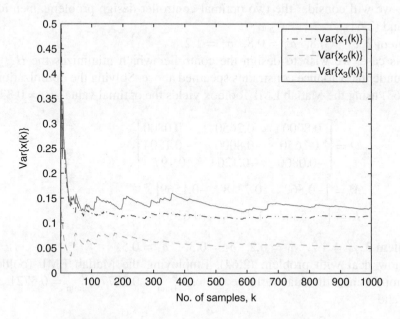

**Figure 2.12**    The state variance evolution for Case 2

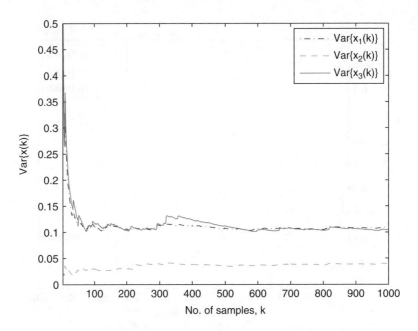

**Figure 2.13**    The state variance evolution for Case 3

Next, we will consider the two optimal controller design problems mentioned in (2.63) and (2.64).

**Problem 2.** $\sigma_1^2 = 0.8$, $\sigma_2^2 = 0.8$, $\sigma_3^2 = 1.2$.

In this case, we wish to design the controller which minimizes the $H_\infty$ performance under the variance constraints specified above. Solving the optimization problem (2.63) using the Matlab LMI Toolbox yields the optimal value $\gamma_{opt}^2 = 0.8329$ and

$$Q = \begin{bmatrix} 0.8000 & 0.2650 & -0.0800 \\ 0.2650 & 0.8000 & -0.1204 \\ -0.0800 & -0.1204 & 0.197 \end{bmatrix},$$

$$M = \begin{bmatrix} -0.8627 & 0.7318 & -0.1569 \end{bmatrix},$$

$$K = \begin{bmatrix} -1.5863 & 1.3474 & -0.6166 \end{bmatrix}.$$

**Problem 3.** $\gamma^2 = 1.2$, $\delta_1 = 0.2$, $\delta_2 = 0.3$, $\delta_3 = 0.5$.

We now deal with problem (2.64). Employing the Matlab LMI Toolbox, we obtain minimum individual variance values $\sigma_{1min}^2 = 0.2977, \sigma_{2min}^2 = 0.5721, \sigma_{3min}^2 = 0.1464$ and

$$Q = \begin{bmatrix} 0.2977 & 0.1595 & 0.0158 \\ 0.1595 & 0.5721 & -0.0988 \\ 0.0158 & -0.0988 & 0.1464 \end{bmatrix},$$

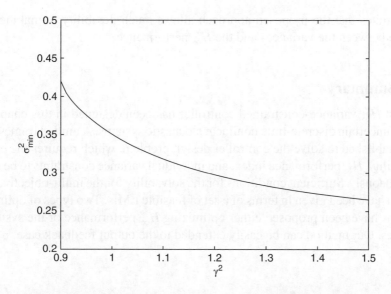

**Figure 2.14**   The $H_\infty$ performance $\gamma^2$ versus the optimal value $\sigma^2_{min}$

$$M = \begin{bmatrix} -0.2635 & 0.4431 & -0.1866 \end{bmatrix},$$
$$K = \begin{bmatrix} -1.4668 & 1.1213 & -0.3599 \end{bmatrix}.$$

Through Problem 2 and Problem 3, we can see that within the LMI framework developed in this chapter, there is some tradeoff that can be used for satisfying specific performance requirements. Figure 2.14 shows that if the value of the $H_\infty$ performance constraint is allowed to be increased, then the steady-state variances can be further reduced. Also, the $H_\infty$ performance will be improved if the variance constraints become more relaxed. Hence, the proposed approach allows much flexibility in making a compromise between the variances and the $H_\infty$ performance, while the essential multiple objectives can all be achieved simultaneously.

The simulation results are shown in Figures 2.1 to 2.14. Figure 2.1 illustrates the signal of noise $\omega(k)$, which obeys the Gaussian distribution. Figures 2.2, 2.4, and 2.6 represent the system open-loop responses for three different cases respectively, while Figures 2.3, 2.5, and 2.7 are the corresponding closed-loop responses. We can see clearly that by the state feedback control law proposed in this chapter, the original unstable systems have been stabilized. Figures 2.8, 2.9, and 2.10 represent the stochastic nonlinearities of three different cases discussed in this chapter. We can see from Figures 2.11, 2.12, and 2.13 that all the steady-state variances of different cases stay below the pre-specified upper bounds, which means that the design requirements have been achieved. Figure 2.14 shows the relationship between the $H_\infty$ performance $\gamma^2$ and the optimal value of system variance upper bounds $\sigma^2_{min}$. We can see clearly that as $\sigma^2_{min}$ increases, $\gamma^2$ will decrease. Such a result is reasonable and logical, which demonstrates that the proposed approach allows much flexibility in making a compromise between the variances and the $H_\infty$ performance.

## 2.5   Summary

A robust $H_\infty$ variance-constrained controller has been designed in this chapter for a class of uncertain discrete-time nonlinear stochastic systems. A unified framework has been established to solve the controller design problem, which requires the exponential stability, $H_\infty$ performance index, and individual variance constraints to be satisfied simultaneously. Sufficient conditions for the solvability for the multi-objective control problem have been given in terms of a set of feasible LMIs. Two types of optimization problems have been proposed either optimizing $H_\infty$ performance or the system state variances. Our method can be easily extended to the output feedback case.

# 3

# Robust Mixed $H_2/H_\infty$ Filtering

The filtering problem has been playing an important role in signal processing and control engineering. Among various filtering schemes, the celebrated Kalman filtering (also known as $H_2$ filtering) approach minimizes the $H_2$ norm of the estimation error, under the assumptions that an exact model is available and the noise processes have exactly known statistical properties. However, this is seldom the case in practical applications. As an alternative, the $H_\infty$ filtering method has been proposed, which provides an upper bound for the worst-case estimation error without the need for knowledge of noise statistics. The $H_\infty$ filter has also been proven to be more robust than the traditional Kalman filter, when model uncertainties exist in the system. However, there is no provision in $H_\infty$ filtering to ensure that the variance of the state estimation error lies within acceptable bounds. In this respect, it is natural to consider combining the performance requirements of the Kalman filter and the $H_\infty$ filter into a mixed $H_2/H_\infty$ filtering problem.

For deterministic systems, the mixed $H_2/H_\infty$ filtering problems have been extensively studied. Various methods have been proposed to solve the mixed $H_2/H_\infty$ filtering problems, such as the algebraic equation approach, time-domain Nash game approach, and convex optimization approach. It should be pointed out that, results of $H_2/H_\infty$ filtering for *nonlinear* systems are relatively few. On the other hand, as for the stochastic setting, recently there has been growing research interest in the stochastic filtering problems with or without $H_\infty$ performance constraints. The main reason is that stochastic modeling has been applied in many practical systems, such as image processing, communication systems, biological systems, and aerospace systems. However, the filtering problem for *nonlinear* stochastic systems with or without $H_2/H_\infty$ performance constraints has received little attention, which is still open and remains challenging. It is, therefore, our intention in this chapter to cope with the robust $H_2/H_\infty$ filtering problem for a class of nonlinear stochastic systems.

Motivated by the above discussion, in this chapter we will fully investigate the mixed $H_2/H_\infty$ filtering problem for systems with deterministic uncertainties and

*Variance-Constrained Multi-Objective Stochastic Control and Filtering*, First Edition.
Lifeng Ma, Zidong Wang and Yuming Bo.
© 2015 John Wiley & Sons, Ltd. Published 2015 by John Wiley & Sons, Ltd.

stochastic nonlinearities, where the nonlinearities are characterized by statistical means. Our aim is to design a filter such that, for all admissible stochastic nonlinearities and deterministic uncertainties, the overall filtering process is exponentially mean-square quadratically stable, the $H_2$ filtering performance is achieved, and the prescribed disturbance attenuation level is guaranteed in an $H_\infty$ sense. New formulas are derived for exponential mean-square quadratic stability, $H_2$ performance, and $H_\infty$ performance. In particular, due to the introduction of the stochastic nonlinearities, a new lemma about the relations between the stability of the system and the non-negative definite solution to matrix inequality is developed for deriving our $H_2$ performance, which is actually the extension of results for linear systems. The solution to the $H_2/H_\infty$ filtering problem is enforced within a unified LMI framework. In order to demonstrate the flexibility of the proposed framework, we will examine two types of optimization problems that optimize either the $H_2$ performance or the $H_\infty$ performance, and a numerical example is provided to illustrate the design procedures and performances of the proposed method.

The remainder of this chapter is organized as follows: In Section 3.1, a class of uncertain discrete-time nonlinear stochastic systems is described and the robust $H_2/H_\infty$ filtering problem for the systems is formulated. The solution to the robust $H_2/H_\infty$ filtering problem is derived based on the notion of exponential mean-square quadratic stability, characterizations of the $H_2$ performance, and $H_\infty$ performance in Section 3.2. An LMI algorithm is developed in Section 3.3 for designing the $H_2/H_\infty$ filter for the systems with stochastic nonlinearities and deterministic norm-bounded parameter uncertainties. An illustrative example is presented in Section 3.4 to demonstrate the applicability of the method and some concluding remarks are provided in Section 3.5.

## 3.1   System Description and Problem Formulation

Consider the following class of discrete-time systems with stochastic nonlinearities and deterministic norm-bounded parameter uncertainties:

$$\begin{cases} x(k+1) = (A + H_1FE)x(k) + f(x(k)) + B_1w(k), \\ y(k) = (C + H_2FE)x(k) + g(x(k)) + D_{11}w(k), \\ z_2(k) = L_2x(k), \\ z_\infty(k) = L_\infty x(k), \end{cases} \tag{3.1}$$

where $x(k) \in \mathbb{R}^n$ is the state, $y(k) \in \mathbb{R}^m$ is the measured output, $z_{2k} \in \mathbb{R}^{p1}$ is a combination of the states to be estimated (with respect to $H_2$-norm constraints), $z_\infty(k) \in \mathbb{R}^{p2}$ is another combination of the states to be estimated (with respect to $H_\infty$-norm constraints), $w(k) \in \mathbb{R}^r$ is a zero mean Gaussian white noise sequence with covariance $\Phi > 0$, and $A, B_1, C, D_{11}, L_2, L_\infty, H_1, H_2$ and $E$ are known real matrices with appropriate dimensions.

The matrix $F \in \mathbb{R}^{i \times j}$ represents the deterministic norm-bounded parameter uncertainties satisfying

$$FF^T \le I. \tag{3.2}$$

The deterministic uncertainty $F$ is said to be admissible if the condition (3.2) is met.

The functions $f(x(k)): \mathbb{R}^n \to \mathbb{R}^n$ and $g(x(k)): \mathbb{R}^n \to \mathbb{R}^m$ are stochastic nonlinear functions of the states, which are assumed to have the following first moments for all $x(k)$:

$$\mathbb{E}\left\{ \begin{bmatrix} f(x(k)) \\ g(x(k)) \end{bmatrix} \Big| x(k) \right\} = 0, \tag{3.3}$$

with the covariance given by

$$\mathbb{E}\left\{ \begin{bmatrix} f(x(k)) \\ g(x(k)) \end{bmatrix} \begin{bmatrix} f(x_j)^T & g(x_j)^T \end{bmatrix} \Big| x(k) \right\} = 0, \qquad k \ne j, \tag{3.4}$$

and

$$\mathbb{E}\left\{ \begin{bmatrix} f(x(k)) \\ g(x(k)) \end{bmatrix} \begin{bmatrix} f(x(k))^T & g(x(k))^T \end{bmatrix} \Big| x(k) \right\} = \sum_{i=1}^q \begin{bmatrix} \pi_{1i} \\ \pi_{2i} \end{bmatrix} \begin{bmatrix} \pi_{1i} \\ \pi_{2i} \end{bmatrix}^T x(k)^T \Gamma_i x(k), \tag{3.5}$$

where $\pi_{1i} \in \mathbb{R}^{n \times 1}$ and $\pi_{2i} \in \mathbb{R}^{m \times 1}$ ($i = 1,...,q$) are known column vectors with compatible dimensions of $f(x(k))$ and $g(x(k))$, and $\Gamma_i$ ($i = 1,...,q$) are known positive definite matrices with appropriate dimensions.

Now consider the following filter for the system (3.1):

$$\begin{cases} \hat{x}(k + 1) = \hat{A}\hat{x}(k) + \hat{K}y(k), \\ \hat{z}_2(k) = \hat{L}_2\hat{x}(k), \\ \hat{z}_\infty(k) = \hat{L}_\infty\hat{x}(k), \end{cases} \tag{3.6}$$

where $\hat{x}(k)$ is the state estimate, $\hat{z}_2(k)$ is an estimate for $z_2(k)$, and $\hat{z}_\infty(k)$ is an estimate for $z_\infty(k)$, and $\hat{A}, \hat{K}, \hat{L}_2$ and $\hat{L}_\infty$ are the filter parameters to be determined.

The augmented system is described as follows by combining (3.6) with (3.1):

$$\begin{cases} \tilde{x}(k + 1) = \tilde{A}\tilde{x}(k) + B_e h(x(k)) + \tilde{B}w(k), \\ e_2(k) \triangleq z_2(k) - \hat{z}_2(k) = L_2 x(k) - \hat{L}_2\hat{x}(k) = C_2\tilde{x}(k), \\ e_\infty(k) \triangleq z_\infty(k) - \hat{z}_{\infty k} = L_\infty x(k) - \hat{L}_\infty\hat{x}(k) = C_\infty\tilde{x}(k), \end{cases} \tag{3.7}$$

where

$$\tilde{x}(k) = \begin{bmatrix} x(k) \\ \hat{x}(k) \end{bmatrix}, \quad h(x(k)) = \begin{bmatrix} f(x(k)) \\ g(x(k)) \end{bmatrix}, \tag{3.8}$$

$$\tilde{A} = \bar{A} + \Delta A, \quad \bar{A} = \begin{bmatrix} A & 0 \\ \hat{K}C & \hat{A} \end{bmatrix}, \tag{3.9}$$

$$\Delta A = \begin{bmatrix} H_1 \\ \hat{K}H_2 \end{bmatrix} F \begin{bmatrix} E & 0 \end{bmatrix} \triangleq H_e F E_e, \tag{3.10}$$

$$B_e = \begin{bmatrix} I & 0 \\ 0 & \hat{K} \end{bmatrix}, \quad \tilde{B} = \begin{bmatrix} B_1 \\ \hat{K}D_{11} \end{bmatrix}, \tag{3.11}$$

$$C_\infty = \begin{bmatrix} L_\infty & -\hat{L}_\infty \end{bmatrix}, \quad C_2 = \begin{bmatrix} L_2 & -\hat{L}_2 \end{bmatrix}. \tag{3.12}$$

Since the augmented system (3.7) contains both deterministic and stochastic entries, we need to introduce the notion of stochastic stability (in the mean-square sense) for the augmented system (3.7).

**Definition 3.1.1** *The system* (3.7) *is said to be exponentially mean-square quadratically stable if with* $w(k) = 0$ *there exist constants* $\alpha \geq 1$ *and* $\tau \in (0, 1)$ *such that*

$$\mathbb{E}\{\|\tilde{x}(k)\|^2\} \leq \alpha \tau^k \mathbb{E}\{\|\tilde{x}_0\|^2\}, \quad \forall \tilde{x}_0 \in \mathbb{R}^n, \quad k \in \mathbb{I}^+, \tag{3.13}$$

*for all uncertainties* $F$ *satisfying the condition* (3.2).

In this chapter, our objective is to design the filter (3.6) for the system (3.1) such that, for all stochastic nonlinearities and all admissible deterministic uncertainties, the augmented system (3.7) is exponentially mean-square quadratically stable, the estimation error $e_2(k)$ satisfies the $H_2$ performance constraint, and the estimation error $e_\infty(k)$ satisfies the $H_\infty$ performance constraint. More specifically, we aim to design the filter (3.6) such that the following requirements are satisfied simultaneously:

(R1) For a given $\beta > 0$, the system (3.7) is exponentially mean-square quadratically stable and

$$J_2 = \lim_{k \to \infty} \mathbb{E}\{\|e_2(k)\|^2\} < \beta. \tag{3.14}$$

(R2) For a given $\gamma > 0$, the system (3.7) is exponentially mean-square quadratically stable and

$$\sum_{k=0}^{\infty} \mathbb{E}\{\|e_\infty(k)\|^2\} < \gamma^2 \sum_{k=0}^{\infty} \mathbb{E}\{\|w(k)\|^2\}, \tag{3.15}$$

for all nonzero $w(k)$ under zero initial condition.

The above design problem will be referred to as the robust $H_2/H_\infty$ filtering problem.

## 3.2 Algebraic Characterizations for Robust $H_2/H_\infty$ Filtering

In this section, we will present the algebraic characterizations for robust $H_2$ filtering and robust $H_\infty$ filtering. Under the framework of a common Lyapunov function, the solution to the robust $H_2/H_\infty$ filtering problem is provided for the system (3.7).

### 3.2.1 Robust $H_2$ Filtering

Before proceeding, we denote

$$\tilde{\Gamma}_i \triangleq \begin{bmatrix} \Gamma_i & 0 \\ 0 & 0 \end{bmatrix}, \quad \Pi_i \triangleq \begin{bmatrix} \pi_{1i} \\ \pi_{2i} \end{bmatrix} \begin{bmatrix} \pi_{1i} \\ \pi_{2i} \end{bmatrix}^{\mathrm{T}}. \tag{3.16}$$

In order to derive the characterization guaranteeing the robust $H_2$ filtering performance, we need the following technical results.

**Lemma 3.2.1** *Let* $V(\tilde{x}(k)) = \tilde{x}(k)^{\mathrm{T}} P \tilde{x}(k)$ *be a Lyapunov functional, where* $P > 0$. *If there exist positive real scalars* $\lambda, \mu, \nu,$ *and* $0 < \psi < 1$ *such that*

$$\mu \|\tilde{x}(k)\|^2 \le V(\tilde{x}(k)) \le \nu \|\tilde{x}(k)\|^2 \tag{3.17}$$

*and*

$$\mathbb{E}\{V(\tilde{x}(k+1))|\tilde{x}(k)\} - V(\tilde{x}(k)) \le \lambda - \psi V(\tilde{x}(k)), \tag{3.18}$$

*then the sequence* $\tilde{x}(k)$ *satisfies*

$$\mathbb{E}\{\|\tilde{x}(k)\|^2\} \le \frac{\nu}{\mu} \|\tilde{x}_0\|^2 (1 - \psi)^k + \frac{\lambda}{\mu\psi}. \tag{3.19}$$

**Lemma 3.2.2** *If the system* (3.7) *is exponentially mean-square quadratically stable, then*

$$\rho\{\tilde{A} \otimes \tilde{A} + \sum_{i=1}^{q} \mathrm{st}(B_e \Pi_i B_e^{\mathrm{T}}) \mathrm{st}^{\mathrm{T}}(\tilde{\Gamma}_i)\} < 1 \tag{3.20}$$

*or, equivalently,*

$$\rho\{\tilde{A}^{\mathrm{T}} \otimes \tilde{A}^{\mathrm{T}} + \sum_{i=1}^{q} \mathrm{st}(\tilde{\Gamma}_i) \mathrm{st}^{\mathrm{T}}(B_e \Pi_i B_e^{\mathrm{T}})\} < 1. \tag{3.21}$$

We are now ready to give the stability conditions.

**Theorem 3.2.3** *The system* (3.7) *is exponentially mean-square quadratically stable if, for all admissible uncertainties, there exists a positive definite matrix P satisfying*

$$\tilde{A}^{\mathrm{T}}P\tilde{A} - P + \sum_{i=1}^{q}[\tilde{\Gamma}_i\mathrm{tr}(B_e\Pi_iB_e^{\mathrm{T}}P)] < 0. \tag{3.22}$$

*Proof.* Define the Lyapunov functional $V(\tilde{x}(k)) = \tilde{x}(k)^{\mathrm{T}}P\tilde{x}(k)$, where $P > 0$ is the solution to (3.22). By using the super-Martingale property for the system (3.7) with $w(k) = 0$, we obtain

$$\mathbb{E}\{V(\tilde{x}(k+1))|\tilde{x}(k)\} - V(\tilde{x}(k))$$

$$= \tilde{x}(k)^{\mathrm{T}}\tilde{A}^{\mathrm{T}}P\tilde{A}\tilde{x}(k) + \mathbb{E}\{h^{\mathrm{T}}(x(k))B_e^{\mathrm{T}}PB_eh(x(k))\} - \tilde{x}(k)^{\mathrm{T}}P\tilde{x}(k)$$

$$= \tilde{x}(k)^{\mathrm{T}}(\tilde{A}^{\mathrm{T}}P\tilde{A} - P)\tilde{x}(k) + \mathrm{tr}[\mathbb{E}\{B_eh(x(k))h(x(k))^{\mathrm{T}}B_e^{\mathrm{T}}\}P]$$

$$= \tilde{x}(k)^{\mathrm{T}}(\tilde{A}^{\mathrm{T}}P\tilde{A} - P)\tilde{x}(k) + \mathrm{tr}[B_e\sum_{i=1}^{q}\{\Pi_ix(k)^{\mathrm{T}}\Gamma_ix(k)\}B_e^{\mathrm{T}}P]$$

$$= \tilde{x}(k)^{\mathrm{T}}(\tilde{A}^{\mathrm{T}}P\tilde{A} - P)\tilde{x}(k) + \sum_{i=1}^{q}[x(k)^{\mathrm{T}}\Gamma_ix(k)\mathrm{tr}[B_e\Pi_iB_e^{\mathrm{T}}P]]$$

$$= \tilde{x}(k)^{\mathrm{T}}(\tilde{A}^{\mathrm{T}}P\tilde{A} - P)\tilde{x}(k) + \sum_{i=1}^{q}[\tilde{x}(k)^{\mathrm{T}}\begin{bmatrix}\Gamma_i & 0\\ 0 & 0\end{bmatrix}\tilde{x}(k)\mathrm{tr}[B_e\Pi_iB_e^{\mathrm{T}}P]]$$

$$= \tilde{x}(k)^{\mathrm{T}}(\tilde{A}^{\mathrm{T}}P\tilde{A} - P)\tilde{x}(k) + \tilde{x}(k)^{\mathrm{T}}\sum_{i=1}^{q}[\tilde{\Gamma}_i\mathrm{tr}[B_e\Pi_iB_e^{\mathrm{T}}P]]\tilde{x}(k)$$

$$= \tilde{x}(k)^{\mathrm{T}}\{\tilde{A}^{\mathrm{T}}P\tilde{A} - P + \sum_{i=1}^{q}[\tilde{\Gamma}_i\mathrm{tr}[B_e\Pi_iB_e^{\mathrm{T}}P]]\}\tilde{x}(k). \tag{3.23}$$

We know from (3.21) that there must exist a sufficiently small scalar $\alpha$ satisfying $0 < \alpha < \lambda_{\max}(P)$ and

$$\tilde{A}^{\mathrm{T}}P\tilde{A} - P + \sum_{i=1}^{q}\tilde{\Gamma}_i\mathrm{tr}[B_e\Pi_iB_e^{\mathrm{T}}P] < -\alpha I. \tag{3.24}$$

Therefore, we obtain

$$\mathbb{E}\{V(\tilde{x}(k+1))|\tilde{x}(k)\} - V(\tilde{x}(k)) \le -\alpha\tilde{x}(k)^{\mathrm{T}}\tilde{x}(k) \le -\frac{\alpha}{\lambda_{\max}(P)}V(\tilde{x}(k)) \tag{3.25}$$

and the proof follows immediately from Lemma 3.2.1. □

Let us proceed to compute the $H_2$ performance $J_2$ that is used in the constraint (3.14). Define the state variance by

$$Q(k) \triangleq \mathbb{E}\{\tilde{x}(k)\tilde{x}(k)^{\mathrm{T}}\}$$

$$= \mathbb{E}\left\{\begin{bmatrix} \tilde{x}_1(k) & \tilde{x}_2(k) & \cdots & \tilde{x}_n(k)\end{bmatrix}\begin{bmatrix} \tilde{x}_1(k) & \tilde{x}_2(k) & \cdots & \tilde{x}_n k\end{bmatrix}^{\mathrm{T}}\right\}. \tag{3.26}$$

The Lyapunov-type equation that governs the evolution of the state variance matrix $Q(k)$ can be derived from the system (3.7) as follows:

$$Q(k+1) = \tilde{A}Q(k)\tilde{A}^{\mathrm{T}} + \mathbb{E}[B_e h(x(k))h(x(k)^{\mathrm{T}})B_e^{\mathrm{T}}] + \tilde{B}\Phi\tilde{B}^{\mathrm{T}}$$

$$= \tilde{A}Q(k)\tilde{A}^{\mathrm{T}} + \mathbb{E}\left\{B_e\begin{bmatrix} f(x(k))\\ g(x(k))\end{bmatrix}\begin{bmatrix} f(x(k))\\ g(x(k))\end{bmatrix}^{\mathrm{T}}B_e^{\mathrm{T}}\right\} + \tilde{B}R\Phi\tilde{B}^{\mathrm{T}}$$

$$= \tilde{A}Q(k)\tilde{A}^{\mathrm{T}} + \mathbb{E}\left[B_e\sum_{i=1}^{q}(\Pi_i x(k)^{\mathrm{T}}\Gamma_i x(k))B_e^{\mathrm{T}}\right] + \tilde{B}\Phi\tilde{B}^{\mathrm{T}}$$

$$= \tilde{A}Q(k)\tilde{A}^{\mathrm{T}} + \sum_{i=1}^{q}\left\{B_e\Pi_i\mathbb{E}(\tilde{x}(k)^{\mathrm{T}}\begin{bmatrix} \Gamma_i & 0\\ 0 & 0\end{bmatrix}\tilde{x}(k))B_e^{\mathrm{T}}\right\} + \tilde{B}\Phi\tilde{B}^{\mathrm{T}}$$

$$= \tilde{A}Q(k)\tilde{A}^{\mathrm{T}} + \sum_{i=1}^{q}\{B_e\Pi_i B_e^{\mathrm{T}}\mathrm{tr}[\mathbb{E}(\tilde{x}(k)\tilde{x}(k)^{\mathrm{T}})\tilde{\Gamma}_i]\} + \tilde{B}\Phi\tilde{B}^{\mathrm{T}}$$

$$= \tilde{A}Q(k)\tilde{A}^{\mathrm{T}} + \sum_{i=1}^{q}[B_e\Pi_i B_e^{\mathrm{T}}\mathrm{tr}(Q(k)\tilde{\Gamma}_i)] + \tilde{B}\Phi\tilde{B}^{\mathrm{T}}. \tag{3.27}$$

Rewrite (3.27) in the form of a stack matrix as follows:

$$\mathrm{st}(Q(k+1)) = \Psi\mathrm{st}(Q(k)) + \mathrm{st}(\tilde{B}\Phi\tilde{B}^{\mathrm{T}}), \tag{3.28}$$

where

$$\Psi \triangleq \tilde{A}\otimes\tilde{A} + \sum_{i=1}^{q}\mathrm{st}(B_e\Pi_i B_e^{\mathrm{T}})\mathrm{st}^{\mathrm{T}}(\tilde{\Gamma}_i). \tag{3.29}$$

If the system (3.7) is exponentially mean-square quadratically stable, it then follows from Lemma 3.2.2 that $\rho(\Psi) < 1$ and $Q(k)$ in (3.28) will converge to $Q$ when $k \to \infty$, i.e.,

$$Q = \lim_{k\to\infty} Q(k). \tag{3.30}$$

Therefore, the $H_2$ performance can be written by

$$
\begin{aligned}
J_2 &= \lim_{k \to \infty} \mathbb{E}\{\|e_2(k)\|^2\} \\
&= \lim_{k \to \infty} \mathbb{E}\{\tilde{x}(k)^\mathrm{T} C_2^\mathrm{T} C_2 \tilde{x}(k)\} \\
&= \lim_{k \to \infty} \mathbb{E}\{\mathrm{tr}[C_2 \tilde{x}(k)\tilde{x}(k)^\mathrm{T} C_2^\mathrm{T}]\} \\
&= \lim_{k \to \infty} \mathrm{tr}[C_2 Q(k) C_2^\mathrm{T}] \\
&= \mathrm{tr}[C_2 Q C_2^\mathrm{T}].
\end{aligned}
\tag{3.31}
$$

Now suppose that there exists a symmetric matrix $\hat{P}(k)$ such that the following backward recursion is satisfied:

$$
\hat{P}(k) = \tilde{A}^\mathrm{T} \hat{P}(k+1)\tilde{A} + \sum_{i=1}^{q} \tilde{\Gamma}_i \mathrm{tr}[B_e \Pi_i B_e^\mathrm{T} \hat{P}(k+1)] + C_2^\mathrm{T} C_2.
\tag{3.32}
$$

We rearrange (3.32) in the form of a stack matrix as follows:

$$
\mathrm{st}(P\hat{(}k)) = \Lambda \mathrm{st}(\hat{P}(k+1)) + \mathrm{st}(C_2^\mathrm{T} C_2),
\tag{3.33}
$$

where

$$
\Lambda \triangleq \tilde{A}^\mathrm{T} \otimes \tilde{A}^\mathrm{T} + \sum_{i=1}^{q} \mathrm{st}(\tilde{\Gamma}_i)\mathrm{st}^\mathrm{T}(B_e \Pi_i B_e^\mathrm{T}).
\tag{3.34}
$$

If the system (3.7) is exponentially mean-square quadratically stable, it follows from Lemma 3.2.2 that $\rho(\Lambda) < 1$ and $\hat{P}(k)$ in (3.33) will converge to $\hat{P}$ when $k \to \infty$, i.e.,

$$
\hat{P} = \lim_{k \to \infty} \hat{P}(k).
\tag{3.35}
$$

Hence, in the steady state, (3.32) becomes

$$
\hat{P} = \tilde{A}^\mathrm{T} \hat{P}\tilde{A} + \sum_{i=1}^{q} \tilde{\Gamma}_i \mathrm{tr}[B_e \Pi_i B_e^\mathrm{T} \hat{P}] + C_2^\mathrm{T} C_2.
\tag{3.36}
$$

Summarizing the above results, we obtain the following theorem, which gives an alternative to compute the $H_2$ performance.

**Theorem 3.2.4** *If the system (3.7) is exponentially mean-square quadratically stable, the $H_2$ performance can be expressed in terms of $\hat{P}$ as follows:*

$$
J_2 = \mathrm{tr}[\Phi \tilde{B}^\mathrm{T} \hat{P} \tilde{B}],
\tag{3.37}
$$

*where $\hat{P} > 0$ is the solution to (3.36).*

*Proof.* Note that

$$\lim_{k\to\infty} \text{tr}\{Q(k+1)\hat{P}(k+1) - Q(k)\hat{P}(k)\}$$

$$= \lim_{k\to\infty} \text{tr}\{[\tilde{A}Q(k)\tilde{A}^T + \sum_{i=1}^{q} B_e\Pi_i B_e^T \text{tr}(Q(k)\tilde{\Gamma}_i) + \tilde{B}\Phi\tilde{B}^T]\hat{P}(k+1)$$

$$-Q(k)[\tilde{A}^T\hat{P}(k+1)\tilde{A} + \sum_{i=1}^{q} \tilde{\Gamma}_i\text{tr}(B_e\Pi_i B_e^T\hat{P}(k+1)) + C_2^T C_2]\}$$

$$= 0. \tag{3.38}$$

Using the properties of $\text{tr}[AB] = \text{tr}[BA]$ and $\text{tr}[A+B] = \text{tr}[A] + \text{tr}[B]$ to (3.38), we have

$$\text{tr}[C_2 Q C_2^T] = \text{tr}[\Phi\tilde{B}^T\hat{P}\tilde{B}]. \tag{3.39}$$

This completes the proof. □

**Remark 3.1** *We use (3.34) to compute the $H_2$ performance instead of (3.29). The reason is that the $H_2$ filtering performance and the $H_\infty$ filtering performance need to be characterized as a similar structure so that the solution to the addressed $H_2/H_\infty$ filtering problem can be obtained by using a unified LMI approach. Although we have already discussed the $H_2$ filtering performance and $H_\infty$ filtering performance, we still need to put in much effort because it is not a trivial task to merge two performance objectives into a unified LMI framework. One will see in the next subsection that the structure of (3.37) is similar to that for the $H_\infty$ filtering performance.*

Notice that the system model in (3.1) involves parameter uncertainties, and hence the exact $H_2$ performance (3.37) cannot be obtained by simply solving the equation (3.36). One way to deal with this problem is to provide an upper bound for the actual $H_2$ performance.

Suppose that there exists a positive definite matrix $P$ such that the following matrix inequality is satisfied:

$$\tilde{A}^T P\tilde{A} - P + \sum_{i=1}^{q} \tilde{\Gamma}_i\text{tr}[B_e\Pi_i B_e^T P] + C_2^T C_2 < 0. \tag{3.40}$$

Before proving that the solution $P > 0$ to (3.40) is an upper bound for $\hat{P}$ in Theorem 3.2.6, we develop the following lemma, which is very important to reveal the relationship between the stochastic stability and the non-negative-definite solution to a matrix inequality.

**Lemma 3.2.5** *Consider the system*

$$\xi(k+1) = M\xi(k) + B_e f(x(k)), \tag{3.41}$$

*where* $\xi(k) = \begin{bmatrix} x(k) \\ \hat{x}(k) \end{bmatrix} \in \mathbb{R}^{2n}$, $x(k) \in \mathbb{R}^n$, $\mathbb{E}\{f(x(k))|x(k)\} = 0$ *and* $\mathbb{E}\{f(x(k))f^{\mathrm{T}}(x(k))|x(k)\} = \sum_{i=1}^{q} \Pi_i(x(k)^{\mathrm{T}}\Xi_i x(k))$. *Here,* $\Pi_i = \pi_i \pi_i^{\mathrm{T}}$, $\pi_i$ $(i = 1,...,q)$ *are column vectors,* $\Xi_i$ $(i = 1,...,q)$ *are known positive definite matrices with appropriate dimensions. If the system (3.41) is exponentially mean-square stable, and there exists a symmetric matrix* $Y$ *satisfying*

$$M^{\mathrm{T}}YM - Y + \sum_{i=1}^{q} \tilde{\Xi}_i \mathrm{tr}[B_e\Pi_iB_e^{\mathrm{T}}Y] < 0 \tag{3.42}$$

*where*

$$\tilde{\Xi}_i = \begin{bmatrix} \Xi_i & 0 \\ 0 & 0 \end{bmatrix}, \tag{3.43}$$

*then* $Y \geq 0$.

*Proof.* It follows from (3.42) that

$$M^{\mathrm{T}}YM - Y + \sum_{i=1}^{q} \tilde{\Xi}_i \mathrm{tr}[B_e\Pi_iB_e^{\mathrm{T}}Y] = -\Sigma, \tag{3.44}$$

for some $\Sigma > 0$. Define a functional $W(\xi(k)) = \xi(k)^{\mathrm{T}}Y\xi(k)$. Applying the super-Martingale property to system (3.41) yields

$$\mathbb{E}\{W(\xi(k+1))|\xi(k)\} - W(\xi(k))$$
$$= \xi(k)^{\mathrm{T}}(M^{\mathrm{T}}YM - Y)\xi(k) + \mathbb{E}\{f^{\mathrm{T}}(x(k))B_e^{\mathrm{T}}YB_ef(x(k))\}$$
$$= \xi(k)^{\mathrm{T}}(M^{\mathrm{T}}YM - Y)\xi(k) + x(k)^{\mathrm{T}}\Xi_i x(k)\mathrm{tr}(B_e\Pi_iB_e^{\mathrm{T}}Y)$$
$$= \xi(k)^{\mathrm{T}}[M^{\mathrm{T}}YM - Y + \sum_{i=1}^{q} \tilde{\Xi}_i \mathrm{tr}[B_e\Pi_iB_e^{\mathrm{T}}Y]]\xi(k)$$
$$= -\xi(k)^{\mathrm{T}}\Sigma\xi(k). \tag{3.45}$$

Summing (3.45) from 0 to $n$ with respect to $k$, we obtain

$$\mathbb{E}\{\xi_n^{\mathrm{T}}Y\xi_n\} - \xi_0^{\mathrm{T}}Y\xi_0 = -\sum_{k=0}^{n} \xi(k)^{\mathrm{T}}\Sigma\xi(k). \tag{3.46}$$

Let $n \to \infty$ in (3.46); it then follows from the exponential mean-square stability of the system (3.41) and the fact

$$\lim_{n\to\infty} \mathbb{E}\{\xi_n^{\mathrm{T}}Y\xi_n\} \leq \|Y\| \lim_{n\to\infty} \mathbb{E}\{\xi_n^{\mathrm{T}}\xi_n\}$$

that $\lim\limits_{n\to\infty} \mathbb{E}\{\xi_n^{\mathrm{T}} Y \xi_n\} = 0$. Hence, we have from (3.46) that

$$\xi_0^{\mathrm{T}} Y \xi_0 = \sum_{k=0}^{\infty} \xi(k)^{\mathrm{T}} \Sigma \xi(k) \geq 0. \tag{3.47}$$

Since (3.47) holds for any nonzero initial state $\xi_0$, we arrive at the conclusion that $Y \geq 0$. □

**Remark 3.2** *For linear time-invariant stochastic systems, it is well known that the stability of the system implies the existence of a positive definite solution to a Lyapunov matrix equation. Lemma 3.2.5 provides a "nonlinear version" of such a property, which would help establish the upper bound for the $H_2$ performance.*

Now we are ready to give the upper bound for $\hat{P}$. Comparing (3.36) to (3.40), we obtain the following main result in this subsection.

**Theorem 3.2.6** *If there exists a positive definite matrix $P$ satisfying (3.40), then the system (3.7) is exponentially mean-square quadratically stable, and*

$$\hat{P} \leq P, \tag{3.48}$$

$$\mathrm{tr}[\Phi \tilde{B}^{\mathrm{T}} \hat{P} \tilde{B}] \leq \mathrm{tr}[\Phi \tilde{B}^{\mathrm{T}} P \tilde{B}], \tag{3.49}$$

*where $\hat{P}$ satisfies (3.36).*

*Proof.* It is obvious that (3.40) implies (3.22). Hence it follows directly from Theorem 3.2.3 that the system (3.7) is exponentially mean-square quadratically stable and that, $\hat{P}$ exists and meets (3.36). Subtracting (3.40) from (3.36) yields

$$\tilde{A}^{\mathrm{T}}(P - \hat{P})\tilde{A} - (P - \hat{P}) + \sum_{i=1}^{q} \tilde{\Gamma}_i \mathrm{tr}[B_e \Pi_i B_e^{\mathrm{T}}(P - \hat{P})] < 0, \tag{3.50}$$

which indicates from Lemma 3.2.5 that $P - \hat{P} \geq 0$. Also, (3.48) implies (3.49). This completes the proof. □

The corollary given below follows immediately from Theorem 3.2.6 and (3.14).

**Corollary 3.2.7** *If there exists a positive definite matrix $P$ satisfying (3.40) and $\mathrm{tr}[\Phi \tilde{B}^{\mathrm{T}} P \tilde{B}] < \beta$, then the system (3.7) is exponentially mean-square quadratically stable and (3.14) is satisfied for some $\beta$.*

## 3.2.2   Robust $H_\infty$ Filtering

In this subsection, we present the algebraic characterization for the $H_\infty$ filtering performance in (R2). Note that the expression defined in (3.15) is used to describe the $H_\infty$ filtering performance of the stochastic system, where the expectation operator is utilized on both the filtering error and the disturbance input.

In the following theorem, the sufficient conditions are provided for establishing the $H_\infty$-norm performance.

**Theorem 3.2.8** *For a given $\gamma > 0$, the system (3.7) is exponentially mean-square quadratically stable and achieves the $H_\infty$-norm constraint (3.15) for all nonzero $w(k)$ if there exists a positive definite matrix $P$ satisfying*

$$\begin{bmatrix} \tilde{A}^T P \tilde{A} - P + \sum_{i=1}^{q} \tilde{\Gamma}_i \mathrm{tr}[B_e \Pi_i B_e^T P] + C_\infty^T C_\infty & \tilde{A}^T P \tilde{B} \\ \tilde{B}^T P \tilde{A} & \tilde{B}^T P \tilde{B} - \gamma^2 I \end{bmatrix} < 0, \qquad (3.51)$$

*for all admissible uncertainties.*

*Proof.* It is obvious that (3.51) implies (3.22); hence it follows from Theorem 3.2.3 that the system (3.7) is exponentially mean-square quadratically stable.

Next, for any nonzero $w(k)$, it follows from (3.51) that

$$\mathbb{E}\{V(\tilde{x}(k+1))|\tilde{x}(k)\} - \mathbb{E}\{V(\tilde{x}(k))\} + \mathbb{E}\{e_\infty(k)^T e_\infty(k)\} - \gamma^2 \mathbb{E}\{w(k)^T w(k)\}$$

$$= \mathbb{E}\{\tilde{x}(k)^T[\tilde{A}^T P \tilde{A} - P + \sum_{i=1}^{q} \tilde{\Gamma}_i \mathrm{tr}[B_e \Pi_i B_e^T P]]\tilde{x}(k) + \tilde{x}(k)^T \tilde{A}^T P \tilde{B} w(k) + w(k)^T \tilde{B}^T P \tilde{A} \tilde{x}(k)$$

$$+ w(k)^T \tilde{B}^T P \tilde{B} w(k) + \tilde{x}(k)^T C_\infty^T C_\infty \tilde{x}(k) - \gamma^2 w(k)^T w(k)\}$$

$$= \mathbb{E}\{\tilde{x}(k)^T[\tilde{A}^T P \tilde{A} - P + \sum_{i=1}^{q} \tilde{\Gamma}_i \mathrm{tr}[B_e \Pi_i B_e^T P] + C_\infty^T C_\infty]\tilde{x}(k)$$

$$+ \tilde{x}(k)^T \tilde{A}^T P \tilde{B} w(k) + w(k)^T \tilde{B}^T P \tilde{A} \tilde{x}(k) + w(k)^T (\tilde{B}^T P \tilde{B} - \gamma^2 I) w(k)\}$$

$$= \mathbb{E} \left\{ \begin{bmatrix} \tilde{x}(k) \\ w(k) \end{bmatrix}^T \begin{bmatrix} \tilde{A}^T P \tilde{A} - P + \sum_{i=1}^{q} \tilde{\Gamma}_i \mathrm{tr}[B_e \Pi_i B_e^T P] + C_\infty^T C_\infty & \tilde{A}^T P \tilde{B} \\ \tilde{B}^T P \tilde{A} & \tilde{B}^T P \tilde{B} - \gamma^2 I \end{bmatrix} \right.$$

$$\left. \times \begin{bmatrix} \tilde{x}(k) \\ w(k) \end{bmatrix} \right\} < 0. \qquad (3.52)$$

Now, summing (3.52) from 0 to $\infty$ with respect to $k$ yields

$$\sum_{k=0}^{\infty} [\mathbb{E}\{V(\tilde{x}(k+1))|\tilde{x}(k)\} - \mathbb{E}\{V(\tilde{x}(k))\} + \mathbb{E}\{e_\infty(k)^T e_{\infty k}\} - \gamma^2 \mathbb{E}\{w(k)^T w(k)\}] < 0,$$

$$(3.53)$$

i.e.,

$$\sum_{k=0}^{\infty} \mathbb{E}\{\|e_\infty(k)\|^2\} < \gamma^2 \sum_{k=0}^{\infty} \mathbb{E}\{\|w(k)\|^2\} + \mathbb{E}\{V(\tilde{x}_0)\} - \mathbb{E}\{V(\tilde{x}_\infty)\}. \quad (3.54)$$

Since $\tilde{x}_0 = 0$ and the system (3.7) is exponentially mean-square quadratically stable, it is straightforward to see that

$$\sum_{k=0}^{\infty} \mathbb{E}\{\|e_\infty(k)\|^2\} < \gamma^2 \sum_{k=0}^{\infty} \mathbb{E}\{\|w(k)\|^2\}. \quad (3.55)$$

This ends the proof. □

So far, the solutions to the $H_2$ filtering problem $(R1)$ and $H_\infty$ filtering problem $(R2)$ have been developed separately. In the next section, we will focus on the mixed $H_2/H_\infty$ filtering problem.

## 3.3 Robust $H_2/H_\infty$ Filter Design Techniques

In this section, we will present the solution to the robust $H_2/H_\infty$ filtering design problem for discrete-time systems with *stochastic* nonlinearities and *deterministic* norm-bounded parameter uncertainty, i.e., we design the filter that satisfies the performance requirements $(R1)$ and $(R2)$ simultaneously.

According to the results obtained in the previous section and the conditions (3.22), (3.40), and (3.51), we conclude that (3.40) and (3.51) imply (3.22). Hence (3.22) becomes redundant and is not involved in the filter design. Therefore, in order to achieve our design objectives $(R1)$ and $(R2)$, we can cast the original robust $H_2/H_\infty$ filtering problem as follows:

**Problem** (Pa). Design the filter (3.6) such that there exists a positive definite matrix $P$ satisfying the following inequalities:

$$\mathrm{tr}[\Phi \tilde{B}^{\mathrm{T}} P \tilde{B}] < \beta, \quad (3.56)$$

$$\tilde{A}^{\mathrm{T}} P \tilde{A} - P + \sum_{i=1}^{q} \tilde{\Gamma}_i \mathrm{tr}[B_e \Pi_i B_e^{\mathrm{T}} P] + C_2^{\mathrm{T}} C_2 < 0, \quad (3.57)$$

$$\begin{bmatrix} \tilde{A}^{\mathrm{T}} P \tilde{A} - P + \sum_{i=1}^{q} \tilde{\Gamma}_i \mathrm{tr}[B_e \Pi_i B_e^{\mathrm{T}} P] + C_\infty^{\mathrm{T}} C_\infty & \tilde{A}^{\mathrm{T}} P \tilde{B} \\ \tilde{B}^{\mathrm{T}} P \tilde{A} & \tilde{B}^{\mathrm{T}} P \tilde{B} - \gamma^2 I \end{bmatrix} < 0. \quad (3.58)$$

The problem (Pa) is to find the filter (3.6) to ensure that (3.56) to (3.58) are satisfied for all admissible uncertainties. Note that, at this stage, such a problem is not solved yet, since the matrix trace terms and the uncertainty $F$ are involved in (3.56) to (3.58), which make the problem very complicated. Our goal is therefore to transform (3.56) to (3.58) into LMIs, in order to obtain solutions to the aforementioned filtering problem.

In order to transform problem (Pa) into a convex optimization problem, we first deal with the matrix trace terms in (3.56) to (3.58) by introducing new variables. The following theorem presents a sufficient condition for solving the problem (Pa).

**Theorem 3.3.1** *Given $\gamma > 0$ and $\beta > 0$. If there exist positive definite matrices $P > 0$, $\Theta > 0$, and positive scalars $\alpha_i > 0$ ($i = 1, \dots, q$) such that the following matrix inequalities*

$$\mathrm{tr}[\Theta] < \beta, \tag{3.59}$$

$$\begin{bmatrix} -\Theta & \Phi^{\frac{1}{2}}\tilde{B}^{\mathrm{T}}P \\ P\tilde{B}\Phi^{\frac{1}{2}} & -P \end{bmatrix} < 0, \tag{3.60}$$

$$\begin{bmatrix} -\alpha_i & [\pi_{1i}^{\mathrm{T}} \quad \pi_{2i}^{\mathrm{T}}] B_e^{\mathrm{T}} P \\ PB_e \begin{bmatrix} \pi_{1i} \\ \pi_{2i} \end{bmatrix} & -P \end{bmatrix} < 0 \quad (i = 1, \dots, q), \tag{3.61}$$

$$\begin{bmatrix} -P & \tilde{A}^{\mathrm{T}} & \alpha_1\tilde{\Gamma}_1^{\frac{1}{2}} & \cdots & \alpha_q\tilde{\Gamma}_q^{\frac{1}{2}} & C_2^{\mathrm{T}} \\ \tilde{A} & -P^{-1} & 0 & \cdots & 0 & 0 \\ \alpha_1\tilde{\Gamma}_1^{\frac{1}{2}} & 0 & -\alpha_1 I & \cdots & 0 & 0 \\ \cdots & \cdots & \cdots & \cdots & \cdots & \cdots \\ \alpha_q\tilde{\Gamma}_q^{\frac{1}{2}} & 0 & 0 & \cdots & -\alpha_q I & 0 \\ C_2 & 0 & 0 & \cdots & 0 & -I \end{bmatrix} < 0, \tag{3.62}$$

$$\begin{bmatrix} -P & \tilde{A}^{\mathrm{T}} & \alpha_1\tilde{\Gamma}_1^{\frac{1}{2}} & \cdots & \alpha_q\tilde{\Gamma}_q^{\frac{1}{2}} & C_\infty^{\mathrm{T}} & 0 \\ \tilde{A} & -P^{-1} & 0 & \cdots & 0 & 0 & \tilde{B} \\ \alpha_1\tilde{\Gamma}_1^{\frac{1}{2}} & 0 & -\alpha_1 I & \cdots & 0 & 0 & 0 \\ \cdots & \cdots & \cdots & \cdots & \cdots & \cdots & \cdots \\ \alpha_q\tilde{\Gamma}_q^{\frac{1}{2}} & 0 & 0 & \cdots & -\alpha_q I & 0 & 0 \\ C_\infty & 0 & 0 & \cdots & 0 & -I & 0 \\ 0 & \tilde{B}^{\mathrm{T}} & 0 & \cdots & 0 & 0 & -\gamma^2 I \end{bmatrix} < 0, \tag{3.63}$$

*hold, then (3.56) to (3.58) are satisfied.*

*Proof.* We define new variables $\alpha_i > 0$ ($i = 1, \dots, q$) satisfying

$$\mathrm{tr}[B_e \Pi_i B_e^{\mathrm{T}} P] < \alpha_i \quad (i = 1, \dots, q), \tag{3.64}$$

i.e.,

$$\mathrm{tr}[B_e \begin{bmatrix} \pi_{1i} \\ \pi_{2i} \end{bmatrix} \begin{bmatrix} \pi_{1i} \\ \pi_{2i} \end{bmatrix}^{\mathrm{T}} B_e^{\mathrm{T}} P] < \alpha_i \quad (i = 1, \dots, q). \tag{3.65}$$

Using the properties of trace and the Schur Complement Lemma, we obtain

$$\text{tr}\left[B_e \begin{bmatrix} \pi_{1i} \\ \pi_{2i} \end{bmatrix} \begin{bmatrix} \pi_{1i} \\ \pi_{2i} \end{bmatrix}^{\mathrm{T}} B_e^{\mathrm{T}} P\right] = \begin{bmatrix} \pi_{1i}^{\mathrm{T}} & \pi_{2i}^{\mathrm{T}} \end{bmatrix} B_e^{\mathrm{T}} P B_e \begin{bmatrix} \pi_{1i} \\ \pi_{2i} \end{bmatrix} < \alpha_i \tag{3.66}$$

$$\Longleftrightarrow \begin{bmatrix} -\alpha_i & \begin{bmatrix} \pi_{1i}^{\mathrm{T}} & \pi_{2i}^{\mathrm{T}} \end{bmatrix} B_e^{\mathrm{T}} \\ B_e \begin{bmatrix} \pi_{1i} \\ \pi_{2i} \end{bmatrix} & -P^{-1} \end{bmatrix} < 0 \quad (i = 1, \dots, q), \tag{3.67}$$

which is equivalent to (3.61).

Next, let us prove that (3.62) is equivalent to

$$\tilde{A}^{\mathrm{T}} P \tilde{A} - P + \sum_{i=1}^{q} \alpha_i \tilde{\Gamma}_i + C_2^{\mathrm{T}} C_2 < 0. \tag{3.68}$$

By using the Schur Complement Lemma again in (3.68), we have

$$\begin{bmatrix} -P + \sum_{i=1}^{q} \alpha_i \tilde{\Gamma}_i + C_2^{\mathrm{T}} C_2 & \tilde{A}^{\mathrm{T}} \\ \tilde{A} & -P^{-1} \end{bmatrix} < 0 \tag{3.69}$$

$$\Longleftrightarrow$$

$$\begin{bmatrix} -P + C_2^{\mathrm{T}} C_2 & \tilde{A}^{\mathrm{T}} & \tilde{\Gamma}_1^{\frac{1}{2}} & \cdots & \tilde{\Gamma}_q^{\frac{1}{2}} \\ \tilde{A} & -P^{-1} & 0 & \cdots & 0 \\ \tilde{\Gamma}_1^{\frac{1}{2}} & 0 & -\alpha_1^{-1} I & \cdots & 0 \\ \cdots & \cdots & \cdots & \cdots & \cdots \\ \tilde{\Gamma}_q^{\frac{1}{2}} & 0 & 0 & \cdots & -\alpha_q^{-1} I \end{bmatrix} < 0, \tag{3.70}$$

which is equivalent to (3.62). Moreover, it follows from (3.64) and (3.68) that

$$\tilde{A}^{\mathrm{T}} P \tilde{A} - P + \sum_{i=1}^{q} \tilde{\Gamma}_i \text{tr}[B_e \Pi_i B_e^{\mathrm{T}} P] + C_2^{\mathrm{T}} C_2$$

$$< \tilde{A}^{\mathrm{T}} P \tilde{A} - P + \sum_{i=1}^{q} \alpha_i \tilde{\Gamma}_i + C_2^{\mathrm{T}} C_2 < 0, \tag{3.71}$$

which indicates (3.57).

Similarly, by using the Schur Complement Lemma, (3.63) results in (3.58). At last, we need to prove that (3.59) and (3.60) imply (3.56). Since (3.60) is equivalent to

$$\Phi^{\frac{1}{2}} \tilde{B}^{\mathrm{T}} P \tilde{B} \Phi^{\frac{1}{2}} < \Theta, \tag{3.72}$$

it follows from (3.59) and (3.72) that

$$\text{tr}[\Phi \tilde{B}^{\mathrm{T}} P \tilde{B}] = \text{tr}[\Phi^{\frac{1}{2}} \tilde{B}^{\mathrm{T}} P \tilde{B} \Phi^{\frac{1}{2}}] < \text{tr}[\Theta] < \beta. \tag{3.73}$$

This completes the proof. $\qquad \square$

In the following, we will carry on to "eliminate" the uncertainty $F$ contained in (3.62) and (3.63) using the $S$-procedure technique. Then, a convex optimization problem will be formulated, and the robust $H_2/H_\infty$ filter can be designed by using an LMI approach.

**Theorem 3.3.2** *Given $\gamma > 0$ and $\beta > 0$. If there exist positive definite matrices $S > 0$, $R > 0$, and $\Theta > 0$, real matrices $Q_i$ $(i = 1, 2, 3, 4)$, positive scalars $\alpha_i > 0$ $(i = 1, \ldots, q)$ and $\varepsilon_i > 0$ $(i = 1, 2)$ such that the following linear matrix inequalities are feasible:*

$$
\begin{bmatrix} 1 & 0 & \cdots & 0 \end{bmatrix} \Theta \begin{bmatrix} 1 \\ 0 \\ \vdots \\ 0 \end{bmatrix} + \cdots + \begin{bmatrix} 0 & \cdots & 0 & 1 \end{bmatrix} \Theta \begin{bmatrix} 0 \\ 0 \\ \vdots \\ 1 \end{bmatrix} < \beta, \tag{3.74}
$$

$$
\begin{bmatrix} -\Theta & \Phi^{\frac{1}{2}} B_1^{\mathrm{T}} & \Phi^{\frac{1}{2}}(B_1^{\mathrm{T}} R + D_{11}^{\mathrm{T}} Q_2^{\mathrm{T}}) \\ B_1 \Phi^{\frac{1}{2}} & -S & -S \\ (RB_1 + Q_2 D_{11})\Phi^{\frac{1}{2}} & -S & -R \end{bmatrix} < 0, \tag{3.75}
$$

$$
\begin{bmatrix} -\alpha_i & \pi_{1i}^{\mathrm{T}} & \pi_{1i}^{\mathrm{T}} R + \pi_{2i}^{\mathrm{T}} Q_2^{\mathrm{T}}) \\ \pi_{1i} & -S & -S \\ R\pi_{1i} + Q_2 \pi_{2i} & -S & -R \end{bmatrix} < 0 \quad (i = 1, \ldots, q), \tag{3.76}
$$

$$
\begin{bmatrix} G & G_{11}^{\mathrm{T}} \\ G_{11} & G_{22} \end{bmatrix} < 0, \tag{3.77}
$$

$$
\begin{bmatrix} G & G_{33}^{\mathrm{T}} \\ G_{33} & G_{44} \end{bmatrix} < 0, \tag{3.78}
$$

*where*

$$
G = \begin{bmatrix} -S & -S & A^{\mathrm{T}} S & A^{\mathrm{T}} R + C^{\mathrm{T}} Q_2^{\mathrm{T}} + Q_1^{\mathrm{T}} \\ -S & -R & A^{\mathrm{T}} S & A^{\mathrm{T}} R + C^{\mathrm{T}} Q_2^{\mathrm{T}} \\ SA & SA & -S & -S \\ RA + Q_2 C + Q_1 & RA + Q_2 C & -S & -R \\ \Gamma_1^{\frac{1}{2}} & \Gamma_1^{\frac{1}{2}} & 0 & 0 \\ 0 & 0 & 0 & 0 \\ \cdots & \cdots & \cdots & \cdots \\ \Gamma_q^{\frac{1}{2}} & \Gamma_q^{\frac{1}{2}} & 0 & 0 \\ 0 & 0 & 0 & 0 \end{bmatrix}
$$

$$\begin{bmatrix} \Gamma_1^{\frac{1}{2}} & 0 & \cdots & \Gamma_q^{\frac{1}{2}} & 0 \\ \Gamma_1^{\frac{1}{2}} & 0 & \cdots & \Gamma_q^{\frac{1}{2}} & 0 \\ 0 & 0 & \cdots & 0 & 0 \\ 0 & 0 & \cdots & 0 & 0 \\ -\alpha_1 I & 0 & \cdots & 0 & 0 \\ 0 & -\alpha_1 I & \cdots & 0 & 0 \\ \cdots & \cdots & & \cdots & \cdots \\ 0 & 0 & \cdots & -\alpha_q I & 0 \\ 0 & 0 & \cdots & 0 & -\alpha_q I \end{bmatrix}, \tag{3.79}$$

$$G_{11} = \begin{bmatrix} L_2 - Q_3 & L_2 & 0 & 0 & 0 & 0 & \cdots & 0 & 0 \\ 0 & 0 & H_1^T S & H_1^T R + H_2^T Q_2^T & 0 & 0 & \cdots & 0 & 0 \\ \varepsilon_1 E & \varepsilon_1 E & 0 & 0 & 0 & 0 & \cdots & 0 & 0 \end{bmatrix}, \tag{3.80}$$

$$G_{22} = \begin{bmatrix} -I & 0 & 0 \\ 0 & -\varepsilon_1 I & 0 \\ 0 & 0 & -\varepsilon_1 I \end{bmatrix}, \tag{3.81}$$

$$G_{33} = \begin{bmatrix} L_\infty - Q_4 & L_\infty & 0 & 0 & 0 & 0 & \cdots & 0 & 0 \\ 0 & 0 & B_1^T S & B_1^T R + D_{11}^T Q_2^T & 0 & 0 & \cdots & 0 & 0 \\ 0 & 0 & H_1^T S & H_1^T R + H_2^T Q_2^T & 0 & 0 & \cdots & 0 & 0 \\ \varepsilon_2 E & \varepsilon_2 E & 0 & 0 & 0 & 0 & \cdots & 0 & 0 \end{bmatrix}, \tag{3.82}$$

$$G_{44} = \begin{bmatrix} -I & 0 & 0 & 0 \\ 0 & -\gamma^2 I & 0 & 0 \\ 0 & 0 & -\varepsilon_2 I & 0 \\ 0 & 0 & 0 & -\varepsilon_2 I \end{bmatrix}, \tag{3.83}$$

then there exists a filter of the form (3.6) such that the requirements (R1) and (R2) are satisfied for all stochastic nonlinearities and all admissible deterministic uncertainties. Moreover, if LMIs (3.74) to (3.78) are feasible, the desired filter parameters can be determined by

$$\hat{A} = X_{12}^{-1} Q_1 (S - R)^{-1} X_{12},$$

$$\hat{K} = X_{12}^{-1} Q_2,$$

$$\hat{L}_2 = Q_3 (S - R)^{-1} X_{12},$$

$$\hat{L}_\infty = Q_4 (S - R)^{-1} X_{12}, \tag{3.84}$$

where the matrix $X_{12}$ comes from the factorization $I - RS^{-1} = X_{12} Y_{12}^T < 0.$

*Proof.* Rewrite the condition in the following form:

$$M + HFE + E^T F^T H^T < 0, \tag{3.85}$$

where

$$M = \begin{bmatrix} -P & \bar{A}^T & \alpha_1 \tilde{\Gamma}_1^{\frac{1}{2}} & \cdots & \alpha_q \tilde{\Gamma}_q^{\frac{1}{2}} & C_2^T \\ \bar{A} & -P^{-1} & 0 & \cdots & 0 & 0 \\ \alpha_1 \tilde{\Gamma}_1^{\frac{1}{2}} & 0 & -\alpha_1 I & \cdots & 0 & 0 \\ \cdots & \cdots & \cdots & \cdots & \cdots & \cdots \\ \alpha_q \tilde{\Gamma}_q^{\frac{1}{2}} & 0 & 0 & \cdots & -\alpha_q I & 0 \\ C_2 & 0 & 0 & \cdots & 0 & -I \end{bmatrix},$$

$$H = \begin{bmatrix} 0 & H_e^T & 0 & \cdots & 0 & 0 \end{bmatrix}^T,$$

$$E = \begin{bmatrix} E_e & 0 & 0 & \cdots & 0 & 0 \end{bmatrix}.$$

Applying Lemma 2.3.1 to (3.85), it follows that (3.85) holds if and only if there exists a positive scalar parameter $\varepsilon_1$ such that the following LMI holds:

$$\begin{bmatrix} -P & \bar{A}^T & \alpha_1 \tilde{\Gamma}_1^{\frac{1}{2}} & \cdots & \alpha_q \tilde{\Gamma}_q^{\frac{1}{2}} & C_2^T & 0 & \varepsilon_1 E_e^T \\ \bar{A} & -P^{-1} & 0 & \cdots & 0 & 0 & H_e & 0 \\ \alpha_1 \tilde{\Gamma}_1^{\frac{1}{2}} & 0 & -\alpha_1 I & 0 & 0 & 0 & 0 & 0 \\ \cdots & \cdots & \cdots & \cdots & \cdots & \cdots & \cdots & \cdots \\ \alpha_q \tilde{\Gamma}_q^{\frac{1}{2}} & 0 & 0 & \cdots & -\alpha_q I & 0 & 0 & 0 \\ C_2 & 0 & 0 & \cdots & 0 & -I & 0 & 0 \\ 0 & H_e^T & 0 & \cdots & 0 & 0 & -\varepsilon_1 I & 0 \\ \varepsilon_1 E_e & 0 & 0 & \cdots & 0 & 0 & 0 & -\varepsilon_1 I \end{bmatrix} < 0. \tag{3.86}$$

Applying the congruence transformation $\text{diag}\{I, P, I, \ldots, I, I, I, I\}$ to (3.86), we obtain

$$\begin{bmatrix} -P & \bar{A}^T P & \alpha_1 \tilde{\Gamma}_1^{\frac{1}{2}} & \cdots & \alpha_q \tilde{\Gamma}_q^{\frac{1}{2}} & C_2^T & 0 & \varepsilon_1 E_e^T \\ P\bar{A} & -P & 0 & \cdots & 0 & 0 & PH_e & 0 \\ \alpha_1 \tilde{\Gamma}_1^{\frac{1}{2}} & 0 & -\alpha_1 I & 0 & 0 & 0 & 0 & 0 \\ \cdots & \cdots & \cdots & \cdots & \cdots & \cdots & \cdots & \cdots \\ \alpha_q \tilde{\Gamma}_q^{\frac{1}{2}} & 0 & 0 & \cdots & -\alpha_q I & 0 & 0 & 0 \\ C_2 & 0 & 0 & \cdots & 0 & -I & 0 & 0 \\ 0 & H_e^T P & 0 & \cdots & 0 & 0 & -\varepsilon_1 I & 0 \\ \varepsilon_1 E_e & 0 & 0 & \cdots & 0 & 0 & 0 & -\varepsilon_1 I \end{bmatrix} < 0. \tag{3.87}$$

Recall that our goal is to derive the expression of the filter parameters from (3.6). To do this, we partition $P$ and $P^{-1}$ as

$$P = \begin{bmatrix} R & X_{12} \\ X_{12}^T & X_{22} \end{bmatrix}, \quad P^{-1} = \begin{bmatrix} S^{-1} & Y_{12} \\ Y_{12}^T & Y_{22} \end{bmatrix}, \tag{3.88}$$

where the partitioning of $P$ and $P^{-1}$ is compatible with that of $\bar{A}$ defined in (3.10), i.e., $R \in R^{n \times n}$, $X_{12} \in R^{n \times n}$, $X_{22} \in R^{n \times n}$, $S \in R^{n \times n}$, $Y_{12} \in R^{n \times n}$, and $Y_{22} \in R^{n \times n}$. Define

$$T_1 = \begin{bmatrix} S^{-1} & I \\ Y_{12}^T & 0 \end{bmatrix}, \quad T_2 = \begin{bmatrix} I & R \\ 0 & X_{12}^T \end{bmatrix}, \tag{3.89}$$

which imply that $PT_1 = T_2$ and $T_1^T P T_1 = T_1^T T_2$.

Again, define the change of filter parameters as follows:

$$Q_1 = X_{12} \hat{A} Y_{12}^T S, \quad Q_2 = X_{12} \hat{K}, \quad Q_3 = \hat{L}_2 Y_{12}^T S, \quad Q_4 = \hat{L}_\infty Y_{12}^T S. \tag{3.90}$$

Further applying the congruence transformations $\text{diag}\{T_1, T_1, I, \dots, I, I, I, I\}$ to (3.89), we obtain

$$\begin{bmatrix} J_{11} & J_{12}^T \\ J_{12} & J_{22} \end{bmatrix} < 0, \tag{3.91}$$

where

$$J_{11} = \begin{bmatrix}
-S^{-1} & -I & S^{-1}A^T & S^{-1}(A^T R + C^T Q_2^T + Q_1^T) \\
-I & -R & A^T & A^T R + C^T Q_2^T \\
AS^{-1} & A & -S^{-1} & -I \\
(RA + Q_2 C + Q_1)S^{-1} & RA + Q_2 C & -I & -R \\
\Gamma_1^{\frac{1}{2}} S^{-1} & \Gamma_1^{\frac{1}{2}} & 0 & 0 \\
0 & 0 & 0 & 0 \\
\cdots & \cdots & \cdots & \cdots \\
\Gamma_q^{\frac{1}{2}} S^{-1} & \Gamma_q^{\frac{1}{2}} & 0 & 0 \\
0 & 0 & 0 & 0
\end{bmatrix}$$

$$\begin{matrix}
S^{-1}\Gamma_1^{\frac{1}{2}} & 0 & \cdots & S^{-1}\Gamma_q^{\frac{1}{2}} & 0 \\
\Gamma_1^{\frac{1}{2}} & 0 & \cdots & \Gamma_q^{\frac{1}{2}} & 0 \\
0 & 0 & \cdots & 0 & 0 \\
0 & 0 & \cdots & 0 & 0 \\
-\alpha_1 I & 0 & \cdots & 0 & 0 \\
0 & -\alpha_1 I & \cdots & 0 & 0 \\
\cdots & \cdots & \cdots & \cdots & \cdots \\
0 & 0 & \cdots & -\alpha_q I & 0 \\
0 & 0 & \cdots & 0 & -\alpha_q I
\end{matrix}, \tag{3.92}$$

$$J_{12} = \begin{bmatrix} (L_2 - Q_3)S^{-1} & L_2 & 0 & 0 & 0 & 0 & \cdots & 0 & 0 \\ 0 & 0 & H_1^T S & H_1^T R + H_2^T Q_2^T & 0 & 0 & \cdots & 0 & 0 \\ \varepsilon_1 E S^{-1} & \varepsilon_1 E & 0 & 0 & 0 & 0 & \cdots & 0 & 0 \end{bmatrix}, \quad (3.93)$$

$$J_{22} = \begin{bmatrix} -I & 0 & 0 \\ 0 & -\varepsilon_1 I & 0 \\ 0 & 0 & -\varepsilon_1 I \end{bmatrix}. \quad (3.94)$$

Also, performing the congruence transformation $\mathrm{diag}\{S, I, S, I, I, I, \ldots, I, I, I, I\}$ to (3.86), we obtain (3.77).

Similarly, the condition (3.63) can be written in the following form:

$$M + HFE + E^T F^T H^T < 0, \quad (3.95)$$

where

$$M = \begin{bmatrix} -P & \bar{A}^T & \alpha_1 \tilde{\Gamma}_1^{\frac{1}{2}} & \cdots & \alpha_q \tilde{\Gamma}_q^{\frac{1}{2}} & C_\infty^T & 0 \\ \bar{A} & -P^{-1} & 0 & \cdots & 0 & 0 & \tilde{B} \\ \alpha_1 \tilde{\Gamma}_1^{\frac{1}{2}} & 0 & -\alpha_1 I & \cdots & 0 & 0 & 0 \\ \cdots & \cdots & \cdots & \cdots & \cdots & \cdots & \cdots \\ \alpha_q \tilde{\Gamma}_q^{\frac{1}{2}} & 0 & 0 & \cdots & -\alpha_q I & 0 & 0 \\ C_\infty & 0 & 0 & \cdots & 0 & -I & 0 \\ 0 & \tilde{B}^T & 0 & \cdots & 0 & 0 & -\gamma^2 I \end{bmatrix},$$

$$H = \begin{bmatrix} 0 & H_e^T & 0 & \cdots & 0 & 0 & 0 \end{bmatrix}^T,$$

$$E = \begin{bmatrix} E_e & 0 & 0 & \cdots & 0 & 0 & 0 \end{bmatrix}.$$

By applying Lemma 2.3.1 to (3.95) to eliminate the uncertainty $F$, we know that (3.95) holds if and only if there exists a positive scalar parameter $\varepsilon_2$ such that the following LMI holds:

$$\begin{bmatrix} -P & \bar{A}^T & \alpha_1 \tilde{\Gamma}_1^{\frac{1}{2}} & \cdots & \alpha_q \tilde{\Gamma}_q^{\frac{1}{2}} & C_\infty^T & 0 & 0 & \varepsilon_2 E_e^T \\ \bar{A} & -P^{-1} & 0 & \cdots & 0 & 0 & \tilde{B} & H_e & 0 \\ \alpha_1 \tilde{\Gamma}_1^{\frac{1}{2}} & 0 & -\alpha_1 I & \cdots & 0 & 0 & 0 & 0 & 0 \\ \cdots & \cdots & \cdots & \cdots & \cdots & \cdots & \cdots & & \cdots \\ \alpha_q \tilde{\Gamma}_q^{\frac{1}{2}} & 0 & 0 & \cdots & -\alpha_q I & 0 & 0 & 0 & 0 \\ C_\infty & 0 & 0 & \cdots & 0 & -I & 0 & 0 & 0 \\ 0 & \tilde{B}^T & 0 & \cdots & 0 & 0 & -\gamma^2 I & 0 & 0 \\ 0 & H_e^T & 0 & \cdots & 0 & 0 & 0 & -\varepsilon_2 I & 0 \\ \varepsilon_2 E_e & 0 & 0 & \cdots & 0 & 0 & 0 & 0 & -\varepsilon_2 I \end{bmatrix} < 0. \quad (3.96)$$

Performing three congruence transformations $\mathrm{diag}\{I,P,I,\ldots,I\}$, $\mathrm{diag}\{T_1,T_1,$ $I,\ldots,I,I\}$, and $\mathrm{diag}\{S,I,S,I,I,I,\ldots,I,I\}$ to (3.96), we get (3.78). Moreover, (3.75) and (3.76) are obtained from (3.60) and (3.61) by applying the congruence transformations $\mathrm{diag}\{I,T_1\}$ and using the property of trace, and (3.74) is derived from (3.59).

Furthermore, if the LMIs (3.74) to (3.78) are feasible, then we have $\begin{bmatrix} -S & -S \\ -S & -R \end{bmatrix} <$ $0$, i.e., $\begin{bmatrix} S^{-1} & I \\ I & R \end{bmatrix} > 0$. It follows directly from $XX^{-1} = I$ that $I - RS^{-1} = X_{12}Y_{12}^{\mathrm{T}} < 0$. Hence, one can always find square and nonsingular $X_{12}$ and $Y_{12}$. Therefore, (3.84) is obtained from (3.90), which concludes the proof. $\qquad\square$

**Remark 3.3** *The addressed robust $H_2/H_\infty$ filter can be obtained by solving the LMIs (3.74) to (3.78) in Theorem 3.3.2. Note that the feasibility of LMIs can be checked efficiently via the interior point method.*

Up to now, the filter has been designed to satisfy the requirements (R1) and (R2). As a by-product, the results in Theorem 3.3.2 also suggest the following two optimization problems:

(P1) *The optimal $H_\infty$ filtering problem with $H_2$ performance constraints for uncertain nonlinear stochastic systems:*

$$\min_{S>0,R>0,\Theta>0,Q_1,Q_2,Q_3,Q_4,\alpha_1,\cdots,\alpha_q,\varepsilon_1,\varepsilon_2} \gamma \quad \text{subject to (3.74) to (3.78),}$$

for some given $\beta$.

(P2) *The optimal $H_2$ filtering problem with $H_\infty$ performance constraints for uncertain nonlinear stochastic systems:*

$$\min_{S>0,R>0,\Theta>0,Q_1,Q_2,Q_3,Q_4,\alpha_1,\cdots,\alpha_q,\varepsilon_1,\varepsilon_2} \beta \quad \text{subject to (3.65) to (3.69)}$$

for some given $\gamma$.

On the other hand, in view of (3.84), we make the linear transformation on the state estimate

$$\bar{x}(k) = X_{12}\hat{x}(k), \tag{3.97}$$

and then obtain a new representation form of the filter as follows:

$$\begin{cases} \bar{x}(k+1) = \check{A}\bar{x}(k) + \check{K}y(k), \\ \hat{z}_2(k) = \check{L}_2\bar{x}(k), \\ \hat{z}_\infty(k) = \check{L}_\infty\bar{x}(k), \end{cases} \tag{3.98}$$

where

$$\check{A} = Q_1(S - R)^{-1},$$

$$\check{K} = Q_2,$$

$$\check{L}_2 = Q_3(S - R)^{-1},$$

$$\check{L}_\infty = Q_4(S - R)^{-1}. \tag{3.99}$$

We can now see from (3.98) that the filter parameters can be obtained directly by solving LMIs (3.74) to (3.78) without solving $I - RS^{-1} = X_{12}Y_{12}^{\mathrm{T}}$ for $X_{12}$ in (3.84).

**Remark 3.4** *In many engineering applications, the performance constraints are often specified a priori. For example, in Theorem 3.3.2, the filter is designed after $H_\infty$ performance and $H_2$ performance are prescribed. In fact, however, we can obtain an improved performance by the optimization method. The aim of problem (P1) is to exploit the design freedom to meet the optimal $H_\infty$ performance under a prescribed $\beta$, while the purpose of the problem (P2) is to search an optimal solution among all solutions achieving the $H_2$ performance under a prescribed $\gamma^2$. These are certainly attractive because the addressed multi-objective problems can be solved while a local optimal performance can also be achieved, and the computation is efficient by using the Matlab LMI Toolbox.*

## 3.4   An Illustrative Example

Consider a discrete-time system described by (3.1) with stochastic nonlinearities and deterministic norm-bounded parameter uncertainties as follows:

$$A = \begin{bmatrix} -0.5 & 0 & -0.1 \\ 0.6 & 0.3 & 0.2 \\ 0.1 & 0.4 & 0.1 \end{bmatrix}, \qquad B_1 = \begin{bmatrix} 0.3 \\ 0 \\ 0.2 \end{bmatrix},$$

$$C = \begin{bmatrix} 1 & -0.6 & 2 \end{bmatrix}, \qquad D_{11} = 1,$$

$$L_2 = \begin{bmatrix} 1 & 0 & 2 \end{bmatrix}, \qquad L_\infty = \begin{bmatrix} 1 & 0.3 & 0.5 \end{bmatrix},$$

$$H_1 = \begin{bmatrix} 0.5 \\ 0.6 \\ 0 \end{bmatrix}, \qquad H_2 = 0.6, \qquad E = \begin{bmatrix} 0.8 & 0 & 0 \end{bmatrix},$$

where $w(k)$ is a zero mean Gaussian white noise sequence with covariance $\Phi = 1$. The deterministic uncertainty $F$ satisfies the condition (3.2), and the stochastic nonlinear functions $f(x(k))$ and $g(x(k))$ satisfy the following assumptions:

$$\mathbb{E}\{f(x(k))|x(k)\} = 0,$$

$$\mathbb{E}\{g(x(k))|x(k)\} = 0,$$

$$\mathbb{E}\{f(x(k))f(x(k))^{\mathrm{T}}|x(k)\} = \begin{bmatrix} 1 \\ 0 \\ 0 \end{bmatrix}\begin{bmatrix} 1 \\ 0 \\ 0 \end{bmatrix}^{\mathrm{T}} x(k)^{\mathrm{T}}\begin{bmatrix} 0.5 & 0 & 0 \\ 0 & 0.8 & 0 \\ 0 & 0 & 0.6 \end{bmatrix}x(k)$$

$$+ \begin{bmatrix} 0.1 \\ 0 \\ 0 \end{bmatrix}\begin{bmatrix} 0.1 \\ 0 \\ 0 \end{bmatrix}^{\mathrm{T}} x(k)^{\mathrm{T}}\begin{bmatrix} 1 & 0 & 0 \\ 0 & 0.5 & 0 \\ 0 & 0 & 0.8 \end{bmatrix}x(k),$$

$$\mathbb{E}\{g(x(k))g(x(k))^{\mathrm{T}}|x(k)\} = x(k)^{\mathrm{T}}\begin{bmatrix} 0.5 & 0 & 0 \\ 0 & 0.8 & 0 \\ 0 & 0 & 0.6 \end{bmatrix}x(k) + 0.1x(k)^{\mathrm{T}}\begin{bmatrix} 1 & 0 & 0 \\ 0 & 0.5 & 0 \\ 0 & 0 & 0.8 \end{bmatrix}x(k).$$

Now let us examine the following three cases.

*Case 1*: $\gamma^2 = 6$, $\beta = 5$.

This case is exactly concerned with the addressed robust $H_2/H_\infty$ filtering problem and hence can be tackled by using Theorem 3.3.2 with $q = 2$. In fact, there are many solutions for this case. We provide one solution by employing the Matlab LMI toolbox, given by

$$\breve{A} = \begin{bmatrix} -0.1110 & -0.0086 & 0.0758 \\ -0.0056 & 0.2307 & -0.0638 \\ -0.5967 & 1.9788 & 0.2797 \end{bmatrix}, \quad \breve{K} = \begin{bmatrix} 0.6632 \\ -0.8490 \\ -0.0875 \end{bmatrix},$$

$$\breve{L}_2 = \begin{bmatrix} -0.7268 & 0.0730 & -0.1260 \end{bmatrix},$$

$$\breve{L}_\infty = \begin{bmatrix} -0.9257 & 0.0696 & -0.0311 \end{bmatrix}.$$

*Case 2*: $\beta = 5$.

In this case, we wish to design the filter that minimizes the $H_\infty$ performance under the $H_2$ performance constraints. That is, we want to solve the problem (P1). Solving the optimization problem (3.97) using the LMI toolbox yields the minimum value $\gamma^2_{\min} = 5.8003$, and

$$\breve{A} = \begin{bmatrix} -0.1013 & -0.0319 & 0.0558 \\ 0.1081 & 0.2080 & -0.0457 \\ -0.6725 & 1.9790 & 0.2148 \end{bmatrix}, \quad \breve{K} = \begin{bmatrix} 0.6655 \\ -0.8767 \\ -0.4613 \end{bmatrix},$$

$$\breve{L}_2 = \begin{bmatrix} -0.7344 & 0.0843 & -0.1176 \end{bmatrix},$$

$$\breve{L}_\infty = \begin{bmatrix} -0.9134 & 0.0804 & -0.0453 \end{bmatrix}.$$

***Case 3***: $\gamma^2 = 6$.

We now deal with the problem ($P2$). Solving the optimization problem (3.97), we obtain the minimum $H_2$ performance $\beta_{\min} = 3.6776$, and

$$\check{A} = \begin{bmatrix} -0.2006 & -0.0161 & 0.0674 \\ 0.1929 & 0.2332 & -0.0077 \\ -0.6979 & 1.9733 & 0.0835 \end{bmatrix}, \quad \check{K} = \begin{bmatrix} 0.6159 \\ -0.3776 \\ -0.7593 \end{bmatrix},$$

$$\check{L}_2 = \begin{bmatrix} -0.6454 & 0.0977 & -0.1511 \end{bmatrix},$$

$$\check{L}_\infty = \begin{bmatrix} -0.8115 & 0.0700 & -0.0551 \end{bmatrix}.$$

The results show that the designed system can satisfy the $H_2$ filtering performance and the $H_\infty$ disturbance rejection performance simultaneously. In Case 2, in order to achieve a better disturbance rejection performance, the optimization algorithm ($P1$) is employed to obtain the optimal solution. Similarly, to get a better $H_2$ filtering performance, the optimization algorithm ($P2$) is applied to obtain the optimal solution in Case 3.

**Remark 3.5** *Within the LMI framework developed in this chapter, we can show that there are some trade offs that can be used for satisfying specific performance requirements. For example, the $H_\infty$ performance will be improved if the $H_2$ performance constraints become more relaxed (larger). Also, if the value of the $H_\infty$ performance constraint is allowed to be increased, then the $H_2$ performance can be further reduced. Hence, the proposed approach allows much flexibility in making compromises between the $H_2$ performance and the $H_\infty$ performance, while the essential multiple objectives can all be met simultaneously.*

## 3.5 Summary

A robust $H_2/H_\infty$ filter has been designed in this chapter for a class of uncertain discrete time nonlinear stochastic systems. A key technology is used to convert the matrix trace terms into linear matrix inequalities and to eliminate the uncertainty in the matrix inequalities. The filter is obtained under a unified flexible LMI framework. Sufficient conditions for the solvability of the $H_2/H_\infty$ filtering problem are given in terms of a set of feasible LMIs. Two types of optimization problems are proposed by either optimizing the $H_2$ performance or the $H_\infty$ performance. Our method can be extended to robust $H_2/H_\infty$ output control.

# 4

# Robust Variance-Constrained Filtering with Missing Measurements

For several decades, filtering techniques have been playing an important role in many branches of signal processing such as target tracking. A number of filtering approaches, including Kalman filtering, $H_\infty$ filtering, and robust filtering, have been proposed in the literature, most of which are under the assumption that the measurements always contain true signals corrupted by the noises. However, in real-world applications, the measurements may contain missing measurements (or incomplete observations) due to various reasons, such as high maneuverability of the tracked targets, sensor temporal failures or network congestion.

On the other hand, it is quite common in practical engineering that, for a class of filtering problems such as the tracking of a maneuvering target, the performance objectives are naturally described as the upper bounds on the error variances of estimation. This gives rise to the so-called variance-constrained filtering problem, which has been motivated from the well-known covariance control theory. Note that the variance-constrained filtering or control theory has been extensively investigated in a variety of practical situations. The key point of the covariance control theory is that the specified variance constraints may not be minimal, but should meet certain given engineering requirements. Therefore, after assigning to the filtering error dynamics a specified variance upper bound, there remains much freedom, which can be used to attempt to directly achieve other desired performance requirements, but the traditional optimal (robust) Kalman filtering methods may not have such an advantage.

Because of its clear engineering insights, in the past few years the filtering problem with missing measurements has received much attention, and many results have been reported in the literature, where the missing data were usually modeled as a binary switching sequence specified by a conditional probability distribution. However, it

*Variance-Constrained Multi-Objective Stochastic Control and Filtering*, First Edition.
Lifeng Ma, Zidong Wang and Yuming Bo.
© 2015 John Wiley & Sons, Ltd. Published 2015 by John Wiley & Sons, Ltd.

should be pointed out that almost all the results concerning variance-constrained filtering have been concerned with linear systems only, and the corresponding literature for nonlinear systems has been very few, due primarily to the difficulty in analyzing the steady-state estimation error covariance for nonlinear systems. Up to now, to the best of the authors' knowledge, *in the presence of probabilistic measurements missing*, the filtering problem for *nonlinear* stochastic systems *with error variance constraints* has not yet been investigated and therefore remains open and challenging.

Motivated by the above discussion, this chapter is concerned with the robust filtering problem for a class of nonlinear stochastic systems with missing measurements and parameter uncertainties. We model the missing measurements by a Bernoulli distributed white sequence with a known conditional probability distribution. The nonlinearities considered in this chapter are expressed by statistical means and could cover several well-studied nonlinearities as special cases. Based on this model, the robust variance-constrained filtering problem is addressed for a class of nonlinear stochastic systems with missing measurements. We aim at designing a filter such that, for all parameter uncertainties and possible measurements missing, (1) the filtering error system is exponentially mean-square stable and (2) the variance of the estimation error for the individual state is not more than prescribed upper bound. It is shown that the solvability of the addressed filtering problem can be expressed as the feasibility of a certain set of LMIs, and the explicit expression of the desired robust filters is also derived. A simulation numerical example is provided to illustrate the usefulness of the proposed design approach.

The main contributions of this chapter are summarized as follows: (1) for the first time, the filtering problem has been solved for the stochastic systems with both stochastic nonlinearities and missing measurements phenomenon; (2) meanwhile, the important error variance performance has been taken into consideration for such kinds of stochastic nonlinear systems with missing measurements.

The rest of this chapter is arranged as follows: Section 4.1 formulates the robust variance-constrained filter design problem for uncertain nonlinear stochastic discrete-time systems. In Section 4.2, the exponential mean-square stability of the filtering error system and the individual variance constraints of the estimation error are analyzed separately. The solution of the robust filter design problem is given in terms of a certain set of LMIs in Section 4.3. In Section 4.4, an illustrative numerical example is provided to show the effectiveness and usefulness of the proposed approach. Section 4.5 gives our summary.

## 4.1 Problem Formulation

Consider the following uncertain discrete-time nonlinear stochastic system:

$$x(k + 1) = (A + \Delta A)x(k) + f(x(k)) + B\omega(k), \tag{4.1}$$

with the measurement equation

$$y(k) = \gamma(k) (Cx(k) + g(x(k))) + D\omega(k), \tag{4.2}$$

where $x(k) \in \mathbb{R}^n$ is the state, $y(k) \in \mathbb{R}^m$ is the measured output, and $A, B, C, D$ are known constant matrices with appropriate dimensions. $\omega(k) \in \mathbb{R}^n$ is a zero mean Gaussian white noise sequence with covariance $W > 0$. $\Delta A$ is a real-valued perturbation matrix that represents parametric uncertainty, being of the following form:

$$\Delta A = HFE, \qquad FF^{\mathrm{T}} \leq I, \tag{4.3}$$

where $H$ and $E$ are known constant matrices with appropriate dimensions. The uncertainties in $\Delta A$ are said to be admissible if (4.3) holds. The stochastic variable $\gamma(k) \in \mathbb{R}$ is a Bernoulli distributed white sequence taking values on 0 and 1 with

$$\mathrm{Prob}\{\gamma(k) = 1\} = \mathbb{E}\{\gamma(k)\} \triangleq \bar{\gamma}, \tag{4.4}$$

where $\bar{\gamma}$ is a known positive constant and $\gamma(k) \in \mathbb{R}$ is assumed to be independent of both $w(k)$ and the system initial state $x_0$. Therefore, we have

$$\mathrm{Prob}\{\gamma(k) = 0\} = 1 - \bar{\gamma},$$

$$\sigma_\gamma^2 \triangleq \mathbb{E}\{(\gamma(k) - \bar{\gamma})^2\} = (1 - \bar{\gamma})\bar{\gamma}. \tag{4.5}$$

**Remark 4.1** *Notice that the parameter uncertainty only enters into the system matrix A. However, it is worth pointing out that, within the same framework to be developed, we can also consider the case when the uncertainties exist in the output equation. The reason why we discuss the system (4.1) to (4.2) is to make our theory more understandable and to avoid unnecessarily complicated notations.*

The nonlinear stochastic functions $f(x(k))$ and $g(x(k))$ are assumed to have the following first moments for all $x(k)$:

$$\mathbb{E}\left\{ \begin{bmatrix} f(x(k)) \\ g(x(k)) \end{bmatrix} \Big| x(k) \right\} = 0, \tag{4.6}$$

with the covariance given by

$$\mathbb{E}\left\{ \begin{bmatrix} f(x(k)) \\ g(x(k)) \end{bmatrix} [f^{\mathrm{T}}(x(j)) \ g^{\mathrm{T}}(x(j))] \Big| x(k) \right\} = 0, \qquad k \neq j, \tag{4.7}$$

and

$$\mathbb{E}\left\{ \begin{bmatrix} f(x(k)) \\ g(x(k)) \end{bmatrix} [f^{\mathrm{T}}(x(k)) \ g^{\mathrm{T}}(x(k))] \Big| x(k) \right\} = \sum_{i=1}^{q} \Pi_i x^{\mathrm{T}}(k) \Gamma_i x(k), \tag{4.8}$$

where $\Pi_i$ and $\Gamma_i$ $(i = 1, 2, \ldots, q)$ are known positive definite matrices with the following structures:

$$\Pi_i = \begin{bmatrix} \pi_{1i} \\ \pi_{2i} \end{bmatrix} \begin{bmatrix} \pi_{1i} \\ \pi_{2i} \end{bmatrix}^{\mathrm{T}}, \qquad \Gamma_i = \theta_i \theta_i^{\mathrm{T}},$$

with $\pi_{1i} \in \mathbb{R}^n$, $\pi_{2i} \in \mathbb{R}^m$, and $\theta_i \in \mathbb{R}^n (i = 1, 2, \ldots, q)$ being known column vectors of appropriate dimensions.

Introduce now a new stochastic sequence

$$\tilde{\gamma}(k) \triangleq \gamma(k) - \bar{\gamma}. \tag{4.9}$$

It is easy to see that $\tilde{\gamma}(k)$ is a scalar zero mean stochastic sequence with variance

$$\sigma_{\tilde{\gamma}}^2 = (1 - \bar{\gamma})\bar{\gamma}. \tag{4.10}$$

Consider the following filter for the system (4.1):

$$\hat{x}(k + 1) = G\hat{x}(k) + K(y(k) - \bar{\gamma} C\hat{x}(k)), \tag{4.11}$$

where $\hat{x}(k)$ stands for the state estimate, and $G$ and $K$ are the filter parameters to be scheduled.

Define the estimation error as

$$e(k) = x(k) - \hat{x}(k), \tag{4.12}$$

and the steady-state estimation error covariance as

$$X_{ee} \triangleq \lim_{k \to \infty} \mathbb{E}\{e(k)e^{\mathrm{T}}(k)\}. \tag{4.13}$$

Then, we obtain the following augmented system

$$z(k + 1) = \tilde{A}z(k) + \tilde{B}h(x(k)) + \tilde{D}\omega(k), \tag{4.14}$$

where

$$z(k) = \begin{bmatrix} x(k) \\ e(k) \end{bmatrix},$$

$$\tilde{A} = \begin{bmatrix} A + \Delta A & 0 \\ A + \Delta A - G - \tilde{\gamma}(k)KC & G - \bar{\gamma}KC \end{bmatrix},$$

$$\tilde{B} = \begin{bmatrix} I & 0 \\ I & -\gamma(k)K \end{bmatrix},$$

$$\tilde{D} = \begin{bmatrix} B \\ B - KD \end{bmatrix},$$

$$h(x(k)) = \begin{bmatrix} f(x(k)) \\ g(x(k)) \end{bmatrix}.$$

**Remark 4.2** *It is mentionable that there is a stochastic variable $\tilde{\gamma}(k)$ involved in $\tilde{A}$, which reflects the characteristic of the missing measurement for the addressed filtering problem, and the augmented system (4.14) is, therefore, essentially a stochastic parameter system. Note that a more general stochastic parameter system is the so-called Markovian jump system, where the stochastic variable $\tilde{\gamma}(k)$ may be an ergodic finite state Markov chain. Markovian jump systems have received much research attention in the past decade; see Refs [135, 136], and the references therein. The robust Kalman filtering problem has recently been studied in Ref. [137] for linear jumping systems by solving two sets of coupled algebraic Riccati equations. Nevertheless, focusing on the particular data missing problem, we let the stochastic variable $\tilde{\gamma}(k)$ be a Bernoulli sequence and, therefore, we are able to obtain more practical solutions. For example, the algorithm developed in this note will not involve solving coupled matrix equations/inequalities. Another motivation is that, instead of the minimum variance filtering, we consider the constrained variance filtering problem here, which should give more design freedom. This will be demonstrated later.*

Before stating our design objective, we introduce the following stability concept for the system (4.14).

**Definition 4.1.1** *The system (4.14) is said to be exponentially mean-square stable if, with $\omega(k) = 0$, there exist constants $\zeta \geq 1$ and $\tau \in (0, 1)$ such that*

$$\mathbb{E}\{\|z(k)\|^2\} \leq \zeta\tau^k\mathbb{E}\{\|z_0\|^2\}, \qquad \forall z_0 \in \mathbb{R}^{2n}, \quad k \in \mathbb{I}^+. \tag{4.15}$$

*for all admissible uncertainties and possible missing measurements.*

In this chapter, our objective is to design the filter (4.11) for the system (4.1) such that, for all admissible uncertainties and possible missing measurements, the following two objectives are satisfied simultaneously:

(R1) The augmented system (4.14) is exponentially mean-square stable.
(R2) The steady-state error variance $X_{ee}$ satisfies

$$X_{ee}^i \leq \sigma_i^2, \qquad i = 1, 2, \dots, n, \tag{4.16}$$

where $X_{ee}^i$ stands for the steady-state variance of the $i$th error state and $\sigma_i^2$ ($i = 1, 2, \dots, n$) denotes the pre-specified steady-state estimation error variance constraint on the $i$th state.

**Remark 4.3** *In engineering practice, the variance upper bounds that represent the control or estimation precision of the system state should be specified according to the actual requirements before the system design. For instance, in the problem of tracking maneuvering targets, the position and velocity of the target are measured* *in every sampling instant. However, due to the high maneuver of the tracked target, it is neither possible nor necessary to track the target in a precise way. Instead, an acceptable compromise is to keep the target within a given "window" as frequently as possible, and such a requirement can be expressed as upper bounds on the estimation error variance. In this sense, before the actual system design, we could specify the error variance upper bound according to the length and width of the required "window".*

## 4.2   Stability and Variance Analysis

Before giving our derivation, we first introduce some useful lemmas.

**Lemma 4.2.1** *[59] Let $V(z(k)) = z^T(k)Pz(k)$ be a Lyapunov functional where $P > 0$. If there exist real scalars $\lambda$, $\mu > 0$, $\nu > 0$, and $0 < \psi < 1$ such that both*

$$\mu \|z(k)\|^2 \leq V(z(k)) \leq \nu \|z(k)\|^2 \tag{4.17}$$

*and*

$$\mathbb{E}\{V(z(k+1))|z(k)\} - V(z(k)) \leq \lambda - \psi V(z(k)) \tag{4.18}$$

*hold, then the process $z(k)$ satisfies*

$$\mathbb{E}\left\{\|z(k)\|^2\right\} \leq \frac{\nu}{\mu}\|z_0\|^2(1-\psi)^k + \frac{\lambda}{\mu\psi}. \tag{4.19}$$

For simplicity of notation, before giving the following lemma, we denote

$$\hat{A} = \begin{bmatrix} A & 0 \\ A-G & G-\bar{\gamma}KC \end{bmatrix}, \qquad \Delta\hat{A} = \begin{bmatrix} \Delta A & 0 \\ \Delta A & 0 \end{bmatrix},$$

$$J = \begin{bmatrix} 0 & 0 \\ \sigma_{\bar{\gamma}}KC & 0 \end{bmatrix}, \qquad \tilde{\Gamma}_i = \begin{bmatrix} \Gamma_i & 0 \\ 0 & 0 \end{bmatrix},$$

$$\tilde{\Pi}_i = \begin{bmatrix} \pi_{1i}\pi_{1i}^T & \pi_{1i}\pi_{1i}^T - \bar{\gamma}\pi_{1i}\pi_{2i}^T K^T \\ * & \pi_{1i}\pi_{1i}^T - \bar{\gamma}(\pi_{1i}\pi_{2i}^T K^T + K\pi_{2i}\pi_{1i}^T) + (\bar{\gamma}^2 + \sigma_{\gamma}^2)K\pi_{2i}\pi_{2i}^T K^T \end{bmatrix}.$$

**Lemma 4.2.2** *Given the filter parameters $G$ and $K$. The following statements are equivalent.*

*1.*

$$\rho\left\{(\hat{A} + \Delta\hat{A})^T \otimes (\hat{A} + \Delta\hat{A})^T + J^T \otimes J^T + \sum_{i=1}^{q} \text{st}(\tilde{\Gamma}_i)\text{st}^T(\tilde{\Pi}_i)\right\} < 1 \tag{4.20}$$

*or*

$$\rho\left\{(\hat{A}+\Delta\hat{A})\otimes(\hat{A}+\Delta\hat{A})+J\otimes J+\sum_{i=1}^{q}\text{st}(\tilde{\Pi}_i)\text{st}^{\mathrm{T}}(\tilde{\Gamma}_i)\right\}<1. \qquad (4.21)$$

2. *There exists a positive definite matrix* $P > 0$ *such that*

$$(\hat{A}+\Delta\hat{A})^{\mathrm{T}}P(\hat{A}+\Delta\hat{A})+J^{\mathrm{T}}PJ-P+\sum_{i=1}^{q}\tilde{\Gamma}_i\text{tr}[P\tilde{\Pi}_i]<0. \qquad (4.22)$$

3. *There exists a positive definite matrix* $Q > 0$ *such that*

$$(\hat{A}+\Delta\hat{A})Q(\hat{A}+\Delta\hat{A})^{\mathrm{T}}+JQJ^{\mathrm{T}}-Q+\sum_{i=1}^{q}\tilde{\Pi}_i\text{tr}[Q\tilde{\Gamma}_i]<0. \qquad (4.23)$$

4. *The system* (4.14) *is exponentially mean-square stable.*

*Proof.* Firstly, it can be noticed that the main difference between this lemma and Theorem 1 of Ref. [133] is that the state matrix of the system (4.14) in this chapter contains stochastic variables $\gamma(k)$ and $\tilde{\gamma}(k)$, which result from the possible measurements missing in the process of output sampling. Hence, we just need to prove the relationship "2 ⇒ 4" in order to demonstrate how we tackle the matrix involving stochastic variables in the derivation; the rest of this lemma can then be easily proved as in Theorem 1 of Ref. [133] using the techniques shown below.

2 ⇒ 4: Define a Lyapunov functional $V(z(k)) = z^{\mathrm{T}}(k)Pz(k)$ where $P > 0$ is the solution to (4.22). Then,

$$\mathbb{E}\{V(z(k+1))|z(k)\} - V(z(k))$$
$$= \mathbb{E}\left\{\left(\tilde{A}z(k)+\tilde{B}h(x(k))\right)^{\mathrm{T}}P(\tilde{A}z(k)+\tilde{B}h(x(k)))|z(k)\right\} - z^{\mathrm{T}}(k)Pz(k)$$
$$= \mathbb{E}\{(z^{\mathrm{T}}(k)\tilde{A}^{\mathrm{T}}P\tilde{A}z(k))|z(k)\}$$
$$\quad + \mathbb{E}\left\{\left(h^{\mathrm{T}}(x(k))\tilde{B}^{\mathrm{T}}P\tilde{B}h(x(k))\right)|z(k)\right\} - z^{\mathrm{T}}(k)Pz(k). \qquad (4.24)$$

Using the statistics of $\omega(k)$, $\gamma(k)$, and $\tilde{\gamma}(k)$, we obtain

$$\mathbb{E}\{(z^{\mathrm{T}}(k)\tilde{A}^{\mathrm{T}}P\tilde{A}z(k))|z(k)\}$$
$$= \mathbb{E}\left\{z^{\mathrm{T}}(k)\left((\hat{A}+\Delta\hat{A})+\tilde{\gamma}(k)\begin{bmatrix}0&0\\-KC&0\end{bmatrix}\right)^{\mathrm{T}}P\right.$$
$$\left.\times\left((\hat{A}+\Delta\hat{A})+\tilde{\gamma}(k)\begin{bmatrix}0&0\\-KC&0\end{bmatrix}\right)z(k)\right\}$$
$$= z^{\mathrm{T}}(k)((\hat{A}+\Delta\hat{A})^{\mathrm{T}}P(\hat{A}+\Delta\hat{A})+J^{\mathrm{T}}PJ)z(k) \qquad (4.25)$$

and

$$
\mathbb{E}\left\{\left(h^{\mathrm{T}}(x(k))\tilde{B}^{\mathrm{T}}P\tilde{B}h(x(k))\right)|z(k)\right\}
$$

$$
=\mathbb{E}\left\{\mathrm{tr}\left(\tilde{B}^{\mathrm{T}}P\tilde{B}h(x(k))h^{\mathrm{T}}(x(k))\right)|z(k)\right\}
$$

$$
=\sum_{i=1}^{q}\mathbb{E}\left\{\mathrm{tr}(\tilde{B}^{\mathrm{T}}P\tilde{B}\Pi_i)z^{\mathrm{T}}(k)\tilde{\Gamma}_i z(k)|z(k)\right\} \tag{4.26}
$$

$$
=z^{\mathrm{T}}(k)\sum_{i=1}^{q}\tilde{\Gamma}_i\mathbb{E}\left\{\mathrm{tr}(P\tilde{B}\Pi_i\tilde{B}^{\mathrm{T}})\right\}z(k).
$$

Rewrite $\tilde{B}$ as

$$
\tilde{B}=\begin{bmatrix} I & 0 \\ I & -\bar{\gamma}K \end{bmatrix}+(\bar{\gamma}-\gamma(k))\begin{bmatrix} 0 & 0 \\ 0 & K \end{bmatrix}. \tag{4.27}
$$

Then, we have

$$
\tilde{B}\Pi_i\tilde{B}^{\mathrm{T}}=\tilde{B}\begin{bmatrix} \pi_{1i} \\ \pi_{2i} \end{bmatrix}\begin{bmatrix} \pi_{1i} \\ \pi_{2i} \end{bmatrix}^{\mathrm{T}}\tilde{B}^{\mathrm{T}}
$$

$$
=\left(\begin{bmatrix} \pi_{1i} \\ \pi_{1i}-\bar{\gamma}K\pi_{2i} \end{bmatrix}+(\bar{\gamma}-\gamma(k))\begin{bmatrix} 0 \\ K\pi_{2i} \end{bmatrix}\right) \tag{4.28}
$$

$$
\times\left(\begin{bmatrix} \pi_{1i} \\ \pi_{1i}-\bar{\gamma}K\pi_{2i} \end{bmatrix}+(\bar{\gamma}-\gamma(k))\begin{bmatrix} 0 \\ K\pi_{2i} \end{bmatrix}\right)^{\mathrm{T}}.
$$

Thus,

$$
\mathbb{E}\{\mathrm{tr}(P\tilde{B}\Pi_i\tilde{B}^{\mathrm{T}})\}
$$

$$
=\mathbb{E}\left\{\mathrm{tr}\left(P\begin{bmatrix} \pi_{1i} \\ \pi_{1i}-\bar{\gamma}K\pi_{2i} \end{bmatrix}\begin{bmatrix} \pi_{1i} \\ \pi_{1i}-\bar{\gamma}K\pi_{2i} \end{bmatrix}^{\mathrm{T}}\right)\right.
$$

$$
\left.+(\bar{\gamma}-\gamma(k))^2\mathrm{tr}\left(P\begin{bmatrix} 0 \\ K\pi_{2i} \end{bmatrix}\begin{bmatrix} 0 \\ K\pi_{2i} \end{bmatrix}^{\mathrm{T}}\right)\right\}
$$

$$
=\mathrm{tr}\left(P\begin{bmatrix} \pi_{1i}\pi_{1i}^{\mathrm{T}} & \pi_{1i}\pi_{1i}^{\mathrm{T}}-\bar{\gamma}\pi_{1i}\pi_{2i}^{\mathrm{T}}K^{\mathrm{T}} \\ * & \pi_{1i}\pi_{1i}^{\mathrm{T}}-\bar{\gamma}(\pi_{1i}\pi_{2i}^{\mathrm{T}}K^{\mathrm{T}}+K\pi_{2i}\pi_{1i}^{\mathrm{T}})+\bar{\gamma}^2K\pi_{2i}\pi_{2i}^{\mathrm{T}}K^{\mathrm{T}} \end{bmatrix}\right) \tag{4.29}
$$

$$
+\sigma_\gamma^2\mathrm{tr}\left(P\begin{bmatrix} 0 & 0 \\ 0 & K\pi_{2i}\pi_{2i}^{\mathrm{T}}K^{\mathrm{T}} \end{bmatrix}\right)
$$

$$
=\mathrm{tr}(P\tilde{\Pi}_i).
$$

Hence,

$$\mathbb{E}\{V(z(k+1))|z(k)\} - V(z(k))$$

$$= z^{\mathrm{T}}(k)\left((\hat{A} + \Delta\hat{A})^{\mathrm{T}}P(\hat{A} + \Delta\hat{A}) + J^{\mathrm{T}}PJ - P + \sum_{i=1}^{q}\tilde{\Gamma}_i\mathrm{tr}[P\tilde{\Pi}_i]\right)z(k). \tag{4.30}$$

It follows from (4.22) that there always exists a sufficiently small scalar $\eta$ satisfying $0 < \eta < \lambda_{\max}(P)$ such that

$$(\hat{A} + \Delta\hat{A})^{\mathrm{T}}P(\hat{A} + \Delta\hat{A}) + J^{\mathrm{T}}PJ - P + \sum_{i=1}^{q}\tilde{\Gamma}_i\mathrm{tr}[P\tilde{\Pi}_i] < -\eta I, \tag{4.31}$$

which means

$$\mathbb{E}\{V(z(k+1))|z(k)\} - V(z(k)) \le -\eta z^{\mathrm{T}}(k)z(k) \le -\frac{\eta}{\lambda_{\max}(P)}V(z(k)). \tag{4.32}$$

Finally, the exponential mean-square stability of (4.14) can be immediately obtained from Lemma 4.2.1. $\qquad\square$

**Lemma 4.2.3** *Given the filter parameters G and K. If the system (4.14) is exponentially mean-square stable and there exists a symmetric matrix Y satisfying*

$$(\hat{A} + \Delta\hat{A})Y(\hat{A} + \Delta\hat{A})^{\mathrm{T}} + JYJ^{\mathrm{T}} - Y + \sum_{i=1}^{q}\tilde{\Pi}_i\mathrm{tr}[Y\tilde{\Gamma}_i] < 0, \tag{4.33}$$

*then Y ≥ 0.*

*Proof.* Lemma 4.2.3 can be easily proved by the Lyapunov method together with Lemma 4.2.2; hence the proof is omitted. $\qquad\square$

Now let us proceed to deal with the error variance constraints. Defining the following second moment for system (4.14) by

$$Q(k) \triangleq \mathbb{E}\{z(k)z^{\mathrm{T}}(k)\}$$

$$= \mathbb{E}\left\{\begin{bmatrix} x(k) \\ e(k) \end{bmatrix}\begin{bmatrix} x(k) \\ e(k) \end{bmatrix}^{\mathrm{T}}\right\}$$

$$\triangleq \begin{bmatrix} X_{xxk} & X_{xek} \\ X_{xek}^{\mathrm{T}} & X_{eek} \end{bmatrix}, \tag{4.34}$$

the evolution of $Q(k)$ can be derived from the system (4.14) as follows:

$$Q(k+1) = (\hat{A} + \Delta\hat{A})Q(k)(\hat{A} + \Delta\hat{A})^{\mathrm{T}} + JQ(k)J^{\mathrm{T}}$$
$$+ \sum_{i=1}^{q} \tilde{\Pi}_i \mathrm{tr}[Q(k)\tilde{\Gamma}_i] + \tilde{D}W\tilde{D}^{\mathrm{T}}. \tag{4.35}$$

Rewrite (4.35) in the form of stack matrices

$$\mathrm{st}(Q(k+1)) = \left[ (\hat{A} + \Delta\hat{A}) \otimes (\hat{A} + \Delta\hat{A}) + J \otimes J \right.$$
$$\left. + \sum_{i=1}^{q} \mathrm{st}(\tilde{\Pi}_i)\mathrm{st}^{\mathrm{T}}(\tilde{\Gamma}_i) \right] \mathrm{st}(Q(k)) + \mathrm{st}(\tilde{D}W\tilde{D}^{\mathrm{T}}). \tag{4.36}$$

If the system (4.14) is exponentially mean-square stable, it then follows from Lemma 4.2.2 that (4.21) holds and hence in the steady-state

$$\hat{Q} \triangleq \lim_{k\to\infty} Q(k) = \begin{bmatrix} X_{xx} & X_{xe} \\ X_{xe}^{\mathrm{T}} & X_{ee} \end{bmatrix} \tag{4.37}$$

exists and satisfies

$$(\hat{A} + \Delta\hat{A})\hat{Q}(\hat{A} + \Delta\hat{A})^{\mathrm{T}} + J\hat{Q}J^{\mathrm{T}} - \hat{Q} + \sum_{i=1}^{q} \tilde{\Pi}_i \mathrm{tr}[\hat{Q}\tilde{\Gamma}_i] + \tilde{D}W\tilde{D}^{\mathrm{T}} = 0. \tag{4.38}$$

Based on the results we have obtained so far concerning the exponential mean-square stability as well as the steady-state variance, we are now ready to cope with the addressed multi-objective filter design problem.

## 4.3   Robust Filter Design

In this section, an LMI method is proposed to design the robust filter with constrained variance for the uncertain nonlinear stochastic system with missing measurements. To start with, a theorem is given that combines exponential mean-square stability and the error variance upper bound constraints.

**Theorem 4.3.1** *Given the filter parameters G and K. If there exits a positive definite matrix Q > 0 such that*

$$(\hat{A} + \Delta\hat{A})Q(\hat{A} + \Delta\hat{A})^{\mathrm{T}} + JQJ^{\mathrm{T}} - Q + \sum_{i=1}^{q} \tilde{\Pi}_i \mathrm{tr}[Q\tilde{\Gamma}_i] + \tilde{D}W\tilde{D}^{\mathrm{T}} < 0 \tag{4.39}$$

*holds, then the system (4.14) is exponentially mean-square stable and the steady-state state covariance satisfies $\hat{Q} \leq Q$.*

*Proof.* It follows from (4.39) that

$$(\hat{A} + \Delta\hat{A})Q(\hat{A} + \Delta\hat{A})^{\mathrm{T}} + JQJ^{\mathrm{T}} - Q + \sum_{i=1}^{q} \tilde{\Pi}_i \mathrm{tr}[Q\tilde{\Gamma}_i] < -\tilde{D}W\tilde{D}^{\mathrm{T}} < 0, \qquad (4.40)$$

which indicates from Lemma 4.2.2 that the system (4.14) is exponentially mean-square stable. Therefore, in the steady-state, the state covariance of (4.14) $\hat{Q}$ exists and satisfies (4.38). Subtracting (4.38) from (4.39), we obtain

$$(\hat{A} + \Delta\hat{A})\tilde{Q}(\hat{A} + \Delta\hat{A})^{\mathrm{T}} + J\tilde{Q}J^{\mathrm{T}} - \tilde{Q} + \sum_{i=1}^{q} \tilde{\Pi}_i \mathrm{tr}[\tilde{Q}\tilde{\Gamma}_i] < 0, \qquad (4.41)$$

where $\tilde{Q} \triangleq Q - \hat{Q}$. From Lemma 4.2.3, we know that $Q - \hat{Q} \geq 0$ and the proof is complete. $\qquad \square$

The following theorem provides an LMI approach to the addressed filter design problem for the uncertain discrete-time nonlinear stochastic system (4.1).

**Theorem 4.3.2** *Given $\sigma_i^2 > 0$ ($i = 1, 2, \ldots, n$). If there exist positive definite matrices $R > 0$, $S > 0$, real matrices $M$, $N$, a positive scalar $\varepsilon$, and positive scalars $\alpha_i > 0$ ($i = 1, 2, \ldots, q$) then for all admissible parameter uncertainties and possible measurement missing, the following set of LMIs:*

$$\begin{bmatrix} -\alpha_i & \alpha_i\theta_i^{\mathrm{T}} \\ \alpha_i\theta_i & -R \end{bmatrix} < 0, \qquad (4.42)$$

$$\begin{bmatrix} \Psi_{11} & \Psi_{12} \\ \Psi_{12}^{\mathrm{T}} & \Psi_{22} \end{bmatrix} < 0, \qquad (4.43)$$

$$\hat{X} - S \leq 0, \qquad (4.44)$$

*where*

$$\Psi_{11} = \begin{bmatrix} -R & 0 & RA & 0 & 0 & 0 \\ * & -S & SA - M & M - \bar{\gamma}NC & \sigma_{\bar{\gamma}}NC & 0 \\ * & * & -R & 0 & 0 & 0 \\ * & * & * & -S & 0 & 0 \\ * & * & * & * & -R & 0 \\ * & * & * & * & * & -S \end{bmatrix},$$

$$\Psi_{12} = \begin{bmatrix} \Psi_{121} & 0 & RB & RH & 0 \\ \Psi_{122} & \Psi_{123} & SB - ND & SH & 0 \\ 0 & 0 & 0 & 0 & \varepsilon E^{\mathrm{T}} \\ 0 & 0 & 0 & 0 & 0 \\ 0 & 0 & 0 & 0 & 0 \\ 0 & 0 & 0 & 0 & 0 \end{bmatrix},$$

$$\Psi_{121} = \begin{bmatrix} R\pi_{11} & R\pi_{12} & \cdots & R\pi_{1q} \end{bmatrix},$$

$$\Psi_{122} = \begin{bmatrix} S\pi_{11} - \bar{\gamma}N\pi_{21} & S\pi_{12} - \bar{\gamma}N\pi_{22} & \cdots & S\pi_{1q} - \bar{\gamma}N\pi_{2q} \end{bmatrix},$$

$$\Psi_{123} = \begin{bmatrix} \sigma_{\gamma}N\pi_{21} & \sigma_{\gamma}N\pi_{22} & \cdots & \sigma_{\gamma}N\pi_{2q} \end{bmatrix},$$

$$\Psi_{22} = \text{diag}\{-\alpha_1 I, \cdots, -\alpha_q I, -\alpha_1 I, \cdots, -\alpha_q I, -W^{-1}, -\varepsilon I, -\varepsilon I\},$$

$$\hat{X} = \text{diag}\{(\sigma_1^2)^{-1}, (\sigma_2^2)^{-1}, \cdots, (\sigma_n^2)^{-1}\},$$

*is feasible, then there exists a filter of the form (4.11) such that the requirements (R1) and (R2) are simultaneously satisfied. Moreover, the desired filter can be determined by*

$$G = S^{-1}M,$$
$$K = S^{-1}N. \tag{4.45}$$

*Proof.* Assume that the matrix $Q$ has a block diagonal form as follows:

$$Q = \begin{bmatrix} R & 0 \\ 0 & S \end{bmatrix}^{-1} > 0, \tag{4.46}$$

where $R > 0$ and $S > 0$ are both $n \times n$ real-valued matrices. Now, we define new variables $\alpha_i > 0$ $(i = 1, 2, \ldots, q)$ satisfying

$$\alpha_i < (\text{tr}[Q\tilde{\Gamma}_i])^{-1}. \tag{4.47}$$

Letting

$$\breve{\theta}_i = \begin{bmatrix} \theta_i \\ 0 \end{bmatrix} \in \mathbb{R}^{2n}, \tag{4.48}$$

we have $\tilde{\Gamma}_i = \breve{\theta}_i \breve{\theta}_i^{\mathrm{T}}$.

Using the property of matrix trace and the Schur Complement Lemma, we have

$$\text{tr}[Q\tilde{\Gamma}_i] < \alpha_i^{-1} \iff \begin{bmatrix} -\alpha_i & \alpha_i \breve{\theta}_i^{\mathrm{T}} \\ \alpha_i \breve{\theta}_i & -Q^{-1} \end{bmatrix} < 0. \tag{4.49}$$

Then, after transformation, (4.49) is equivalent to (4.42).

Next, we prove that (4.43) is equivalent to

$$(\hat{A} + \Delta\hat{A})Q(\hat{A} + \Delta\hat{A})^{\mathrm{T}} + JQJ^{\mathrm{T}} - Q + \sum_{i=1}^{q} \tilde{\Pi}_i \alpha_i^{-1} + \tilde{D}W\tilde{D}^{\mathrm{T}} < 0. \tag{4.50}$$

By the Schur Complement Lemma, (4.50) is equivalent to

$$\begin{bmatrix} -Q + \sum_{i=1}^{q} \tilde{\Pi}_i \alpha_i^{-1} & \hat{A} + \Delta\hat{A} & J & \tilde{D} \\ \hat{A}^{\mathrm{T}} + \Delta\hat{A}^{\mathrm{T}} & -Q^{-1} & 0 & 0 \\ J^{\mathrm{T}} & 0 & -Q^{-1} & 0 \\ \tilde{D}^{\mathrm{T}} & 0 & 0 & -W^{-1} \end{bmatrix} < 0. \tag{4.51}$$

Performing the congruence transformation by $\mathrm{diag}\{Q^{-1},I,I,I\}$, we can see that (4.51) is equivalent to

$$
\begin{bmatrix}
-Q^{-1} + Q^{-1}\left(\displaystyle\sum_{i=1}^{q}\tilde{\Pi}_i\alpha_i^{-1}\right)Q^{-1} & Q^{-1}(\hat{A} + \Delta\hat{A}) & Q^{-1}J & Q^{-1}\tilde{D} \\
(\hat{A}^{\mathrm{T}} + \Delta\hat{A}^{\mathrm{T}})Q^{-1} & -Q^{-1} & 0 & 0 \\
J^{\mathrm{T}}Q^{-1} & 0 & -Q^{-1} & 0 \\
\tilde{D}^{\mathrm{T}}Q^{-1} & 0 & 0 & -W^{-1}
\end{bmatrix} < 0. \qquad (4.52)
$$

Rewrite $\tilde{\Pi}_i$ in the following form:

$$
\tilde{\Pi}_i = \begin{bmatrix} \pi_{1i} \\ \pi_{1i} - \bar{\gamma}K\pi_{2i} \end{bmatrix}\begin{bmatrix} \pi_{1i} \\ \pi_{1i} - \bar{\gamma}K\pi_{2i} \end{bmatrix}^{\mathrm{T}} + \begin{bmatrix} 0 \\ \sigma_\gamma K\pi_{2i} \end{bmatrix}\begin{bmatrix} 0 \\ \sigma_\gamma K\pi_{2i} \end{bmatrix}^{\mathrm{T}}. \qquad (4.53)
$$

Using the Schur Complement Lemma again, after some tedious calculations, we obtain that (4.52) is equivalent to the following matrix inequality:

$$
\Upsilon = \begin{bmatrix} \Upsilon_{11} & \Upsilon_{12} \\ * & \Upsilon_{22} \end{bmatrix} < 0, \qquad (4.54)
$$

where

$$
\Upsilon_{11} = \begin{bmatrix}
-R & 0 & R\check{A} & 0 & 0 & 0 \\
* & -S & S\check{A} - M & M - \bar{\gamma}NC & \sigma_{\bar{\gamma}}NC & 0 \\
* & * & -R & 0 & 0 \\
* & * & * & -S & 0 & 0 \\
* & * & * & * & -R & 0 \\
* & * & * & * & * & -S
\end{bmatrix},
$$

$$
\Upsilon_{12} = \begin{bmatrix}
\Psi_{121} & 0 & RB \\
\Psi_{122} & \Psi_{123} & SB - ND \\
0 & 0 & 0 \\
0 & 0 & 0 \\
0 & 0 & 0 \\
0 & 0 & 0
\end{bmatrix},
$$

$$
\Upsilon_{22} = \mathrm{diag}\{-\alpha_1 I, \cdots, -\alpha_q I, -\alpha_1 I, \cdots, -\alpha_q I, -W^{-1}\},
$$

with

$$
\check{A} = (A + \Delta A),
$$

$$
SG = M,
$$

$$
SK = N.
$$

In order to eliminate the parameter uncertainty that occurs in the system matrix, we rewrite (4.54) as follows:

$$L + \hat{H}F\hat{E} + \hat{E}^{\mathrm{T}}F^{\mathrm{T}}\hat{H}^{\mathrm{T}} < 0, \tag{4.55}$$

where

$$L = \begin{bmatrix} \Psi_{11} & \Upsilon_{12} \\ * & \Upsilon_{22} \end{bmatrix},$$

$$\hat{H} = \begin{bmatrix} H^{\mathrm{T}}R & H^{\mathrm{T}}S & 0 & 0 & 0 & 0 & 0 & \cdots & 0 & 0 & \cdots & 0 & 0 \end{bmatrix}^{\mathrm{T}},$$

$$\hat{E} = \begin{bmatrix} 0 & 0 & E & 0 & 0 & 0 & 0 & \cdots & 0 & 0 & \cdots & 0 & 0 \end{bmatrix}.$$

Now, applying Lemma 2.3.1 to (4.55), we know that (4.43) holds if and only if (4.50) holds. Moreover, taking note of (4.47), we arrive at (4.39) from (4.43). Therefore, according to Theorem 4.3.1, the system (4.14) is exponentially mean-square stable and the steady-state state covariance satisfies

$$\hat{Q} \le Q. \tag{4.56}$$

Since

$$\hat{Q} = \begin{bmatrix} X_{xx} & X_{xe} \\ X_{xe}^{\mathrm{T}} & X_{ee} \end{bmatrix} \quad \text{and} \quad Q = \begin{bmatrix} R^{-1} & 0 \\ 0 & S^{-1} \end{bmatrix}, \tag{4.57}$$

we know that

$$X_{ee} \le S^{-1}. \tag{4.58}$$

Noting that (4.44) implies

$$S^{-1} \le \hat{X}^{-1}, \tag{4.59}$$

the requirement (R2) is also achieved. The proof is now complete.      □

**Remark 4.4** *The robust variance-constrained filtering problem has been solved for a class of nonlinear stochastic systems with missing measurements in terms of the feasibility of the LMIs (4.42) to (4.44) in Theorem 4.3.2. The LMIs can be solved efficiently via the interior point method [138]. Note that LMIs (4.42) to (4.44) are affine in the scalar positive parameters ε and $\alpha_i > 0$. Hence, they can be defined as LMI variables in order to increase the possibility of the solutions and decrease conservatism with respect to the uncertainty F. Note that our main results are based on the LMI conditions. The LMI Control Toolbox implements state-of-the-art interior-point LMI solvers. While these solvers are significantly faster than classical convex optimization algorithms, it should be kept in mind that the complexity of LMI computations remains higher than that of solving, say, a Riccati equation. For instance, problems with a thousand design variables typically take over an hour on today's workstations. However, research on LMI optimization is a very active area in the applied maths, optimization and the operations research community, and substantial speed-ups can be expected in the future.*

**Remark 4.5** *It is worth pointing out that the main results in this chapter can be easily extended to other more complicated systems, such as systems with multiple stochastic data packet losses or stochastic systems with sector-bounded nonlinearity, which is more general than that discussed in this chapter.*

## 4.4   Numerical Example

In this section, we present an illustrative example to demonstrate the effectiveness of the proposed algorithms.

Consider the following discrete uncertain system with stochastic nonlinearities:

$$x(k+1) = \left( \begin{bmatrix} -0.13 & 0.38 & -0.25 \\ 0 & -0.31 & 0.13 \\ 0.12 & 0 & 0.63 \end{bmatrix} + \begin{bmatrix} 0.5 \\ 0.6 \\ 0 \end{bmatrix} F(k) \begin{bmatrix} 0.8 & 0 & 0 \end{bmatrix} \right) x(k)$$

$$+ f(x(k)) + \begin{bmatrix} 0.38 \\ 0 \\ 0.13 \end{bmatrix} \omega(k),$$

$$y(k) = \gamma(k) \left( \begin{bmatrix} 0.12 & -0.75 & 0.25 \end{bmatrix} x(k) + g(x(k)) \right) + \omega(k),$$

where $F(k) = \sin(0.6k)$ is a perturbation matrix satisfying $F(k)F^{\mathrm{T}}(k) \leq I$ and $\omega(k)$ is a zero mean Gaussian white noise process with unity covariance. We now consider the stochastic nonlinearities $f(x(k))$ and $g(x(k))$ in the following three cases:

**Case 1:** $f(x(k))$ and $g(x(k))$ are nonlinearities with multiplicative noise of the following form:

$$f(x(k)) = a_f \sum_{i=1}^{n} \alpha_{fi} x_i \xi_i,$$

$$g(x(k)) = a_g \sum_{i=1}^{n} \alpha_{gi} x_i \xi_i, \qquad i = 1, 2, \ldots, n,$$

where $a_f$ and $a_g$ are known column vectors, $\alpha_{fi}$ and $\alpha_{gi}$ $(i = 1, 2, \ldots, n)$ are known coefficients, and $x_i$ is the $i$th member of $x(k)$. For simplicity of notation, $\xi_i$ is short for $\xi_i(k)$ being zero mean uncorrelated Gaussian white noise processes with unity covariances. In this case, we assume

$$f(x(k)) = \begin{bmatrix} 0.1 \\ 0.2 \\ 0.2 \end{bmatrix} (0.3x_1\xi_1 + 0.4x_2\xi_2 + 0.5x_3\xi_3),$$

$$g(x(k)) = 0.3(0.3x_1\xi_1 + 0.4x_2\xi_2 + 0.5x_3\xi_3).$$

***Case 2:*** $f(x(k))$ and $g(x(k))$ are nonlinearities with the sign of a function:

$$f(x(k)) = b_f \sum_{i=1}^{n} \beta_{fi} \text{ sign}[x_i]x_i\xi_i,$$

$$g(x(k)) = b_g \sum_{i=1}^{n} \beta_{gi} \text{ sign}[x_i]x_i\xi_i, \qquad i = 1, 2, \ldots, n.$$

In this case, we assume

$$f(x(k)) = \begin{bmatrix} 0.1 \\ 0.2 \\ 0.2 \end{bmatrix} (0.3 \text{ sign}[x_1]x_1\xi_1 + 0.4 \text{ sign}[x_2]x_2\xi_2 + 0.5 \text{ sign}[x_3]x_3\xi_3),$$

$$g(x(k)) = 0.3(0.3 \text{ sign}[x_1]x_1\xi_1 + 0.4 \text{ sign}[x_2]x_2\xi_2 + 0.5 \text{ sign}[x_3]x_3\xi_3).$$

***Case 3:*** $f(x(k))$ and $g(x(k))$ are nonlinearities with the following form:

$$f(x(k)) = c_f \sum_{i=1}^{n} \rho_{fi}x_i(\sin(x_i)\xi_i + \cos(x_i)\eta_i),$$

$$g(x(k)) = c_g \sum_{i=1}^{n} \rho_{gi}x_i(\sin(x_i)\xi_i + \cos(x_i)\eta_i),$$

where $\eta_i$ ($i = 1, 2, \ldots, n$) are zero mean Gaussian white noise processes with unity covariances, which are mutually uncorrelated and also uncorrelated with $\xi_i$. In this case, we assume that

$$f(x(k)) = \begin{bmatrix} 0.1 \\ 0.2 \\ 0.2 \end{bmatrix} (0.3x_1(\sin(x_1)\xi_1 + \cos(x_1)\eta_1)$$

$$+ 0.4x_2(\sin(x_2)\xi_2 + \cos(x_2)\eta_2) + 0.5x_3(\sin(x_3)\xi_3 + \cos(x_3)\eta_3)),$$

$$g(x(k)) = 0.3(0.3x_1(\sin(x_1)\xi_1 + \cos(x_1)\eta_1)$$

$$+ 0.4x_2(\sin(x_2)\xi_2 + \cos(x_2)\eta_2) + 0.5x_3(\sin(x_3)\xi_3 + \cos(x_3)\eta_3)).$$

Now we can easily that check all of the above three classes of stochastic nonlinearities satisfy the following equality:

$$\mathbb{E} \left\{ \begin{bmatrix} f(x(k)) \\ g(x(k)) \end{bmatrix} \begin{bmatrix} f^{\mathrm{T}}(x(k)) & g^{\mathrm{T}}(x(k)) \end{bmatrix} | x(k) \right\}$$

$$= \begin{bmatrix} 0.01 & 0.02 & 0.02 & 0.03 \\ 0.02 & 0.04 & 0.04 & 0.06 \\ 0.02 & 0.04 & 0.04 & 0.06 \\ 0.03 & 0.06 & 0.06 & 0.09 \end{bmatrix} x^{\mathrm{T}}(k) \begin{bmatrix} 0.09 & 0 & 0 \\ 0 & 0.16 & 0 \\ 0 & 0 & 0.25 \end{bmatrix} x(k).$$

Let the stochastic variable $\gamma(k) \in \mathbb{R}$ be a Bernoulli distributed white sequence taking values on 0 and 1 with $\text{Prob}\{\gamma(k) = 1\} = \mathbb{E}\{\gamma(k)\} = 0.9$. Hence, $\sigma_\gamma = \sigma_{\bar{\gamma}} = 0.3$.

Choosing $\sigma_1^2 = 0.5, \sigma_2^2 = 0.5, \sigma_3^2 = 0.8$ as the estimation error variances upper bounds, we employ the Matlab Toolbox to find the desired filter parameters by using Theorem 4.3.2 and obtain

$$M = \begin{bmatrix} -0.3152 & -0.0081 & 0.3975 \\ -0.0081 & -0.8295 & -0.1627 \\ 0.3975 & -0.1627 & 2.6246 \end{bmatrix},$$

$$N = \begin{bmatrix} 1.9063 \\ -0.1838 \\ 1.1465 \end{bmatrix},$$

$$R = \begin{bmatrix} 3.5044 & -0.9228 & 0.1288 \\ -0.9228 & 4.3243 & -0.6485 \\ -0.9228 & -0.6485 & 4.3164 \end{bmatrix},$$

$$S = \begin{bmatrix} 3.5715 & -1.1232 & 0.7656 \\ -1.1232 & 3.4382 & -0.4301 \\ 0.7656 & -0.4301 & 4.8642 \end{bmatrix},$$

$$\alpha_1 = 6.1511, \qquad \alpha_2 = 6.0043,$$

$$\alpha_3 = 5.7691, \qquad \varepsilon = 4.2928.$$

Finally, the obtained filter parameters are calculated as follows:

$$G = \begin{bmatrix} -0.1182 & -0.0783 & 0.0019 \\ -0.0288 & -0.2725 & 0.0210 \\ 0.0978 & -0.0452 & 0.5411 \end{bmatrix},$$

$$K = \begin{bmatrix} 0.5444 \\ 0.1448 \\ 0.1628 \end{bmatrix}.$$

The simulation results are shown in Figures 4.1 to 4.13. The actual state responses and their estimates are shown in Figures 4.1 to 4.3 for Case 1, in Figures 4.5 to 4.7 for Case 2, and in Figures 4.9 to 4.11 for Case 3. The filtering error variances of the states $x_1(k)$, $x_2(k)$, and $x_3(k)$ for Case 1 to Case 3 are given in Figures 4.4, 4.8, and 4.12 respectively. We can see clearly that the filtering error variances are constrained below the pre-specified upper bounds, which means that the design requirements are satisfied. We can also see from Figure 4.13 that the variance of the filtering error will become lager when the data missing is severe. Figures 4.14 to 4.16 show the stochastic nonlinearities for three different cases discussed in this chapter.

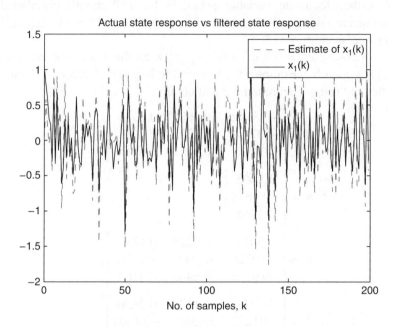

**Figure 4.1**   The actual state $x_1(k)$ and its estimate $\hat{x}_1(k)$ for Case 1

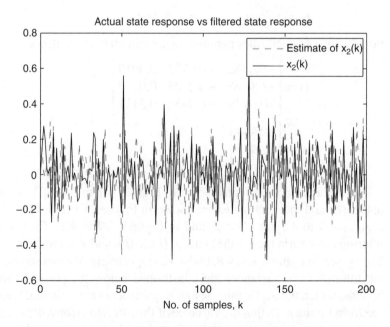

**Figure 4.2**   The actual state $x_2(k)$ and its estimate $\hat{x}_2(k)$ for Case 1

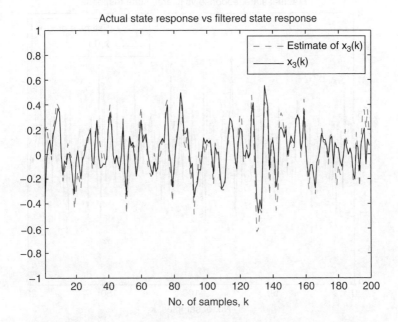

**Figure 4.3**  The actual state $x_3(k)$ and its estimate $\hat{x}_3(k)$ for Case 1

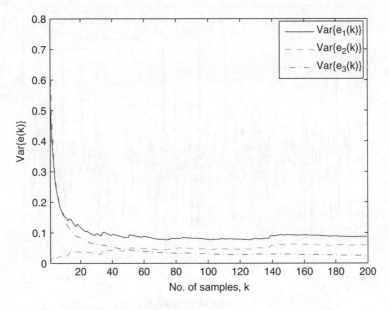

**Figure 4.4**  The filtering error variances for Case 1

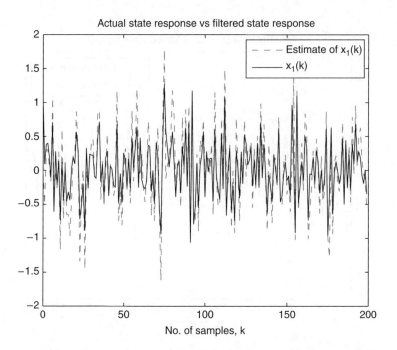

**Figure 4.5**   The actual state $x_1(k)$ and its estimate $\hat{x}_1(k)$ for Case 2

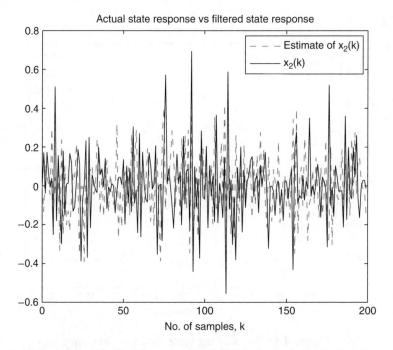

**Figure 4.6**   The actual state $x_2(k)$ and its estimate $\hat{x}_2(k)$ for Case 2

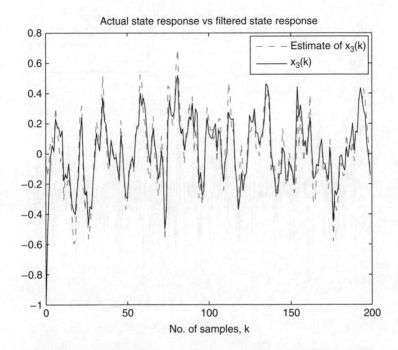

**Figure 4.7** The actual state $x_3(k)$ and its estimate $\hat{x}_3(k)$ for Case 2

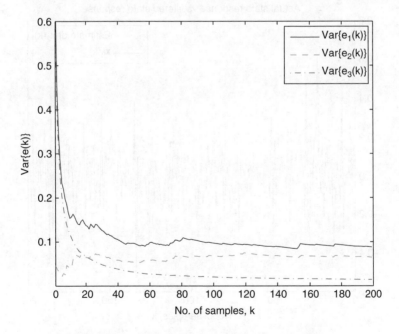

**Figure 4.8** The filtering error variances for Case 2

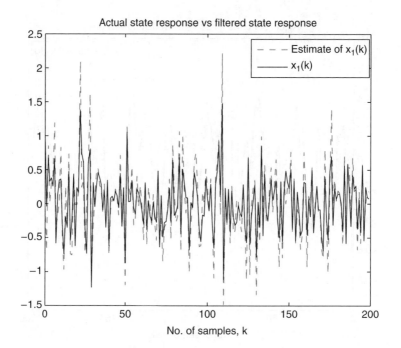

**Figure 4.9**   The actual state $x_1(k)$ and its estimate $\hat{x}_1(k)$ for Case 3

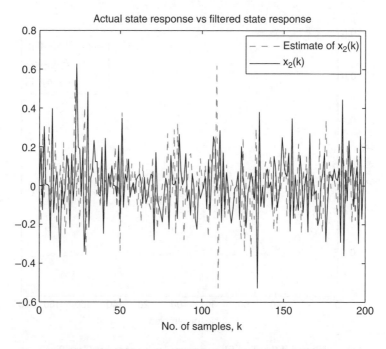

**Figure 4.10**   The actual state $x_2(k)$ and its estimate $\hat{x}_2(k)$ for Case 3

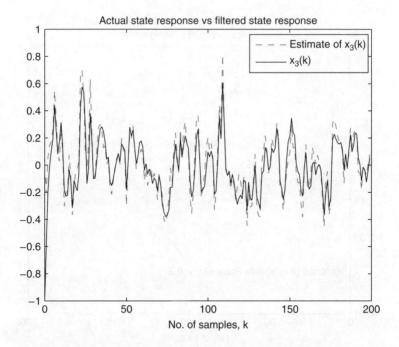

**Figure 4.11** The actual state $x_3(k)$ and its estimate $\hat{x}_3(k)$ for Case 3

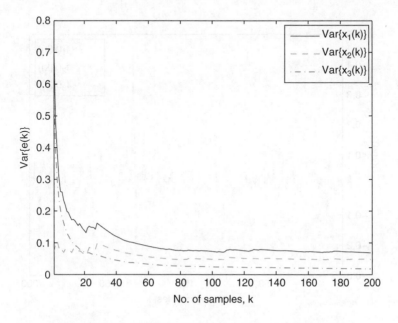

**Figure 4.12** The filtering error variances for Case 3

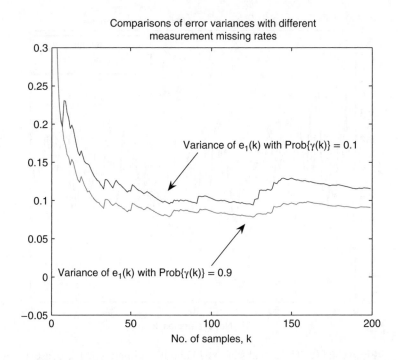

**Figure 4.13**   The comparison between filtering error variance of different data missing rates

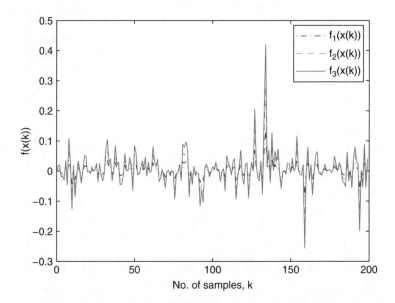

**Figure 4.14**   The stochastic nonlinearity $f(x(k))$ for Case 1

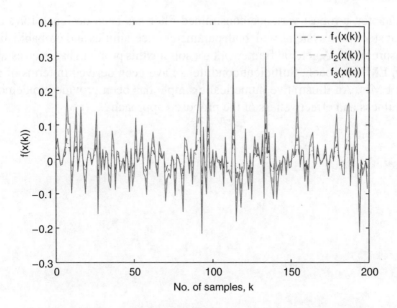

**Figure 4.15**    The stochastic nonlinearity $f(x(k))$ for Case 2

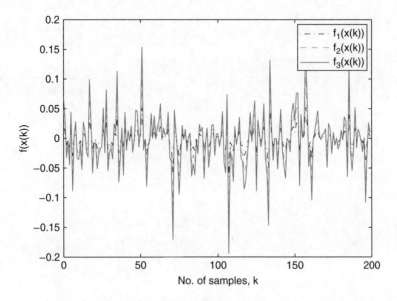

**Figure 4.16**    The stochastic nonlinearity $f(x(k))$ for Case 3

## 4.5   Summary

In this chapter, a robust variance-constrained filter has been designed for a class of nonlinear stochastic systems with both parameter uncertainties and probabilistic missing measurements. A general framework for solving this problem has been established using an LMI approach. Sufficient conditions have been derived in terms of a set of feasible LMIs. An illustrative numerical example has been provided to demonstrate the usefulness and effectiveness of the proposed approach.

# 5

# Robust Fault-Tolerant Control with Variance Constraints

As is well known, in real-world engineering practice, the actuator of a control system might be confronted with different kinds of failures due to a variety of reasons, such as equipment aging, erosion, abrupt changes of working conditions, to name but a few key ones. Those actuator failures may have significant deterioration effects on systems performances and sometimes even lead to instability. This gives rise to the so-called fault-tolerant control problem, which strives to make the system stable and retain acceptable performance under the system faults. Fault-tolerant control has recently received more and more interest as requirements increase toward the reliability of engineering systems. Many results of the control as well as filtering problems for systems against possible actuator/sensor failures have been reported in the literature. Nevertheless, most of the literature has been concerned with system stability only, and other desired performances such as variance constraints have seldom been taken into account. Up to now, to the best of the authors' knowledge, the *variance-constrained fault-tolerant control* problem for stochastic systems with *general nonlinearities* has not yet been thoroughly investigated.

In view of the above observations, in this chapter we aim at designing a state feedback controller for a class of uncertain nonlinear systems such that, for all admissible parameter uncertainties and all possible actuator failures, the closed-loop system is exponentially mean-square stable, and the individual steady-state variance is not more than the pre-specified value. The nonlinearities considered here are characterized by statistical means, which can cover several classes of well-studied nonlinearities, and the actuator failure model adopted here is more practical than the conventional outage case. The solvability of the addressed problem is shown to be converted into a sequential optimization problem subject to LMI constraints.

The main contributions of this chapter can be summarized as follows: (1) the variance-constrained control problem is studied for a class of nonlinear stochastic

*Variance-Constrained Multi-Objective Stochastic Control and Filtering*, First Edition.
Lifeng Ma, Zidong Wang and Yuming Bo.
© 2015 John Wiley & Sons, Ltd. Published 2015 by John Wiley & Sons, Ltd.

systems where the nonlinearities are characterized by statistical means; (2) the challenging issue of obtaining steady-state covariance of such a system is dealt with; (3) multiple objectives are simultaneously considered that include the stability, variance constraints, robustness, as well as the reliability; and (4) the solvability of the addressed problem is shown to be converted into a sequential optimization problem subject to LMI constraints.

The remainder of this chapter is organized as follows: In Section 5.1, the multi-objective fault-tolerant control problem for the nonlinear stochastic system is formulated. In Section 5.2, the exponential mean-square stability of the closed-loop system and the individual variances of the states are analyzed separately. In Section 5.3, an LMI algorithm is developed for the controller design problem. In Section 5.4, an illustrative example is presented to show the effectiveness of the proposed algorithm. In Section 5.5, summary remarks are provided.

## 5.1  Problem Formulation

Consider the following uncertain nonlinear stochastic system:

$$x(k + 1) = (A + \Delta A)x(k) + Bu(k) + D\omega(k) + f(x(k), u(k)), \qquad (5.1)$$

where $x(k) \in \mathbb{R}^n$ is the state, $u(k) \in \mathbb{R}^r$ is the control input, $\omega(k) \in \mathbb{R}^m$ is a zero mean Gaussian white noise sequence with covariance $R$, and $A$, $B$, and $D$ are known real matrices with appropriate dimensions. $\Delta A$ is a real-valued perturbation matrix that represents parametric uncertainty, being of the following form:

$$\Delta A = HFE, \quad FF^{\mathrm{T}} \leq I, \qquad (5.2)$$

where $H$ and $E$ are known constant matrices with appropriate dimensions. The uncertainties in $\Delta A$ are said to be admissible if (5.2) holds.

The nonlinear stochastic function $f(x(k), u(k))$ is assumed to have the following first moments for all $x(k)$ and $u(k)$:

$$\mathbb{E}\left\{f(x(k), u(k)) \mid x(k), u(k)\right\} = 0, \qquad (5.3)$$

$$\mathbb{E}\{f(x(k), u(k))f^{\mathrm{T}}(x(k), u(k)) \mid x(k), u(k)\}$$
$$= \sum_{i=1}^{q} \theta_i \theta_i^{\mathrm{T}}(x^{\mathrm{T}}(k)\Gamma_i x(k) + u^{\mathrm{T}}(k)\Pi_i u(k)), \qquad (5.4)$$

where $\theta_i$ $(i = 1, \ldots, q)$ are known column vectors and $\Gamma_i$ and $\Pi_i$ $(i = 1, \ldots, q)$ are known as positive definite matrices with appropriate dimensions.

**Remark 5.1** *Note that the discrete-time system (5.1) involves parameter uncertainties, stochastic noises and nonlinear disturbances. The motivation for us to consider such a model stems from the following three facts. (1) In practice, it is almost impossible to get an exact mathematical model of a dynamical system due to the complexity* *of the system, the difficulty of measuring various parameters, environmental noises, uncertain and/or time-varying parameters, etc. Indeed, the model of the system to be controlled almost always contains some type of uncertainty. (2) Stochastic modeling has come to play an important role since, in engineering applications, many systems are subject to stochastic disturbances such as aircraft, chemical, or process control systems, and distributed networks. (3) External nonlinear disturbances may result from the linearization process of an originally highly nonlinear plant or may be an external nonlinear input, and therefore exist in many real-world systems. In this chapter, a single degree-of-freedom (SDOF) model simulating a single-storey building, taken from Ref. [139], is used to demonstrate the application potential of the proposed theory, where the control simulates a force applied to the structure using pre-stressed active tendons connected to a servo-controlled hydraulic actuator.*

Now, by introducing the sensor failure model proposed in Ref. [40], we consider the following state feedback controller for system (5.1):

$$u(k) = KG(k)x(k),  \tag{5.5}$$

where $K$ is the feedback gain to be designed and the time-varying sensor fault function matrix $G(k)$ is defined as follows:

$$G(k) = \text{diag}\{g_1(k), g_2(k), \ldots, g_n(k)\},  \tag{5.6}$$

where the matrix elements $g_i(k)$ $(i = 1, 2, \ldots, n)$ are assumed to satisfy the following bounded condition:

$$0 \le g_i^L \le g_i(k) \le g_i^U < \infty,  \tag{5.7}$$

with $g_i^L < 1$ and $g_i^U \ge 1$ being known real constants.

Let $G(k)$ be decomposed as

$$G(k) = G_0 + G_0 M(k),  \tag{5.8}$$

where

$$G_0 = \text{diag}\{g_1(0), g_2(0), \ldots, g_n(0)\},$$

$$M(k) = \text{diag}\{m_1(k), m_2(k), \ldots, m_n(k)\},$$

with

$$g_i(0) = \frac{g_i^L + g_i^U}{2},$$

$$m_i(k) = \frac{g_i(k) - g_i(0)}{g_i(0)}.$$

Denote

$$L = \text{diag}\{l_1, l_2, \ldots, l_n\}, \tag{5.9}$$

where

$$l_i = \frac{g_i^U - g_i^L}{g_i^U + g_i^L}.$$

Setting

$$|M(k)| = \text{diag}\{|m_1(k)|, |m_2(k)|, \ldots, |m_n(k)|\},$$

we can see that

$$G_0 > 0, \quad |M(k)| \leq L \leq I. \tag{5.10}$$

Applying (5.5) to the system (5.1), we obtain the following closed-loop system:

$$x(k + 1) = A_K x(k) + f(x(k), KG(k)x(k)) + D\omega(k), \tag{5.11}$$

where

$$A_K \triangleq A + HFE + BKG(k).$$

Before giving our design goal, we introduce the following stability concept for the system (5.11).

**Definition 5.1.1** *The system* (5.11) *is said to be exponentially mean-square stable if, with* $\omega(k) = 0$, *there exist constants* $\zeta \geq 1$ *and* $\tau \in (0, 1)$ *such that*

$$\mathbb{E}\{\|x(k)\|^2\} \leq \zeta \tau^k \mathbb{E}\{\|x_0\|^2\}, \quad \forall x_0 \in \mathbb{R}^n, \quad k \in \mathbb{I}^+, \tag{5.12}$$

*for all admissible uncertainties and all possible actuator failures.*

This chapter aims to determine the parameter $K$ for the state feedback controller (5.5) such that, for all admissible parameter uncertainties and all possible actuator failures, the following two requirements are simultaneously satisfied:

(R1) The system (5.11) is exponentially mean-square stable.
(R2) The steady-state state variance achieves the following constraint:

$$\text{Var}\{x_i(k)\} \triangleq \lim_{k \to \infty} \mathbb{E}\{x_i(k)x_i^T(k)\} \leq \sigma_i^2, \tag{5.13}$$

where $x(k) = \begin{bmatrix} x_1(k) & x_2(k) & \cdots & x_n(k) \end{bmatrix}^T$ and $\sigma_i^2 > 0$ $(i = 1, 2, \ldots, n)$ are given pre-specified constants.

## 5.2  Stability and Variance Analysis

Before the analysis, we first introduce several useful lemmas.

**Lemma 5.2.1** *[133] Given the state feedback gain matrix K and actuator failure matrix G(k). The following statements are equivalent:*

*1.*
$$\rho(\Psi) < 1, \tag{5.14}$$

*where*
$$\Psi = A_K \otimes A_K + \sum_{i=1}^{q} \mathrm{st}(\theta_i \theta_i^{\mathrm{T}}) \mathrm{st}^{\mathrm{T}}(\Gamma_i + G^{\mathrm{T}}(k) K^{\mathrm{T}} \Pi_i K G(k))$$

*or*
$$\Psi = A_K^{\mathrm{T}} \otimes A_K^{\mathrm{T}} + \sum_{i=1}^{q} \mathrm{st}(\Gamma_i + G^{\mathrm{T}}(k) K^{\mathrm{T}} \Pi_i K G(k)) \mathrm{st}^{\mathrm{T}}(\theta_i \theta_i^{\mathrm{T}}).$$

*2. There exists P > 0 such that*
$$A_K^{\mathrm{T}} P A_K - P + \sum_{i=1}^{q} (\Gamma_i + G^{\mathrm{T}}(k) K^{\mathrm{T}} \Pi_i K G(k)) \mathrm{tr}[\theta_i \theta_i^{\mathrm{T}} P] < 0. \tag{5.15}$$

*3. There exists Q > 0 such that*
$$A_K Q A_K^{\mathrm{T}} - Q + \sum_{i=1}^{q} \theta_i \theta_i^{\mathrm{T}} \mathrm{tr}[Q(\Gamma_i + G^{\mathrm{T}}(k) K^{\mathrm{T}} \Pi_i K G(k))] < 0. \tag{5.16}$$

*4. The system* (5.11) *is exponentially mean-square stable.*

**Lemma 5.2.2** *[59] Consider the system*
$$\xi(k+1) = M\xi(k) + f(\xi(k)), \tag{5.17}$$

*where*
$$\mathbb{E}\{f(\xi(k))|\xi(k)\} = 0,$$

$$\mathbb{E}\{f(\xi(k)) f^{\mathrm{T}}(\xi(k))|\xi(k)\} = \sum_{i=1}^{q} \rho_i \rho_i^{\mathrm{T}}(\xi^{\mathrm{T}}(k) \Omega_i \xi(k)),$$

*with $\rho_i$ ($i = 0, 1, 2, \ldots, q$) being known column vectors and $\Omega_i$ ($i = 0, 1, 2, \ldots, q$) are known positive definite matrices with appropriate dimensions. If the system* (5.17) *is exponentially mean-square stable and there exists a symmetric matrix Y satisfying*

$$M Y M^{\mathrm{T}} - Y + \sum_{i=1}^{q} \rho_i \rho_i^{\mathrm{T}} \mathrm{tr}[Y \Omega_i] < 0, \tag{5.18}$$

*then $Y \geq 0$.*

Now, let us proceed to deal with the state covariance. Define the state covariance of (5.11) by

$$Q(k) \triangleq \mathbb{E}\{x(k)x^{\mathrm{T}}(k)\}$$
$$= \mathbb{E}\left\{ \begin{bmatrix} x_1(k) & x_2(k) \cdots x_n(k) \end{bmatrix} \begin{bmatrix} x_1(k) & x_2(k) \cdots x_n(k) \end{bmatrix}^{\mathrm{T}} \right\}. \qquad (5.19)$$

Then the Lyapunov-type equation that governs the evolution of the state covariance matrix $Q(k)$ can be derived as follows:

$$Q(k+1) = A_K Q(k)A_K^{\mathrm{T}} + DRD^{\mathrm{T}}$$
$$+ \sum_{i=1}^{q} \theta_i \theta_i^{\mathrm{T}} \mathrm{tr}[Q(k)(\Gamma_i + G^{\mathrm{T}}(k)K^{\mathrm{T}}\Pi_i KG(k))]. \qquad (5.20)$$

Rewriting (5.20) in the form of a stack matrix results in

$$\mathrm{st}(Q(k+1)) = \Psi \mathrm{st}(Q(k)) + \mathrm{st}(DRD^{\mathrm{T}}), \qquad (5.21)$$

where

$$\Psi = A_K \otimes A_K + \sum_{i=1}^{q} \mathrm{st}(\theta_i \theta_i^{\mathrm{T}})\mathrm{st}^{\mathrm{T}}(\Gamma_i + G^{\mathrm{T}}(k)K^{\mathrm{T}}\Pi_i KG(k)).$$

If the system (5.11) is exponentially mean-square stable, it follows from Lemma 5.2.1 that $\rho(\Psi) < 1$, which means that $Q(k)$ in (5.21) converges to a constant matrix $\hat{Q}$ when $k \to \infty$, that is

$$\hat{Q} = \lim_{k\to\infty} Q(k), \qquad (5.22)$$

and, in the steady state, (5.20) can be rewritten as

$$A_K \hat{Q} A_K^{\mathrm{T}} - \hat{Q} + \sum_{i=1}^{q} \theta_i \theta_i^{\mathrm{T}} \mathrm{tr}[\hat{Q}(\Gamma_i + G^{\mathrm{T}}(k)K^{\mathrm{T}}\Pi_i KG(k))] + DRD^{\mathrm{T}} = 0. \qquad (5.23)$$

In the next stage, before discussing the controller designing technique, we first provide the following result, which will play a key role in the derivation of our main results.

**Theorem 5.2.3** *Given the state feedback gain matrix $K$ and actuator failure matrix $G(k)$. If there exists a positive definite matrix $Q > 0$ such that for all admissible uncertainties and all possible actuator failures, the following matrix inequality*

$$A_K Q A_K^{\mathrm{T}} - Q + \sum_{i=1}^{q} \theta_i \theta_i^{\mathrm{T}} \mathrm{tr}[Q(\Gamma_i + G^{\mathrm{T}}(k)K^{\mathrm{T}}\Pi_i KG(k))] + DRD^{\mathrm{T}} < 0 \qquad (5.24)$$

*holds, then the system (5.11) is exponentially mean-square stable and $\hat{Q} \le Q$.*

*Proof.* First, it follows from Lemma 5.2.1 and (5.24) that the system (5.11) is exponentially mean-square stable. Therefore, in the steady state, the system state covariance defined by (5.22) exists and satisfies (5.23). Subtracting (5.23) from (5.24) leads to

$$A_K(Q - \hat{Q})A_K^T - (Q - \hat{Q})$$
$$+ \sum_{i=1}^{q} \theta_i \theta_i^T \mathrm{tr}[(Q - \hat{Q})(\Gamma_i + G^T(k)K^T\Pi_i KG(k))] < 0, \tag{5.25}$$

which indicates from Lemma 5.2.2 that $Q - \hat{Q} \geq 0$. The proof is now complete. □

## 5.3   Robust Controller Design

In this section, we present the solution to the robust variance-constrained state feedback controller design problem for the discrete-time nonlinear stochastic systems with deterministic norm-bounded parametric uncertainty and possible actuator failures. In other words, we aim to design the controller that satisfies the requirements (*R*1) and (*R*2) simultaneously.

The following lemma is needed in our derivation.

**Lemma 5.3.1**  *Given $\mathcal{R}_1$ and $\mathcal{R}_2$ with appropriate dimensions. Let*

$$\mathcal{V} = \mathrm{diag}\{v_1, v_2, \dots, v_n\} \tag{5.26}$$

*be a time-varying diagonal matrix of appropriate dimensions with*

$$|\mathcal{V}| = \mathrm{diag}\{|v_1|, |v_2|, \dots, |v_p|\} \tag{5.27}$$

*and $|\mathcal{V}| < \mathcal{U}$, where $\mathcal{U} > 0$ is a known positive definite diagonal matrix. Then, for any scalar $\psi > 0$, it follows that*

$$\mathcal{R}_1 \mathcal{V} \mathcal{R}_2 + (\mathcal{R}_1 \mathcal{V} \mathcal{R}_2)^T \leq \psi \mathcal{R}_1 \mathcal{U} \mathcal{R}_1^T + \psi^{-1} \mathcal{R}_2^T \mathcal{U} \mathcal{R}_2. \tag{5.28}$$

*Proof.*  Firstly denote

$$\bar{\mathcal{V}} \triangleq \mathrm{diag}\left\{ \sqrt{|v_1|}, \sqrt{|v_2|}, \dots, \sqrt{|v_p|} \right\}.$$

This shows that the following inequality is always true:

$$(\sqrt{\psi}R_1\bar{\mathcal{V}} - \frac{1}{\sqrt{\psi}}R_2^T\bar{\mathcal{V}})(\sqrt{\psi}R_1\bar{\mathcal{V}} - \frac{1}{\sqrt{\psi}}R_2^T\bar{\mathcal{V}})^T \geq 0. \tag{5.29}$$

Then it follows that

$$\psi R_1 \bar{\mathcal{V}} \bar{\mathcal{V}} R_1^T + \frac{1}{\sqrt{\psi}} R_2^T \bar{\mathcal{V}} \bar{\mathcal{V}} R_2 - R_1 \bar{\mathcal{V}} \bar{\mathcal{V}} R_2 - R_2^T \bar{\mathcal{V}} \bar{\mathcal{V}} R_1^T \geq 0, \tag{5.30}$$

which results in

$$R_1|\mathcal{V}|R_2 + R_2^{\mathrm{T}}|\mathcal{V}|R_1^{\mathrm{T}} \leq \psi R_1|\mathcal{V}|R_1^{\mathrm{T}} + \frac{1}{\sqrt{\psi}}R_2^{\mathrm{T}}|\mathcal{V}|R_2. \tag{5.31}$$

Noting that $\mathcal{V} < |\mathcal{V}| < \mathcal{U}$, we could obtain directly

$$\begin{aligned} & R_1\mathcal{V}R_2 + R_2^{\mathrm{T}}\mathcal{V}R_1^{\mathrm{T}} \\ \leq & R_1|\mathcal{V}|R_2 + R_2^{\mathrm{T}}|\mathcal{V}|R_1^{\mathrm{T}} \\ \leq & \psi R_1|\mathcal{V}|R_1^{\mathrm{T}} + \frac{1}{\sqrt{\psi}}R_2^{\mathrm{T}}|\mathcal{V}|R_2 \\ \leq & \psi R_1\mathcal{U}R_1^{\mathrm{T}} + \frac{1}{\sqrt{\psi}}R_2^{\mathrm{T}}\mathcal{U}R_2. \end{aligned} \tag{5.32}$$

□

The proof is complete.

The following theorem presents sufficient conditions for the addressed robust controller design problem.

**Theorem 5.3.2** *Given the state variance upper bounds $\sigma_i^2 > 0 \quad (i = 1, 2, \dots, n)$ and the actuator failure matrix $G(k)$. If there exist a positive definite matrix $Q > 0$, a real matrix $N$, a positive scalar $\varepsilon$, and positive scalars $\alpha_i$ and $\beta_i$ $(i = 1, 2, \dots, q)$ such that the following set of matrix inequalities*

$$Q \geq L, \tag{5.33}$$

$$\begin{bmatrix} -\alpha_i & \delta_i^{\mathrm{T}}Q & \pi_i^{\mathrm{T}}N & \pi_i^{\mathrm{T}}N & 0 \\ * & -Q & 0 & 0 & 0 \\ * & * & -Q & 0 & Q \\ * & * & * & -\mu_1^{-1}Q & 0 \\ * & * & * & * & -\mu_1 L^{-1} \end{bmatrix} < 0, \tag{5.34}$$

$$\begin{bmatrix} -Q & AQ + BN & \Upsilon_{13} & \Upsilon_{14} \\ * & -Q & 0 & \Upsilon_{24} \\ * & * & \Upsilon_{33} & 0 \\ * & * & * & \Upsilon_{44} \end{bmatrix} < 0, \tag{5.35}$$

$$\begin{bmatrix} 1 & 0 & 0 & \cdots & 0 \end{bmatrix} Q \begin{bmatrix} 1 & 0 & 0 & \cdots & 0 \end{bmatrix}^{\mathrm{T}} \leq \sigma_1^2,$$
$$\begin{bmatrix} 0 & 1 & 0 & \cdots & 0 \end{bmatrix} Q \begin{bmatrix} 0 & 1 & 0 & \cdots & 0 \end{bmatrix}^{\mathrm{T}} \leq \sigma_2^2,$$
$$\vdots \tag{5.36}$$
$$\begin{bmatrix} 0 & 0 & \cdots & 0 & 1 \end{bmatrix} Q \begin{bmatrix} 0 & 0 & \cdots & 0 & 1 \end{bmatrix}^{\mathrm{T}} \leq \sigma_n^2,$$

$$\alpha_i\beta_i = 1, \quad i = 1, 2, \dots, q, \tag{5.37}$$

*where*

$$\Upsilon_{13} = \begin{bmatrix} \theta_1 & \theta_2 & \cdots & \theta_3 \end{bmatrix},$$

$$\Upsilon_{14} = \begin{bmatrix} D & BN & 0 & \varepsilon H & 0 \end{bmatrix},$$

$$\Upsilon_{24} = \begin{bmatrix} 0 & 0 & Q & 0 & QE^{\mathrm{T}} \end{bmatrix},$$

$$\Upsilon_{33} = \begin{bmatrix} -\beta_1 I & 0 & \cdots & 0 \\ 0 & -\beta_2 I & \cdots & 0 \\ \vdots & \vdots & \ddots & \vdots \\ 0 & 0 & \cdots & -\beta_q I \end{bmatrix},$$

$$\Upsilon_{44} = \begin{bmatrix} -R^{-1} & 0 & 0 & 0 & 0 \\ 0 & -\mu_2^{-1}Q & 0 & 0 & 0 \\ 0 & 0 & -\mu_2 L^{-1} & 0 & 0 \\ 0 & 0 & 0 & -\varepsilon I & 0 \\ 0 & 0 & 0 & 0 & -\varepsilon I \end{bmatrix},$$

*is feasible for some given positive constants $\mu_1$ and $\mu_2$, then there exists a state feedback controller of the form (5.5) such that the system (5.11) is exponentially mean-square stable and the individual steady-state state variance is not more than the pre-specified value (i.e., the two requirements (R1) and (R2) are satisfied simultaneously). Moreover, the desired controller (5.5) can be determined by*

$$K = NQ^{-1}G_0^{-1}. \tag{5.38}$$

*Proof.* To start with, without losing any generality, we assume that $\Gamma_i = \delta_i \delta_i^T$ and $\Pi_i = \pi_i \pi_i^T$, where $\delta_i$ and $\pi_i$ ($i = 1, 2, \ldots, q$) are column vectors with appropriate dimensions.

Using the Schur Complement Lemma, it follows from (5.34) that

$$\begin{bmatrix} -\alpha_i & \delta_i^{\mathrm{T}}Q & \pi_i^{\mathrm{T}}N \\ Q\delta_i & -Q & 0 \\ N^{\mathrm{T}}\pi_i & 0 & -Q \end{bmatrix} + \begin{bmatrix} \mu_1 \pi_i^{\mathrm{T}}NQ^{-1}N^{\mathrm{T}}\pi_i & 0 & 0 \\ 0 & 0 & 0 \\ 0 & 0 & \mu_1^{-1}QLQ \end{bmatrix} < 0. \tag{5.39}$$

Since $N = KG_0Q$ and $Q \geq L$, we have

$$\begin{bmatrix} -\alpha_i & \delta_i^{\mathrm{T}}Q & \pi_i^{\mathrm{T}}N \\ Q\delta_i & -Q & 0 \\ N^{\mathrm{T}}\pi_i & 0 & -Q \end{bmatrix} + \begin{bmatrix} \mu_1 \pi_i^{\mathrm{T}}KG_0LG_0^{\mathrm{T}}K^{\mathrm{T}}\pi_i & 0 & 0 \\ 0 & 0 & 0 \\ 0 & 0 & \mu_1^{-1}QLQ \end{bmatrix} < 0. \tag{5.40}$$

By Lemma 5.3.1, it follows that

$$
\begin{bmatrix} 0 & 0 & \pi_i^{\mathrm{T}} KG_0 M(k)Q \\ 0 & 0 & 0 \\ QM^{\mathrm{T}}(k)G_0^{\mathrm{T}}K^{\mathrm{T}}\pi_i & 0 & 0 \end{bmatrix}
$$

$$
\leq \begin{bmatrix} \mu_1 \pi_i^{\mathrm{T}} KG_0 LG_0^{\mathrm{T}}K^{\mathrm{T}}\pi_i & 0 & 0 \\ 0 & 0 & 0 \\ 0 & 0 & \mu_1^{-1}QLQ \end{bmatrix}.
$$

(5.41)

Hence, applying the Schur Complement Lemma leads to

$$
\begin{bmatrix} -\alpha_i & \delta_i^{\mathrm{T}}Q & \pi_i^{\mathrm{T}}N \\ Q\delta_i & -Q & 0 \\ N^{\mathrm{T}}\pi_i & 0 & -Q \end{bmatrix} + \begin{bmatrix} 0 & 0 & \pi_i^{\mathrm{T}}KG_0M(k)Q \\ 0 & 0 & 0 \\ QM^{\mathrm{T}}(k)G_0^{\mathrm{T}}K^{\mathrm{T}}\pi_i & 0 & 0 \end{bmatrix} < 0
$$

$$
\Longleftrightarrow \begin{bmatrix} -\alpha_i & \delta_i^{\mathrm{T}}Q & \pi_i^{\mathrm{T}}KG(k)Q \\ Q\delta_i & -Q & 0 \\ QG^{\mathrm{T}}(k)K^{\mathrm{T}}\pi_i & 0 & -Q \end{bmatrix} < 0
$$

(5.42)

$$
\Longleftrightarrow \delta_i^{\mathrm{T}}Q\delta_i + \pi_i^{\mathrm{T}}KG(k)QG^{\mathrm{T}}(k)K^{\mathrm{T}}\pi_i < \alpha_i.
$$

By the property of matrix trace, it is easy to see that

$$
\mathrm{tr}(Q\Gamma_i + QG^{\mathrm{T}}(k)K^{\mathrm{T}}\Pi_i KG(k)) < \alpha_i.
$$

(5.43)

Similarly, we can obtain from (5.35) together with (5.33) that

$$
\begin{bmatrix} -Q & (A+BKG(k))Q & \theta_1 & \cdots & \theta_q & D & \varepsilon H & 0 \\ * & -Q & 0 & \cdots & 0 & 0 & 0 & QE^{\mathrm{T}} \\ * & * & -\beta_1 I & \cdots & 0 & 0 & 0 & 0 \\ * & * & * & \cdots & \cdots & \cdots & \cdots & \cdots \\ * & * & * & * & -\beta_q I & 0 & 0 & 0 \\ * & * & * & * & * & -R^{-1} & 0 & 0 \\ * & * & * & * & * & * & -\varepsilon I & 0 \\ * & * & * & * & * & * & * & -\varepsilon I \end{bmatrix} < 0.
$$

(5.44)

Rewrite (5.44) in the following form:

$$
\begin{bmatrix}
-Q & (A+BKG(k))Q & \theta_1 & \cdots & \theta_q & D \\
* & -Q & 0 & \cdots & 0 & 0 \\
* & * & -\beta_1 I & \cdots & 0 & 0 \\
* & * & * & \cdots & \cdots & \cdots \\
* & * & * & * & -\beta_q I & 0 \\
* & * & * & * & * & -R^{-1}
\end{bmatrix}
$$

$$
+ \varepsilon
\begin{bmatrix}
H \\ 0 \\ 0 \\ 0 \\ 0 \\ 0
\end{bmatrix}
\begin{bmatrix} H^{\mathrm{T}} & 0 & 0 & 0 & 0 & 0 \end{bmatrix}
+ \varepsilon^{-1}
\begin{bmatrix}
0 \\ QE^{\mathrm{T}} \\ 0 \\ 0 \\ 0 \\ 0
\end{bmatrix}
\begin{bmatrix} 0 & EQ & 0 & 0 & 0 & 0 \end{bmatrix} < 0. \quad (5.45)
$$

Using Lemma 2.3.1 and noticing the fact that $FF^{\mathrm{T}} \leq I$, it is obvious that

$$
\begin{bmatrix}
-Q & (A+HFE+BKG(k))Q & \theta_1 & \cdots & \theta_q & D \\
* & -Q & 0 & \cdots & 0 & 0 \\
* & * & -\beta_1 I & \cdots & 0 & 0 \\
* & * & * & \cdots & \cdots & \cdots \\
* & * & * & * & -\beta_q I & 0 \\
* & * & * & * & * & -R^{-1}
\end{bmatrix} < 0. \quad (5.46)
$$

Using the Schur Complement Lemma and the constraint $\alpha_i \beta_i = 1$, the matrix inequality (5.46) is equivalent to

$$
-Q + A_K Q A_K^{\mathrm{T}} + \sum_{i=1}^{q} \alpha_i \theta_i \theta_i^{\mathrm{T}} + DRD^{\mathrm{T}} < 0, \quad (5.47)
$$

and we then know from (5.43) that

$$
A_K Q A_K^{\mathrm{T}} - Q + \sum_{i=1}^{q} \theta_i \theta_i^{\mathrm{T}} \mathrm{tr}[Q(\Gamma_i + G^{\mathrm{T}}(k)K^{\mathrm{T}}\Pi_i KG(k))] + DRD^{\mathrm{T}} < 0. \quad (5.48)
$$

Therefore, according to Theorem 5.2.3, the system (5.11) is exponentially mean-square stable and the steady-state state covariance exists while satisfying $\hat{Q} \leq Q$. Next,

$$
\mathrm{Var}\{x_i(k)\} = \begin{bmatrix} 0 & \cdots & 0 & 1 & 0 & \cdots & 0 \end{bmatrix} \hat{Q} \begin{bmatrix} 0 & \cdots & 0 & 1 & 0 & \cdots & 0 \end{bmatrix}^{\mathrm{T}}
$$

$$
\leq \begin{bmatrix} 0 & \cdots & 0 & 1 & 0 & \cdots & 0 \end{bmatrix} Q \begin{bmatrix} 0 & \cdots & 0 & 1 & 0 & \cdots & 0 \end{bmatrix}^{\mathrm{T}}. \quad (5.49)
$$

Hence, the $n$ LMIs in (5.36) indicate that the individual steady-state state variance is not more than the pre-specified value. In other words, the requirements (R1) and (R2) are simultaneously satisfied. The proof is complete. □

**Remark 5.2** *Notice that the obtained conditions in Theorem 5.3.2 are not all of LMI form because of (5.37), which cannot be solved directly using the LMI Toolbox of Matlab. However, with the cone complementarity linearization (CCL) procedure in Ref. [139, 140], we can handle the non-convex feasibility problem by formulating them into some sequential optimization problems subject to LMI constraints.*

Now using the cone complementarity linearization (CCL) procedure, we suggest the following minimization problem involving LMI conditions instead of the original non-convex feasibility problem formulated in Theorem 5.3.2.

***Problem RFCD (robust fault-tolerant controller design)***

$$
\min \sum_{i=1}^{q} \alpha_i \beta_i \text{ subject to (5.33) to (5.36) and}
$$

$$
\begin{bmatrix} \alpha_i & 1 \\ 1 & \beta_i \end{bmatrix} \geq 0, \quad i = 1, 2, \dots, q. \tag{5.50}
$$

According to Refs [139, 142], if the solution of the above minimization problem is $q$, that is, $\min \left( \sum_{i=1}^{q} \alpha_i \beta_i \right) = q$, then the condition in Theorem 5.3.2 is solvable. Although it is still not guaranteed to always find a global optimal solution for the problem above, the proposed minimization problem is much easier to solve than the original non-convex feasibility problem.

***Algorithm RFCD***

*Step 1.* Find a feasible set $(Q^{(0)}, N^{(0)}, \alpha_i^{(0)}, \beta_i^{(0)}, \varepsilon^{(0)}, i = 1, 2, \dots, q)$ satisfying (5.33) to (5.36) and (5.50). Set $d = 0$.

*Step 2.* Solve the following optimization problem:

$$
\min \sum_{i=1}^{q} \left[ \overset{(d)}{\alpha_i} \beta_i + \alpha_i \overset{(d)}{\beta_i} \right]
$$

subject to (5.33) to (5.36) and (5.50),

and denote $f^*$ to be the optimized value.

*Step 3.* Substitute the obtained matrix variables $(Q, N, \alpha_i, \beta_i, \varepsilon)$ into (5.34) and (5.35). If conditions (5.34) and (5.35) are both satisfied, with

$$
|f^* - 2q| < \nu \tag{5.51}
$$

where $\nu$ is a sufficiently small positive scalar, then output the feasible solutions $(Q, N, \alpha_i, \beta_i, \varepsilon)$. EXIT.

*Step 4.* If $d > Num$, where $Num$ is the maximum number of iterations allowed, EXIT.

*Step 5.* Set $d = d + 1$, $\left( Q^{(d)}, N^{(d)}, \alpha_i^{(d)}, \beta_i^{(d)}, \varepsilon^{(d)} \right) = (Q, N, \alpha_i, \beta_i, \varepsilon)$, and go to *Step 2*.

**Remark 5.3** *The algorithm RFCD suggests an LMI approach to solve the multi-objective fault-tolerant controller design problem. The Matlab LMI control toolbox implements state-of-the-art interior-point LMI solvers. While these solvers are significantly faster than classical convex optimization algorithms, it should be kept in mind that the complexity of LMI computations remains higher than that of solving, say, a Riccati equation. For instance, problems with a thousand design variables typically take over an hour on workstations of today. However, research on LMI optimization is a very active area in the applied math, optimization, and the operations research community, and substantial speed-ups can be expected in the future.*

## 5.4 Numerical Example

In this section, we present an illustrative example to demonstrate the effectiveness of the proposed algorithms.

A single degree-of-freedom (SDOF) model simulating a single-storey building, taken from Ref. [139], is illustrated in Figure 5.1. The control simulates a force applied to the structure using pre-stressed active tendons connected to a servo-controlled hydraulic actuator.

The equation of motion of this system is given by

$$m\ddot{X}(t) + c\dot{X}(t) + kX(t) = -(4k_c \cos(\alpha))u(t) - m\omega(t), \qquad (5.52)$$

where $m$, $c$, and $k$ are the mass, damping, and stiffness values respectively of the SDOF structure and $\ddot{X}(t)$, $\dot{X}(t)$, and $X(t)$ are the acceleration, velocity, and position processes respectively associated with the first floor of the structure. Additionally, $k_c$ and $\alpha$ are constants associated with the structure of the controller, $u(t)$ is the position of the

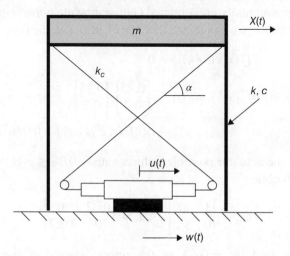

**Figure 5.1** Single degree-of-freedom structure with active tendon control

controlling actuator, and $\omega(t)$ is the ground acceleration, modeled as a Gaussian white noise process. The parameters of the structure are $m = 16.19\,\text{s}^2/\text{in}$, $c = 9.02\,\text{s}/\text{in}$, $k = 7934/\text{in}$, $k_c = 2124/\text{in}$, $\alpha = 36°$, $\mathbb{E}\{\omega(t)\} = 0$, and $\mathbb{E}\{\omega(t)\omega^T(\tau)\} = \delta(t - \tau)$.

In order to use Matlab to analyze the system performance, firstly we discretize system (5.52) into the corresponding discrete-time counterpart:

$$x(k + 1) = \begin{bmatrix} 0.9764 & 0.0099 \\ -4.7034 & 0.9710 \end{bmatrix} x(k) + \begin{bmatrix} -0.0202 \\ -4.0293 \end{bmatrix} u(k)$$

$$+ \begin{bmatrix} 0 \\ -0.0099 \end{bmatrix} \omega(k).$$

It is worth mentioning that, in Ref. [139], the system parameters such as the $k$, $k_c$, and $c$ are constant. However, in the real world, these parameters are usually time-varying since the circumstances are changeable. Taking this into account, we propose the following nonlinear model that is of more practical significance:

$$x(k + 1) = \begin{bmatrix} 0.9764 & 0.0099 \\ -4.7034 & 0.9710 \end{bmatrix} x(k) + \begin{bmatrix} -0.0202 \\ -4.0293 \end{bmatrix} u(k)$$

$$+ \begin{bmatrix} 0 \\ -0.0099 \end{bmatrix} \omega(k) + f(x(k), u(k)),$$

where $f(x(k))$ is a nonlinear term reflecting the perturbations acting on the system state and control input, which are caused by random parameters variation and modeling errors. Let the nonlinear function be of the following form:

$$f(x(k), u(k)) = \begin{bmatrix} 0.25 \\ 0.3 \end{bmatrix} (0.1x_1(k)\xi(k) + 0.1x_2(k)\eta(k) + 0.2u(k)\gamma(k)),$$

where $x_i(k)$ $(i = 1, 2)$ stands for the $i$th element of the system state, and $\xi(k)$, $\eta(k)$, and $\gamma(k)$ represent three mutually uncorrelated Gaussian white noise sequences with unity covariances. Therefore, we can check that $f(x(k), u(k))$ satisfies

$$\mathbb{E}\{f(x(k), u(k))\} = 0,$$

$$\mathbb{E}\{f(x(k), u(k))f^T(x(k), u(k))\} = \begin{bmatrix} 0.25 \\ 0.3 \end{bmatrix} \begin{bmatrix} 0.25 \\ 0.3 \end{bmatrix}^T$$

$$\times \{0.01x^T(k)x(k) + 0.04u^T(k)u(k)\}.$$

Assuming that the actuator possible failures satisfy $0.98 \le g_1(k) \le 1.02$ and $0.4 \le g_2(k) \le 2$, we can obtain

$$G_0 = \begin{bmatrix} 1 & 0 \\ 0 & 1.2 \end{bmatrix}, \quad L = \begin{bmatrix} 0.02 & 0 \\ 0 & 0.67 \end{bmatrix}.$$

Choosing $\sigma_1^2 = 1.44$, $\sigma_2^2 = 0.64$ as the upper bounds of the state variances, we employ the Matlab LMI Toolbox. Setting $v = 0.000001$ and by solving

Problem RFCD with help from Algorithm RFCD, we obtain a feasible solution as follows:

$$Q = \begin{bmatrix} 0.0206 & -0.0208 \\ -0.0208 & 0.7996 \end{bmatrix},$$

$$N = \begin{bmatrix} -0.0223 & 0.1138 \end{bmatrix},$$

$$\alpha_1 = 0.0104, \quad \alpha_2 = 0.0099,$$

$$\beta_1 = 95.8830, \quad \beta_2 = 100.9796.$$

The desired state feedback control law is

$$K = \begin{bmatrix} -0.9665 & 0.0976 \end{bmatrix}.$$

The simulation results are shown in Figures 5.2 to 5.4, which have clearly shown that the designed control law stabilizes the unstable open-loop system and also makes the steady-state variance meet the given upper bound constraints. Figure 5.2 represents the state responses of the open-loop without control. It is clearly seen that the original system is unstable. Figure 5.3 shows the corresponding state responses of closed-loop systems with the state feedback fault-tolerant control scheme, from which we can see that the system trajectories are driven to the stable origin quickly. Figure 5.4 illustrates the evolution of the system steady-state variances. Figures 5.5 to 5.7 show the control input, external noise sequence $\omega(k)$, and the stochastic nonlinearity $f(x(k), u(k))$

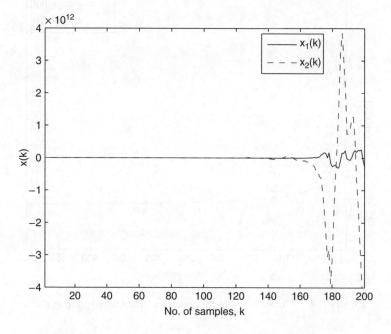

**Figure 5.2**  The state responses of the uncontrolled system

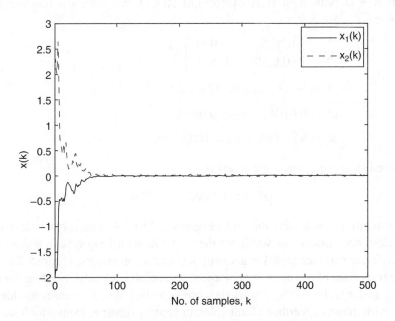

**Figure 5.3**    The response of $x(k)$ of the controlled system

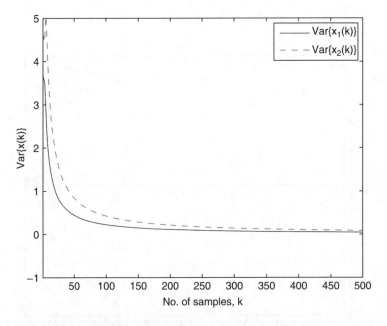

**Figure 5.4**    The steady-state variances of the closed-loop system

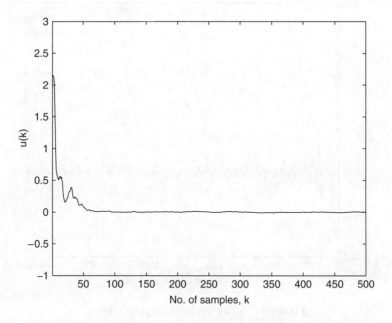

**Figure 5.5**   The control input

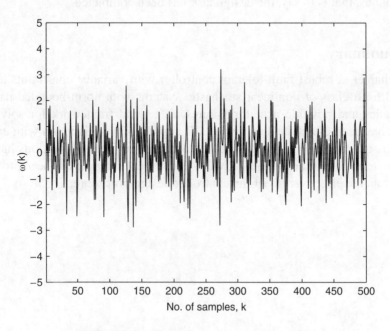

**Figure 5.6**   The white noise sequence $\omega(k)$

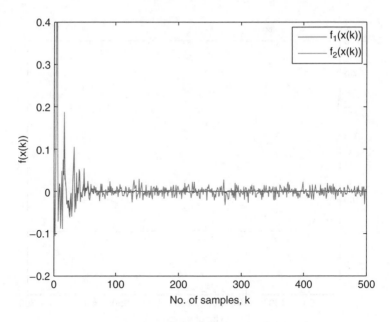

**Figure 5.7**   The nonlinearity $f(x(k), u(k))$

respectively. Obviously, both of the variances are constrained below the pre-specified upper bounds, that is to say, the design task has been completed.

## 5.5   Summary

In this chapter, a robust fault-tolerant controller with variance constraints has been designed for a class of nonlinear stochastic systems with norm-bounded parameter uncertainties and possible actuator failures. A general framework for solving this problem has been established using an LMI approach in conjunction with exponential mean-square stability and variance constraints. Sufficient conditions have been derived in terms of a set of feasible LMIs. A numerical example has been provided to demonstrate the effectiveness of the controller design procedure.

# 6

# Robust $H_2$ Sliding Mode Control

In the past two decades, the sliding mode control (SMC) problem of continuous-time systems has been extensively studied and the corresponding results have been widely applied in various fields, due primarily to the strong robustness of the sliding motion on the sliding surface against model uncertainties, parameter variations, and external disturbances. Since most control strategies are implemented digitally, the SMC problem in the discrete-time setting has recently started to gain some research interests and a number of results have been reported in the literature.

On the other hand, as it is now well recognized that stochastic systems are suitable to be used to model physical systems, financial systems, ecological systems, as well as social systems, the stochastic control problem has drawn much research attention and many analysis as well as synthesis methodologies have been developed. In recent years, the SMC problem for stochastic systems has stirred considerable research interests and a series of papers have been published. However, it should be pointed out that the existing literature on the discrete-time SMC problem has been concerned with deterministic linear systems with or without matched nonlinear disturbances. In terms of the importance of both the discrete-time system and stochastic system in practice, it seems quite promising to investigate the SMC method for discrete stochastic systems. Unfortunately, up to now, the SMC problem for discrete stochastic systems has not yet attracted much attention despite its practical significance, due primarily to the mathematical difficulties. Up to now, to the best of the authors' knowledge, the robust $H_2$ SMC control problem has not been studied for uncertain discrete stochastic systems with either matched or unmatched nonlinearities, and the purpose of this chapter is therefore to shorten such a gap.

In this chapter, the sliding mode control problem for uncertain nonlinear discrete-time stochastic systems with $H_2$ performance constraints is discussed. The nonlinearities considered are of both the matched and unmatched forms, which can encompass several classes of well-studied nonlinearities. A new discrete-time switching function is proposed and then a sufficient condition is derived, which

*Variance-Constrained Multi-Objective Stochastic Control and Filtering*, First Edition.
Lifeng Ma, Zidong Wang and Yuming Bo.

ensures both the exponential mean-square stability and the $H_2$ performance in the sliding surface. Such a condition is expressed as the feasibility of a set of linear matrix inequalities (LMIs) with an equality constraint. An improved discrete-time reaching condition is established to obtain the required SMC controller capable of driving the state trajectories onto the specified switching surface, along which a non-increasing zigzag type sliding motion then results. A simulation example is given to show the effectiveness of the proposed method.

The main contributions can be summarized as follows: (1) for the addressed nonlinear discrete stochastic systems, a new form of discrete switching function is proposed so that we can handle both the matched and unmatched external nonlinearities; (2) the $H_2$ performance is considered, for the first time, for SMC problems with the hope to better evaluate the performances of stochastic systems.

The rest of this chapter is arranged as follows: Section 6.1 formulates the uncertain nonlinear stochastic discrete system to be studied. In Section 6.2, a novel switching function is first designed and then an LMI-based sufficient condition is given to obtain the parameters of the proposed switching function so that both the exponential mean-square stability and the $H_2$ index can be achieved in the sliding surface. An SMC law is then synthesized in Section 6.3 to drive the state trajectories onto the specified surface with probability 1. In Section 6.4, an illustrative numerical example is provided to show the effectiveness and usefulness of the proposed approach. Section 6.5 gives our summary.

## 6.1   The System Model

Consider the following uncertain nonlinear discrete-time stochastic system:

$$\begin{cases} x(k+1) = (A + \Delta A)x(k) + B(u(k) + f(x(k))) + D_1 g(x(k)) + D_2 \omega(k), \\ \quad y(k) = Lx(k) + h(x(k)), \end{cases} \tag{6.1}$$

where $x(k) \in \mathbb{R}^n$ is the state, $u(k) \in \mathbb{R}^m$ is the input, $y(k) \in \mathbb{R}^r$ is a combination of the states to be controlled, and $\omega(k) \in \mathbb{R}^l$ is a zero mean Gaussian white noise process with covariance $W = VV^T > 0$. $A, B, D_1, D_2$ and $L$ are known constant matrices with appropriate dimensions.

The matrix $\Delta A$ is real-valued, which represents the norm-bounded parameter uncertainty as follows:

$$\Delta A = HFE, \tag{6.2}$$

where $H$ and $E$ are known real constant matrices that characterize how the deterministic uncertain parameter in $F$ enters the nominal matrix $A$ with

$$F^T F \leq I. \tag{6.3}$$

The parameter uncertainty $\Delta A$ is said to be admissible if both (6.2) and (6.3) are satisfied. In this chapter, $\Delta A$ is assumed to be bounded in the Euclidean norm.

The unknown nonlinear function $f(x(k))$ represents the *matched* nonlinear external disturbance. The functions $g(x(k))$ and $h(x(k))$ are *unmatched* stochastic nonlinear functions of the states, which have the following first moment for all $x(k)$:

$$\mathbb{E}\left\{\left.\begin{bmatrix} g(x(k)) \\ h(x(k)) \end{bmatrix}\right| x(k)\right\} = 0, \tag{6.4}$$

with the covariance given by

$$\mathbb{E}\left\{\left.\begin{bmatrix} g(x(k)) \\ h(x(k)) \end{bmatrix}\begin{bmatrix} g^{\mathrm{T}}(x(j)) & h^{\mathrm{T}}(x(j)) \end{bmatrix}\right| x(k)\right\} = 0, \qquad k \neq j, \tag{6.5}$$

$$\mathbb{E}\left\{\left.\begin{bmatrix} g(x(k)) \\ h(x(k)) \end{bmatrix}\begin{bmatrix} g^{\mathrm{T}}(x(k)) & h^{\mathrm{T}}(x(k)) \end{bmatrix}\right| x(k)\right\} = \sum_{i=1}^{q} \begin{bmatrix} \Omega_i & \Xi_i \\ \Xi_i^{\mathrm{T}} & \Sigma_i \end{bmatrix} x^{\mathrm{T}}(k)\Gamma_i x(k), \tag{6.6}$$

where $\Omega_i$, $\Xi_i$, $\Sigma_i$ and $\Gamma_i$ $(i = 1, 2, \ldots, q)$ are known positive definite matrices with appropriate dimensions.

The objective of this chapter is to design an SMC law such that the desired control performance is obtained for the resultant closed-loop stochastic system despite parameter uncertainty and unmatched external disturbances.

## 6.2   Robust $H_2$ Sliding Mode Control

In this section, a new switching function is first presented for the uncertain nonlinear discrete-time stochastic system (6.1). Then, two theorems are given to determine the parameters appearing in the proposed switching function so that the specified performance requirements can be satisfied. It is shown that the controller designing problem in the sliding motion can be solved if a set of LMIs with an equality constraint is feasible.

### 6.2.1   Switching Surface

As the first step of the design procedure, in this work the following switching function is chosen:

$$s(k) = Gx(k) - G(A + BK)x(k - 1), \tag{6.7}$$

where $G$ is designed so that $GB$ is nonsingular and $GD = 0$ with $D$ defined by $D = [D_1 \; D_2]$. Both the parameters $G$ and $K$ are to be determined later to achieve the prescribed performances.

The ideal quasi-sliding mode satisfies

$$s(k + 1) = s(k) = 0. \tag{6.8}$$

Then, one can obtain from (6.1), (6.7), and (6.8) that

$$s(k + 1) = Gx(k + 1) - G(A + BK)x(k)$$
$$= G(\Delta A - BK)x(k) + GB(u(k) + f(x(k)))$$
$$= 0. \tag{6.9}$$

Solving the above equation for $u(k)$, the equivalent control law of the sliding motion is given by

$$u_{eq}(k) = (GB)^{-1}G(BK - \Delta A)x(k) - f(x(k)). \tag{6.10}$$

Substituting (6.10) into (6.1) yields

$$\begin{cases} x(k + 1) = A_K x(k) + D_1 g(x(k)) + D_2 \omega(k), \\ \quad y(k) = Lx(k) + h(x(k)), \end{cases} \tag{6.11}$$

where $A_K \triangleq A + \Delta A + B(GB)^{-1}G(BK - \Delta A)$. Obviously, the system (6.11) gives the sliding mode dynamics of (6.1) on the specified switching surface $s(k + 1) = 0$.

Before stating our main objective, we introduce the following stability concept for the system (6.11).

**Definition 6.2.1** *The system (6.11) is said to be exponentially mean-square stable if, with $\omega(k) = 0$, there exist constants $\eta \geq 1$ and $\zeta \in (0, 1)$ such that*

$$\mathbb{E}\{\|x(k)\|^2\} \leq \eta\zeta^k \mathbb{E}\{\|x_0\|^2\}, \qquad \forall x_0 \in \mathbb{R}^n, k \in \mathbb{I}^+,$$

*for all admissible uncertainties.*

Our objective in this chapter is to design an SMC law such that the state trajectories in (6.1) are globally driven onto (with probability 1) the specified sliding surface. Moreover, the following two requirements are simultaneously satisfied after the system state trajectories enter into the sliding motion:

(R1) The sliding motion dynamics (6.11) is exponentially mean-square stable.
(R2) For a given constant $\gamma > 0$, the following $H_2$ performance index is satisfied:

$$J \triangleq \lim_{k \to \infty} \mathbb{E}\{\|y(k)\|^2\} < \gamma.$$

## 6.2.2 Performances of the Sliding Motion

We now analyze the exponential mean-square stability and the $H_2$ performance constraint of the sliding motion described by (6.11) via an LMI method. First of all, we introduce the following lemma, which will be used in this chapter.

**Lemma 6.2.2** *For any real vectors a, b and matrix $Q > 0$ of compatible dimensions, we have*

$$a^\mathrm{T}b + b^\mathrm{T}a \leq a^\mathrm{T}Qa + b^\mathrm{T}Q^{-1}b. \tag{6.12}$$

*Proof.* Since $Q$ is symmetric, we can always find an invertible matrix $T$ such that

$$T^{-1}QT = \Lambda, \tag{6.13}$$

where

$$\Lambda = \text{diag}\{\lambda_1, \lambda_2, \ldots, \lambda_n\} \tag{6.14}$$

with $\lambda_i$ $(i = 1, 2, \ldots)$ being the eigenvalues of $Q$. Obviously, from (6.13), the following transformation is also true:

$$T^{-1}Q^{-1}T = \Lambda^{-1}. \tag{6.15}$$

Moreover, since $Q > 0$, it can be seen that $\lambda_i > 0$. We could define

$$\sqrt{\Lambda} \triangleq \text{diag}\{\sqrt{\lambda_1}, \sqrt{\lambda_2}, \ldots, \sqrt{\lambda_n}\}. \tag{6.16}$$

Therefore, we could select

$$\sqrt{Q} = T\sqrt{\Lambda}T^{-1} \tag{6.17}$$

and

$$\sqrt{Q^{-1}} = T\sqrt{\Lambda^{-1}}T^{-1}. \tag{6.18}$$

Next, it is easy to see that

$$(a^{\text{T}}\sqrt{Q} - b^{\text{T}}\sqrt{Q^{-1}})(a^{\text{T}}\sqrt{Q} - b^{\text{T}}\sqrt{Q^{-1}})^{\text{T}} \geq 0. \tag{6.19}$$

It follows that

$$(a^{\text{T}}\sqrt{Q} - b^{T}\sqrt{Q^{-1}})(\sqrt{Q}a - \sqrt{Q^{-1}}b) \geq 0. \tag{6.20}$$

Expanding the left side of the above inequality, we can obtain

$$a^{\text{T}}\sqrt{Q}\sqrt{Q}a + b^{T}\sqrt{Q^{-1}}\sqrt{Q^{-1}}b - a^{\text{T}}\sqrt{Q}\sqrt{Q^{-1}}b - b^{T}\sqrt{Q^{-1}}\sqrt{Q}a \geq 0, \tag{6.21}$$

which means

$$a^{\text{T}}Qa + b^{T}Q^{-1}b \geq a^{\text{T}}b + b^{\text{T}}a. \tag{6.22}$$

The proof is complete.                                                              □

The following lemma is useful in the derivation of the main results.

**Lemma 6.2.3** *Considering system (6.11) with given $A_K$. The following statements are equivalent:*

*1.*

$$\rho(\Upsilon) < 1 \tag{6.23}$$

*where* $\Upsilon \triangleq A_K \otimes A_K + \sum_{i=1}^{q} \text{st}(D_1\Omega_i D_1^{\text{T}})\text{st}^{\text{T}}(\Gamma_i)$.

*2. There exists $P > 0$ such that*

$$A_K^{\mathrm{T}} P A_K - P + \sum_{i=1}^{q} \Gamma_i \mathrm{tr}[P D_1 \Omega_i D_1^{\mathrm{T}}] < 0. \tag{6.24}$$

*3. The system (6.11) is exponentially mean-square stable.*

**Lemma 6.2.4** *Consider the system*

$$\xi(k+1) = M\xi(k) + g(\xi(k)), \tag{6.25}$$

*where $\mathbb{E}\{g(\xi(k))|\xi(k)\} = 0$ and $\mathbb{E}\{g(\xi(k))g^{\mathrm{T}}(\xi(k))|\xi(k)\} = \sum_{i=1}^{q} \Omega_i \xi^{\mathrm{T}}(k) \Gamma_i \xi(k)$, with $\Omega_i$ and $\Gamma_i$ $(i = 1, 2, \ldots, q)$ being defined in (6.6). If the system (6.25) is exponentially mean-square stable and there exists a symmetric matrix $Y$ satisfying*

$$M^{\mathrm{T}} Y M - Y + \sum_{i=1}^{q} \Gamma_i \mathrm{tr}[Y \Omega_i] < 0, \tag{6.26}$$

*then $Y \geq 0$.*

*Proof.* This lemma can be easily proved by using the Lyapunov method combined with Lemma 6.2.3; hence the proof is omitted. □

Define the state covariance of system (6.11) by

$$\hat{X}(k) \triangleq \mathbb{E}\{x(k) x^{\mathrm{T}}(k)\}$$

$$= \mathbb{E}\left\{ \begin{bmatrix} x_1(k) & x_2(k) & \cdots & x_n(k) \end{bmatrix} \begin{bmatrix} x_1(k) & x_2(k) & \cdots & x_n(k) \end{bmatrix}^{\mathrm{T}} \right\}. \tag{6.27}$$

Then the Lyapunov-type equation that governs the evolution of $\hat{X}(k)$ can be derived from (6.11) as follows:

$$\hat{X}(k+1) = A_K \hat{X}(k) A_K^{\mathrm{T}} + \sum_{i=1}^{q} D_1 \Omega_i D_1^{\mathrm{T}} \mathrm{tr}[\hat{X}(k) \Gamma_i] + D_2 W D_2^{\mathrm{T}}, \tag{6.28}$$

which can be rewritten in terms of the stack matrix by

$$\mathrm{st}(\hat{X}(k+1)) = \Upsilon \mathrm{st}(\hat{X}(k)) + \mathrm{st}(D_2 W D_2^{\mathrm{T}}), \tag{6.29}$$

where $\Upsilon$ is defined in (6.23).

If the system (6.11) is exponentially mean-square stable, it then follows from Lemma 6.2.3 that $\rho(\Upsilon) < 1$, which indicates that $\hat{X}(k)$ will converge to a constant positive definite matrix $\hat{X}$ when $k \to \infty$ and, in the steady state, (6.28) becomes

$$\hat{X} = A_K \hat{X} A_K^{\mathrm{T}} + \sum_{i=1}^{q} D_1 \Omega_i D_1^{\mathrm{T}} \mathrm{tr}[\hat{X} \Gamma_i] + D_2 W D_2^{T}. \tag{6.30}$$

Therefore, the $H_2$ performance defined in ($R2$) can be expressed by

$$J = \lim_{k \to \infty} \mathbb{E}\{\|y(k)\|^2\} = \lim_{k \to \infty} \left\{ \text{tr}[L\hat{X}(k)L^T] + \sum_{i=1}^{q} \text{tr}[\Sigma_i]\text{tr}[\Gamma_i\hat{X}(k)] \right\}$$

$$= \text{tr}[L\hat{X}L^T] + \sum_{i=1}^{q} \text{tr}[\Sigma_i]\text{tr}[\Gamma_i\hat{X}]. \tag{6.31}$$

Let us now derive an equivalent expression of the $H_2$ performance (6.31). Consider the following recursion:

$$\hat{Q}(k) = A_K^T\hat{Q}(k+1)A_K + \sum_{i=1}^{q} \Gamma_i\text{tr}[D_1\Omega_iD_1^T\hat{Q}(k+1)] + L^TL + \sum_{i=1}^{q} \text{tr}[\Sigma_i]\Gamma_i. \tag{6.32}$$

Rewrite (6.32) in the stack matrix form by

$$\text{st}(\hat{Q}(k)) = \Upsilon^T\text{st}(\hat{Q}(k+1)) + \text{st}(L^TL + \sum_{i=1}^{q} \text{tr}[\Sigma_i]\Gamma_i). \tag{6.33}$$

Similar to what we have done before, if the system (6.11) is exponentially mean-square stable, then according to Lemma 6.2.3 and the fact that $\rho(\Upsilon^T) = \rho(\Upsilon) < 1$, one can see that $\hat{Q}(k)$ in (6.33) will converge to a constant matrix $\hat{Q}$ when $k \to \infty$. Therefore, in the steady state, (6.32) can be written as

$$\hat{Q} = A_K^T\hat{Q}A_K + \sum_{i=1}^{q} \Gamma_i\text{tr}[D_1\Omega_iD_1^T\hat{Q}] + L^TL + \sum_{i=1}^{q} \text{tr}[\Sigma_i]\Gamma_i. \tag{6.34}$$

We are now ready to present a theorem that gives an upper bound on the $H_2$ performance $J$.

**Theorem 6.2.5** *Consider the system (6.11) with given $A_K$. If there exists a positive definite matrix $Q$ satisfying*

$$A_K^TQA_K - Q + \sum_{i=1}^{q} \Gamma_i\text{tr}[D_1\Omega_iD_1^TQ] + L^TL + \sum_{i=1}^{q} \text{tr}[\Sigma_i]\Gamma_i < 0, \tag{6.35}$$

*then the system (6.11) is exponentially mean-square stable and $J \leq \text{tr}[D_2WD_2^TQ]$.*

*Proof.* First, from (6.35), it is easy to see that

$$A_K^TQA_K - Q + \sum_{i=1}^{q} \Gamma_i\text{tr}[D_1\Omega_iD_1^TQ] < -L^TL - \sum_{i=1}^{q} \text{tr}[\Sigma_i]\Gamma_i < 0, \tag{6.36}$$

which implies directly from Lemma 6.2.3 that the system (6.11) is exponentially mean-square stable and then both (6.30) and (6.34) are true. Then, we have

$$\lim_{k \to \infty} \mathrm{tr}[\hat{X}(k+1)\hat{Q}(k+1) - \hat{X}(k)\hat{Q}(k)]$$

$$= \lim_{k \to \infty} \mathrm{tr} \left\{ \left[ A_K \hat{X}(k) A_K^{\mathrm{T}} + \left( \sum_{i=1}^{q} D_1 \Omega_i D_1^{\mathrm{T}} \mathrm{tr}[\hat{X}(k)\Gamma_i] \right) + D_2 W D_2^{T} \right] \hat{Q}(k+1) \right.$$

$$\left. -\hat{X}(k) \left[ A_K^{\mathrm{T}} \hat{Q}(k+1) A_K + \sum_{i=1}^{q} \Gamma_i \mathrm{tr}[D_1 \Omega_i D_1^{\mathrm{T}} \hat{Q}(k+1)] + L^{\mathrm{T}} L + \sum_{i=1}^{q} \mathrm{tr}[\Sigma_i]\Gamma_i \right] \right\}$$

$$= 0 \tag{6.37}$$

and the $H_2$ performance index can be written as

$$J = \mathrm{tr}[L\hat{X}L^{\mathrm{T}}] + \sum_{i=1}^{q} \mathrm{tr}[\Sigma_i]\mathrm{tr}[\Gamma_i \hat{X}] = \mathrm{tr}[D_2 W D_2^{\mathrm{T}} \hat{Q}]. \tag{6.38}$$

Subtracting (6.34) from (6.35) results in

$$A_K^{\mathrm{T}}(Q - \hat{Q})A_K - (Q - \hat{Q}) + \sum_{i=1}^{q} \Gamma_i \mathrm{tr}[D_1 \Omega_i D_1^{\mathrm{T}}(Q - \hat{Q})] < 0, \tag{6.39}$$

which indicates from Lemma 6.2.4 that $(Q - \hat{Q}) \geq 0$. We can now conclude that $J \leq \mathrm{tr}[D_2 W D_2^{\mathrm{T}} Q]$ and then the proof is complete. □

Notice that there exist parameter uncertainty in (6.39). In the next stage, we aim to "eliminate" the parameter uncertainty and then propose the design technique for the switching function parameters $G$ and $K$ in order to realize both the requirements ($R1$) and ($R2$).

**Theorem 6.2.6** *Given a positive scalar $\gamma > 0$. If there exist positive definite matrices $Q > 0$, $\Theta > 0$, positive scalars $\varepsilon_1 > 0$, $\varepsilon_2 > 0$, $\alpha_i > 0$ ($i = 1, 2, \ldots, q$), and a real-valued matrix $Z$ satisfying*

$$e_1^{\mathrm{T}}\Theta e_1 + e_2^{\mathrm{T}}\Theta e_2 + \cdots + e_n^{\mathrm{T}}\Theta e_n < \gamma, \tag{6.40}$$

$$\begin{bmatrix} -\Theta & V^{\mathrm{T}}D_2^{\mathrm{T}}Q \\ * & -Q \end{bmatrix} < 0, \tag{6.41}$$

$$\begin{bmatrix} -\alpha_i & \pi_i^{\mathrm{T}}D_1^{\mathrm{T}}Q \\ * & -Q \end{bmatrix} < 0, \tag{6.42}$$

$$\begin{bmatrix} \Pi & \sqrt{2}A^{\mathrm{T}}Q & \sqrt{2}Z^{\mathrm{T}} & 0 & 0 \\ * & -Q & 0 & \sqrt{2}QH & 0 \\ * & * & -B^{\mathrm{T}}QB & 0 & \sqrt{2}B^{\mathrm{T}}QH \\ * & * & * & -\varepsilon_1 I & 0 \\ * & * & * & * & -\varepsilon_2 I \end{bmatrix} < 0, \tag{6.43}$$

$$B^{\mathrm{T}}QD = 0, \tag{6.44}$$

*where*

$$e_j = [\underbrace{0 \; \cdots \; 0}_{j-1} \; 1 \; \underbrace{0 \; \cdots \; 0}_{n-j}] \quad j = 1, 2, \ldots, n,$$

$$\Pi = -Q + \sum_{i=1}^{q} (\alpha_i + \mathrm{tr}[\Sigma_i])\Gamma_i + L^\mathrm{T}L + \varepsilon_1 E^\mathrm{T}E + \varepsilon_2 E^\mathrm{T}E,$$

$$G = B^\mathrm{T}Q,$$

*then the system (6.11) is exponentially mean-square stable and the $H_2$ performance index $J < \gamma$ is satisfied in the sliding motion. Moreover, the parameter $K$ can be determined by $K = (B^\mathrm{T}QB)^{-1}Z$.*

*Proof.* Without loss of any generality, we assume $\Omega_i = \pi_i \pi_i^\mathrm{T}$ $(i = 1, 2, \ldots, q)$. Using the Schur Complement Lemma to (6.42), we have

$$-\alpha_i + \pi_i^\mathrm{T}D_1^\mathrm{T}QD_1\pi_i < 0 \iff \mathrm{tr}[D_1^\mathrm{T}QD_1\Omega_i] < \alpha_i. \tag{6.45}$$

It is easy to see that

$$\begin{aligned}
A_K^\mathrm{T}QA_K &= (A + \Delta A)^\mathrm{T}Q(A + \Delta A) \\
&\quad + (B(GB)^{-1}G(BK - \Delta A))^\mathrm{T}Q(B(GB)^{-1}G(BK - \Delta A)) \\
&\quad + (A + \Delta A)^\mathrm{T}Q(B(GB)^{-1}G(BK - \Delta A)) \\
&\quad + (B(GB)^{-1}G(BK - \Delta A))^\mathrm{T}Q(A + \Delta A). \tag{6.46}
\end{aligned}$$

Since $G = B^\mathrm{T}Q$, we obtain

$$\begin{aligned}
&(B(GB)^{-1}G(BK - \Delta A))^\mathrm{T}Q(B(GB)^{-1}G(BK - \Delta A)) \\
&= (BK - \Delta A)^\mathrm{T}QB(B^\mathrm{T}QB)^{-1}B^\mathrm{T}Q(BK - \Delta A). \tag{6.47}
\end{aligned}$$

Using Lemma 6.2.2, we have

$$\begin{aligned}
&(A + \Delta A)^\mathrm{T}Q(B(GB)^{-1}G(BK - \Delta A)) \\
&+ (B(GB)^{-1}G(BK - \Delta A))^\mathrm{T}Q(A + \Delta A) \\
&\leq (A + \Delta A)^\mathrm{T}Q(A + \Delta A) + (BK - \Delta A)^\mathrm{T}QB(B^\mathrm{T}QB)^{-1}B^\mathrm{T}Q(BK - \Delta A) \tag{6.48}
\end{aligned}$$

and therefore

$$\begin{aligned}
A_K^\mathrm{T}QA_K &\leq 2(A + \Delta A)^\mathrm{T}Q(A + \Delta A) \\
&\quad + 2(BK - \Delta A)^\mathrm{T}QB(B^\mathrm{T}QB)^{-1}B^\mathrm{T}Q(BK - \Delta A). \tag{6.49}
\end{aligned}$$

It follows from (6.45) and (6.49) that

$$\Phi \triangleq A_K^T Q A_K - Q + \sum_{i=1}^{q} \Gamma_i \text{tr}[D_1 \Omega_i D_1^T Q] + L^T L + \sum_{i=1}^{q} \text{tr}[\Sigma_i] \Gamma_i$$

$$< 2(A + \Delta A)^T Q (A + \Delta A)$$

$$+ 2(BK - \Delta A)^T QB(B^T QB)^{-1} B^T Q(BK - \Delta A)$$

$$- Q + \sum_{i=1}^{q} (\alpha_i + \text{tr}[\Sigma_i]) \Gamma_i + L^T L \triangleq \Psi. \tag{6.50}$$

We now proceed to prove that the LMI (6.43) is equivalent to $\Psi < 0$. By using the Schur Complement Lemma twice, $\Psi < 0$ if and only if

$$\begin{bmatrix} \hat{\Pi} & \sqrt{2}(A + \Delta A)^T Q & \sqrt{2}(BK - \Delta A)^T QB \\ * & -Q & 0 \\ * & * & -B^T QB \end{bmatrix} < 0, \tag{6.51}$$

where $\hat{\Pi} = -Q + \sum_{i=1}^{q} (\alpha_i + \text{tr}[\Sigma_i]) \Gamma_i + L^T L$.

In order to eliminate the uncertain parameter $\Delta A$, we can rewrite (6.51) as

$$\begin{bmatrix} \hat{\Pi} & \sqrt{2}A^T Q & \sqrt{2}K^T B^T QB \\ * & -Q & 0 \\ * & * & -B^T QB \end{bmatrix}$$

$$+ \begin{bmatrix} 0 \\ \sqrt{2}QH \\ 0 \end{bmatrix} F \begin{bmatrix} E & 0 & 0 \end{bmatrix} + \begin{bmatrix} E^T \\ 0 \\ 0 \end{bmatrix} F \begin{bmatrix} 0 & \sqrt{2}H^T Q & 0 \end{bmatrix}$$

$$+ \begin{bmatrix} 0 \\ 0 \\ \sqrt{2}B^T QH \end{bmatrix} (-F) \begin{bmatrix} E & 0 & 0 \end{bmatrix} + \begin{bmatrix} E^T \\ 0 \\ 0 \end{bmatrix} (-F) \begin{bmatrix} 0 & 0 & \sqrt{2}H^T QB \end{bmatrix} < 0. \tag{6.52}$$

Using the $S$-procedure twice to (6.52), we can know easily that it is true if and only if there exist two positive scalars $\varepsilon_1$ and $\varepsilon_2$ such that

$$\begin{bmatrix} \hat{\Pi} & \sqrt{2}A^T Q & \sqrt{2}K^T B^T QB \\ * & -Q & 0 \\ * & * & -B^T QB \end{bmatrix} + \varepsilon_1^{-1} \begin{bmatrix} 0 \\ \sqrt{2}QH \\ 0 \end{bmatrix} \begin{bmatrix} 0 & \sqrt{2}H^T Q & 0 \end{bmatrix}$$

$$+ \varepsilon_1 \begin{bmatrix} E^T \\ 0 \\ 0 \end{bmatrix} \begin{bmatrix} E & 0 & 0 \end{bmatrix} + \varepsilon_2^{-1} \begin{bmatrix} 0 \\ 0 \\ \sqrt{2}B^T QH \end{bmatrix} \begin{bmatrix} 0 & 0 & \sqrt{2}H^T QB \end{bmatrix}$$

$$+ \varepsilon_2 \begin{bmatrix} E^T \\ 0 \\ 0 \end{bmatrix} \begin{bmatrix} E & 0 & 0 \end{bmatrix} < 0. \tag{6.53}$$

Letting $Z = B^T QBK$, we know that the LMI (6.43) is equivalent to $\Psi < 0$, and therefore $\Phi < \Psi < 0$. From Theorem 6.2.5, it can be concluded that the system (6.11) is exponentially mean-square stable and the $H_2$ performance $J \leq \text{tr}[D_2 WD_2^T Q]$.

Using the Schur Complement Lemma to (6.41), we have

$$V^T D_2^T QD_2 V < \Theta, \qquad (6.54)$$

and then it follows from (6.40) that

$$J \leq \text{tr}[D_2 WD_2^T Q] = \text{tr}[V^T D_2^T QD_2 V] < \text{tr}[\Theta] < \gamma. \qquad (6.55)$$

Therefore, the requirement (R2) is met as well and the proof is now complete.  □

### 6.2.3 Computational Algorithm

It can be seen that the condition in Theorem 6.2.6 is presented in terms of the feasibility of several LMIs with an equality constraint.

Employing the algorithm proposed in Ref. [107] and noticing that the condition $B^T QD = 0$ is equivalent to $\text{tr}[(B^T QD)^T B^T QD] = 0$, we introduce the condition $(B^T QD)^T B^T QD \leq \beta I$, which, by the Schur Complement Lemma, is equivalent to

$$\begin{bmatrix} -\beta I & D^T QB \\ B^T QD & -I \end{bmatrix} \leq 0. \qquad (6.56)$$

Hence, the original non-convex feasibility problem can be converted into the following minimization problem:

$$\min \beta \quad \text{subject to} \quad (6.40), (6.41), (6.42), (6.43), \text{ and } (6.56). \qquad (6.57)$$

If this infinum equals zero, the solutions will satisfy the LMIs (6.40) to (6.43) and the equality $B^T XD = 0$. Thus, the proposed robust $H_2$ control problem is solvable.

## 6.3 Sliding Mode Controller

In this section, a sliding mode controller is designed such that (1) the trajectory of (6.1), starting from any initial state, is globally driven onto the sliding surface (6.7) in finite time; (2) once the trajectory has crossed the sliding surface for the first time, it will cross the surface again in every successive sampling period, resulting in a zigzag motion along the sliding surface; and (3) the size of each successive zigzagging step is non-increasing and the trajectory stays within a specified band, which is called the quasi-sliding mode band (QSMB) [117]. Moreover, a technique is proposed that is capable of eliminating such a zigzagging step effectively.

In the following, for any vectors

$$a \triangleq \begin{bmatrix} a_1 & a_2 & \cdots & a_m \end{bmatrix} \in \mathbb{R}^m,$$

$$b \triangleq \begin{bmatrix} b_1 & b_2 & \cdots & b_m \end{bmatrix} \in \mathbb{R}^m,$$

by $a \le b$ (respectively $a < b$) we mean $a_i \le b_i$ (respectively $a_i < b_i$), $\forall i \in \{1, 2, \ldots, m\}$.

Now, we extend the reaching condition proposed in Ref. [117] to the following form for the SMC of the uncertain nonlinear discrete-time stochastic system (6.1) with the sliding surface (6.7):

$$
\begin{cases}
\Delta s(k) = s(k+1) - s(k) \\
\qquad\quad \le -\tau\Lambda\mathrm{sgn}[s(k)] - \tau P s(k), \qquad s(k) > 0, \\
\Delta s(k) = s(k+1) - s(k) \\
\qquad\quad \ge -\tau\Lambda\mathrm{sgn}[s(k)] - \tau P s(k), \qquad s(k) < 0,
\end{cases}
\tag{6.58}
$$

where $\tau$ represents the sampling period, $\Lambda \triangleq \mathrm{diag}\{\lambda_1, \lambda_2, \ldots, \lambda_m\} \in \mathbb{R}^{m \times m}$, $P \triangleq \mathrm{diag}\{p_1, p_2, \ldots, p_m\} \in \mathbb{R}^{m \times m}$, and $\lambda_i > 0$, $p_i > 0$ $(i = 1, 2, \ldots, m)$ are properly chosen scalars satisfying $0 < 1 - \tau p_i < 1$, $\forall i \in \{1, 2, \ldots, m\}$.

Since the parameter uncertainty $\Delta A$ and the external disturbance $f(x(k))$ are both assumed to be bounded in the Euclidean norm, $\Delta_a \triangleq G\Delta Ax(k)$ and $\Delta_f \triangleq GBf(x(k))$ are also bounded. Denote $\delta_a^i$ and $\delta_f^i$ as the $i$th element in $\Delta_a$ and $\Delta_f$ respectively. Suppose the upper and lower bounds on $\Delta_a$ and $\Delta_f$ are given as follows:

$$
\begin{aligned}
\delta_{aL}^i &\le \delta_a^i \le \delta_{aU}^i, \\
\delta_{fL}^i &\le \delta_f^i \le \delta_{fU}^i, \qquad i = 1, 2, \ldots, m,
\end{aligned}
\tag{6.59}
$$

where $\delta_{aL}^i$, $\delta_{aU}^i$, $\delta_{fL}^i$, and $\delta_{fU}^i$ are all known constants. Furthermore, we denote

$$
\begin{aligned}
\hat{\Delta}_a &= \begin{bmatrix} \hat{\delta}_a^1 & \hat{\delta}_a^1 & \cdots & \hat{\delta}_a^m \end{bmatrix}^{\mathrm{T}}, \\
\tilde{\Delta}_a &= \mathrm{diag}\{\tilde{\delta}_a^1, \tilde{\delta}_a^2, \cdots, \tilde{\delta}_a^m\}, \\
\hat{\Delta}_f &= \begin{bmatrix} \hat{\delta}_f^1 & \hat{\delta}_f^1 & \cdots & \hat{\delta}_f^m \end{bmatrix}^{\mathrm{T}}, \\
\tilde{\Delta}_f &= \mathrm{diag}\{\tilde{\delta}_f^1, \tilde{\delta}_f^2, \cdots, \tilde{\delta}_f^m\},
\end{aligned}
\tag{6.60}
$$

where

$$
\hat{\delta}_a^i = \frac{\delta_{aU}^i + \delta_{aL}^i}{2},
$$

$$
\tilde{\delta}_a^i = \frac{\delta_{aU}^i - \delta_{aL}^i}{2},
$$

$$
\hat{\delta}_f^i = \frac{\delta_{fU}^i + \delta_{fL}^i}{2},
$$

$$
\tilde{\delta}_f^i = \frac{\delta_{fU}^i - \delta_{fL}^i}{2}.
$$

Now we are ready to give the design technique of the robust SMC controller.

**Theorem 6.3.1** *For the uncertain nonlinear stochastic system (6.1) with the sliding surface (6.7), where $G = B^T Q$ and $Q$, $K$ are the solutions to (6.40) to (6.44). If the SMC law is given as*

$$u(k) = -(GB)^{-1}(\tau \Lambda sgn[s(k)] + \tau Ps(k) + G(A + BK)x(k-1)$$
$$- G(BK + I)x(k) + (\hat{\Delta}_a + \hat{\Delta}_f) + (\tilde{\Delta}_a + \tilde{\Delta}_f)sgn[s(k)]), \qquad (6.61)$$

*then the state trajectories of the system (6.1) are driven onto the sliding surface with probability 1.*

*Proof.* In terms of (6.7) and (6.8), we obtain from (6.61) that

$$\Delta s(k) = s(k+1) - s(k)$$
$$= G(\Delta A - BK)x(k) + GB(u(k) + f(x(k)))$$
$$\quad - Gx(k) + G(A + BK)x(k-1)$$
$$= -\tau \Lambda sgn[s(k)] - \tau Ps(k)$$
$$\quad + G\Delta Ax(k) - (\hat{\Delta}_a + \tilde{\Delta}_a sgn[s(k)])$$
$$\quad + GBf(x(k)) - (\hat{\Delta}_f + \tilde{\Delta}_f sgn[s(k)]). \qquad (6.62)$$

Noticing (6.59) and (6.60), we arrive at the reaching condition (6.58). Therefore, the state trajectories are globally driven onto the specified sliding surface. The proof is complete. □

**Remark 6.1** *We point out that it is not difficult to extend the present results to more general systems that include polytopic parameter uncertainties and constant time delays (or time-varying bounded delays) by using the approach developed in Refs [107, 108] and the same LMI framework can be established. The reason why we discuss the system (6.1) and (6.2) is to make our theory more understandable and to avoid unnecessarily complicated notations.*

## 6.4 Numerical Example

In this section, we present an illustrative example to demonstrate the effectiveness of the proposed algorithms.

Consider a nonlinear stochastic system with the following parameters:

$$A = \begin{bmatrix} 0.32 & -0.20 & 0 \\ 0 & 0.32 & 0.084 \\ 0.04 & -0.30 & -0.20 \end{bmatrix}, \quad B = \begin{bmatrix} 0.04 & 0.12 \\ 0 & -0.08 \\ 0.16 & 0.28 \end{bmatrix},$$

$$D_1 = \begin{bmatrix} 0.12 & 0 & -0.04 \\ 0.08 & 0.12 & 0 \\ 0.16 & 0.20 & -0.04 \end{bmatrix}, \quad D_2 = \begin{bmatrix} 0.08 \\ 0.04 \\ 0 \end{bmatrix}, \quad L = \begin{bmatrix} 0.40 & 0 & -0.40 \end{bmatrix},$$

$$H = \begin{bmatrix} 0.04 \\ 0.08 \\ 0 \end{bmatrix}, \quad F = \sin(0.6k), \quad E = \begin{bmatrix} 0 & 0.04 & 0 \end{bmatrix}.$$

The nonlinear functions $f(x(k))$, $g(x(k))$, and $h(x(k))$ are taken as follows:

$$f(x(k)) = 0.3 \sin(x(k)),$$

$$g(x(k)) = \begin{bmatrix} 0.2 \\ 0.3 \\ 0.5 \end{bmatrix} (0.3x_1(k)\xi_1(k) + 0.4x_2(k)\xi_2(k) + 0.5x_3(k)\xi_3(k)),$$

$$h(x(k)) = 0.05(0.3x_1(k)\xi_1(k) + 0.4x_2(k)\xi_2(k) + 0.5x_3(k)\xi_3(k)),$$

where $x_i(k)$ $(i = 1, 2, 3)$ is the $i$th element of $x(k)$ and $\xi_i(k)$ $(i = 1, 2, 3)$ are zero mean uncorrelated Gaussian white noise processes with unity covariances. We also assume that $\xi_i(k)$ is uncorrelated with $\omega(k)$. Then it can be easily checked that the above class of stochastic nonlinearities satisfies

$$\mathbb{E}\left\{ \begin{bmatrix} g(x(k)) \\ h(x(k)) \end{bmatrix} \right\} = 0,$$

$$\mathbb{E}\left\{ \begin{bmatrix} g(x(k)) \\ h(x(k)) \end{bmatrix} \begin{bmatrix} g(x(k)) \\ h(x(k)) \end{bmatrix}^{\mathrm{T}} \right\} = \begin{bmatrix} 0.2 \\ 0.3 \\ 0.5 \\ 0.05 \end{bmatrix} \begin{bmatrix} 0.2 \\ 0.3 \\ 0.5 \\ 0.05 \end{bmatrix}^{\mathrm{T}} x^{\mathrm{T}}(k)\mathrm{diag}\{0.09, 0.16, 0.25\}x(k).$$

Setting the upper bound of the $H_2$ performance index by $\gamma = 0.001$ and using the Matlab LMI Toolbox, we solve the problem (6.57) to obtain

$$Q = \begin{bmatrix} 0.4356 & -0.3399 & -0.1618 \\ -0.3399 & 0.5365 & 0.0445 \\ -0.1618 & 0.0445 & 0.1926 \end{bmatrix},$$

$$G = \begin{bmatrix} -0.0085 & -0.0065 & 0.0243 \\ 0.0341 & -0.0712 & 0.0309 \end{bmatrix},$$

$$\varepsilon_1 = 1.0625, \qquad \varepsilon_2 = 0.0029.$$

Moreover, it is calculated that $H_2$ performance $J \leq \mathrm{tr}[D_2 W D_2^{\mathrm{T}} Q] = 0.00009 < \gamma$ and $\beta = 2.36 \times 10^{-5}$ (hence the linear constraint $B^{\mathrm{T}} QD = 0$ is satisfied). In order to design the explicit SMC controller, we suppose $G\Delta Ax(k)$ and $GBf(x(k))$ are bounded by the following conditions:

$$\delta^i_{aL} = -\|GH\|\|Ex(k)\|,$$

$$\delta^i_{aU} = \|GH\|\|Ex(k)\|,$$

$$\delta^i_{fL} = -0.3\|GB\sin(x(k))\|,$$

$$\delta^i_{fU} = 0.3\|GB\sin(x(k))\|,$$

where $\delta^i_{aL}$, $\delta^i_{aU}$, $\delta^i_{fL}$, and $\delta^i_{fU}$ ($i = 1, 2$) are defined in (6.59). Choose $\tau = 0.005$ (and $\tau = 0.03$ for comparison), $\lambda_i = 1$, and $p_i = 1$ ($i = 1, 2$). Then it follows from Theorem 6.3.1 that the desired SMC law can be expressed with known parameters. The simulation results are shown in Figures 6.1 to 6.12, which confirm that all the desired performance requirements are well achieved. From the figures, we can see that by implementing the proposed SMC algorithms, the state trajectories are driven onto the pre-specified sliding surface in finite time, subsequently causing a non-increasing zigzag motion. It can also be seen that the larger the parameter $\tau$, the severer the zigzag motion is.

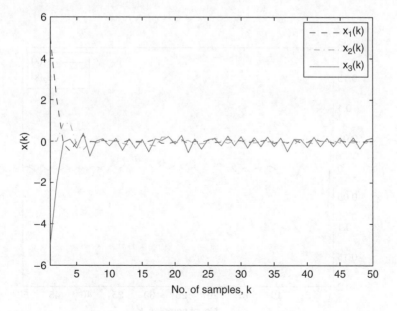

**Figure 6.1** The trajectories of state $x(k)$ ($\tau = 0.005$)

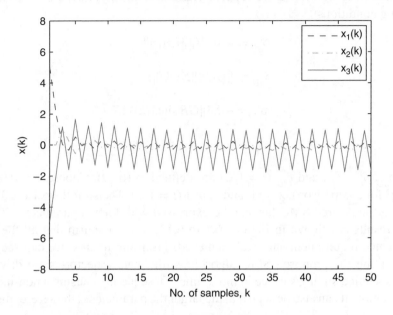

**Figure 6.2** The trajectories of state $x(k)$ ($\tau = 0.03$)

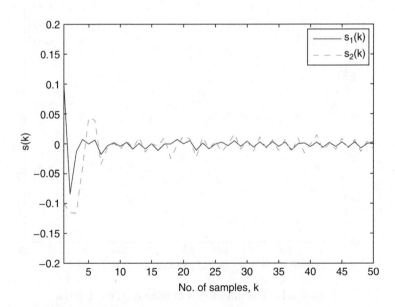

**Figure 6.3** The trajectories of sliding variable $s(k)$ ($\tau = 0.005$)

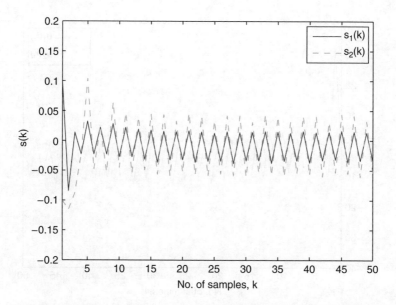

**Figure 6.4**    The trajectories of sliding variable $s(k)$ ($\tau = 0.03$)

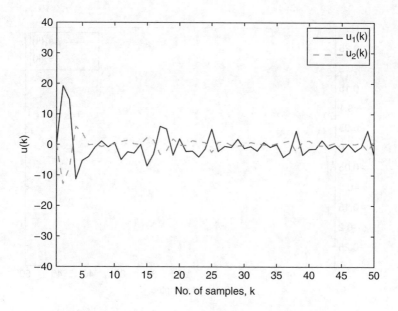

**Figure 6.5**    The control signals $u(k)$ ($\tau = 0.005$)

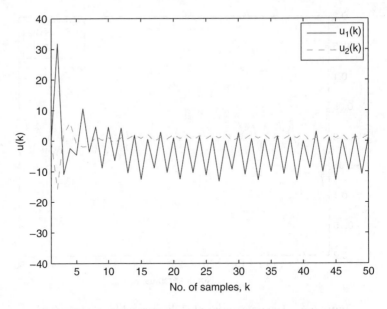

**Figure 6.6**    The control signals $u(k)$ ($\tau = 0.03$)

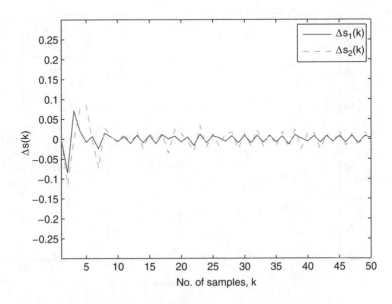

**Figure 6.7**    The signals $\Delta s(k)$ ($\tau = 0.005$)

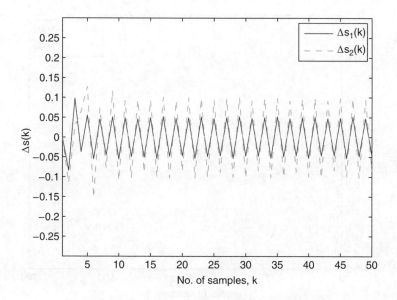

**Figure 6.8** The signals $\Delta s(k)$ ($\tau = 0.03$)

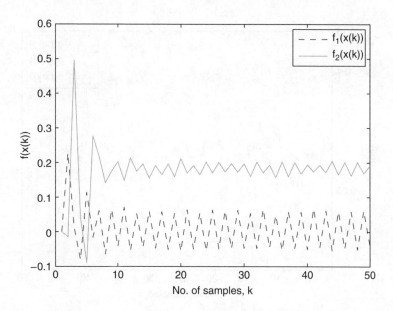

**Figure 6.9** The matched nonlinearity $f(x(k))$ ($\tau = 0.03$)

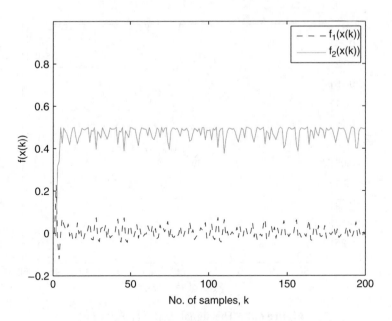

**Figure 6.10**    The matched nonlinearity $f(x(k))$ ($\tau = 0.005$)

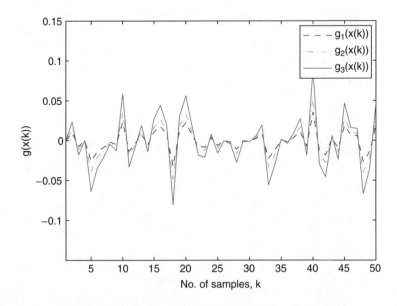

**Figure 6.11**    The matched nonlinearity $g(x(k))$ ($\tau = 0.03$)

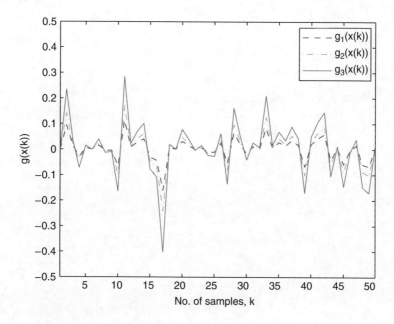

**Figure 6.12** The matched nonlinearity $g(x(k))$ ($\tau = 0.005$)

## 6.5 Summary

In this chapter, a robust SMC design problem for uncertain nonlinear discrete-time stochastic systems with the $H_2$ performance constraint has been studied. A new form of discrete switching function has been used. By means of LMIs, a sufficient condition for the exponential mean-square stability and the $H_2$ performance in the specified sliding surface has been derived. By the improved reaching condition, an SMC controller has been designed that globally drives the state trajectories onto the specified surface with probability 1. It is worth mentioning that, using the proposed method in this chapter, several typical classes of stochastic nonlinearities can be dealt with via SMC method.

# 7

# Variance-Constrained Dissipative Control with Degraded Measurements

It is well recognized that, in real-world engineering practice, the sensors have always been confronted with different kinds of failures due to various reasons, such as the erosion caused by severe circumstance, abrupt changes of working conditions, intense external disturbances, internal component constraints, and aging. The unavoidable sensor failures often occur in a probabilistic way that give rise to the measurements degradation phenomenon. In this case, the sampling output might contain incomplete information. Such a phenomenon has been firstly described in the literature by a binary switching sequence satisfying a conditional probability distribution. Recently, a more general model for missing measurement phenomenon has been proposed where the missing probability for each sensor is governed by an individual random variable satisfying a certain probability distribution over the interval [0, 1]. Such a model is ideal in reflecting multiple missing measurements or multiple measurements degraded due to different failure probabilities of different sensors. So far, all the available results have been concerned with either linear or deterministic systems. When it comes to relatively more complicated systems, such as nonlinear stochastic systems, the phenomenon of multiple degraded measurements has not yet received enough research attention in spite of its clear engineering insight, especially in today's networked environments.

On the other hand, recently the theory of dissipative systems, which plays an important role in system and control areas, has been attracting a great deal of research interest and many results have been reported so far. The dissipative theory serves as a powerful tool in characterizing important system behaviors, such as stability and passivity, and has close connections with bounded real lemma, passivity lemma, and circle criterion. On the other hand, in engineering systems, as was mentioned in previous chapters, it is always desirable for the controlled systems to achieve multiple

*Variance-Constrained Multi-Objective Stochastic Control and Filtering*, First Edition.
Lifeng Ma, Zidong Wang and Yuming Bo.
© 2015 John Wiley & Sons, Ltd. Published 2015 by John Wiley & Sons, Ltd.

performance indices such as stability, robustness, dissipativity, and steady-state variance. Therefore, a seemingly natural research problem arises here: can we handle the robust variance-constrained dissipative control problem for uncertain stochastic systems with general nonlinearities? To the best of the authors' knowledge, such a multi-objective research problem has not received any attention despite its theoretical significance, and this constitutes the main motivation of our current investigation.

This chapter deals with the variance-constrained dissipative control problem for a class of stochastic nonlinear systems with multiple degraded measurements, where the degraded probability for each sensor is governed by an individual random variable satisfying a certain probabilistic distribution over a given interval. The purpose of the problem is to design an observer-based controller such that, for all possible degraded measurements, the closed-loop system is exponentially mean-square stable and strictly dissipative, while the individual steady-state variance is not more than the pre-specified upper bound constraints. A general framework is established so that the required exponential mean-square stability, dissipativity as well as the variance constraints can be easily enforced. A sufficient condition is given for the solvability of the addressed multi-objective control problem, and the desired observer and controller gains are characterized in terms of the solution to a convex optimization problem that can be easily solved by using the semi-definite programming method. Finally, a numerical example is presented to show the effectiveness and applicability of the proposed algorithm.

The main contributions of this chapter can be described as follows: (1) For the first time, the stochastic dissipativity is combined with the steady-state variance in a stochastic control problem, which serve as two important performance requirements for a class of uncertain nonlinear stochastic systems. The tradeoff between these two requirements is investigated in detail by means of a convex optimization approach. (2) Compared with existing literature that uses measurement outputs for dissipative control design, this chapter considers the multiple packet-dropout model, which describes the phenomenon of measurement degradedation that occurs frequently in practical applications (for example, the target tracking problem). Therefore, the system model studied reflects practical engineering systems in a more comprehensive and realistic way.

The rest of the chapter is organized as follows: In Section 7.1, the multi-objective control problem is formulated for a class of nonlinear stochastic systems. In Section 7.2, the stability, dissipativity, and steady-state variance are analyzed one by one. In Section 7.3, an LMI algorithm is developed for controller design. In Section 7.4, an illustrative example is presented to show the effectiveness of the proposed algorithm. In Section 7.5, a summary is provided.

## 7.1 Problem Formulation

Consider the following discrete-time nonlinear stochastic system:

$$\begin{cases} x(k+1) = Ax(k) + B_1u(k) + f(x(k)) + D_1\omega(k), \\ \quad z(k) = Lx(k) + B_2u(k) + D_2\omega(k), \end{cases} \qquad (7.1)$$

with the measurement equation

$$y(k) = \Theta Cx(k) + g(x(k)) + D_3\omega(k)$$

$$= \sum_{j=1}^{m} \theta_j C_j x(k) + g(x(k)) + D_3\omega(k), \tag{7.2}$$

where $x(k) \in \mathbb{R}^n$ is the system state, $u(k) \in \mathbb{R}^p$ is the control input, $z(k) \in \mathbb{R}^r$ is the controlled output, $y(k) \in \mathbb{R}^m$ is the measured output vector, $\omega(k) \in \mathbb{R}^r$ is a zero mean Gaussian white noise sequence with covariance $W > 0$. $A, B_1, B_2, D_1, D_2, D_3, C$, and $L$ are known real constant matrices with appropriate dimensions.

The stochastic matrix $\Theta$ describes the phenomenon of multiple measurement degraded in the process of information retrieval from the sensor output. $\Theta$ is defined as

$$\Theta = \text{diag}\{\theta_1, \theta_2, \ldots, \theta_m\}, \tag{7.3}$$

with $\theta_j (j = 1, 2, \ldots, m)$ being $m$ independent random variables, which are also independent from $\omega(k)$. It is assumed that $\theta_j$ has the probabilistic density function $p_j(s)$ on the interval $[0, 1]$ with mathematical expectation $\bar{\theta}_j$ and variance $\sigma_j^2$. $C_j$ is defined as follows:

$$C_j \triangleq \text{diag}\{0, \underbrace{\ldots, 0, 1, 0,}_{j-1} \underbrace{\ldots, 0\}}_{m-j} C.$$

**Remark 7.1** *It has been illustrated in Ref. [85] that the description of measurements degraded phenomenon given in (7.3) is much more general than those in most previous literature, where the data missing phenomenon is simply modeled by a single Bernoulli sequence [143, 144]. In such a model, when $\theta_j = 1$, it means that the jth sensor is in a good condition; otherwise there might be partial or complete sensor failure. To be specific, when $\theta_j = 0$, the sensor is totally out of work and the measurements are completely missing, while $0 < \theta_j < 1$ means that we could only measure the output signals with reduced gains, namely, degraded measurements. In this sense, the model (7.1) and (7.2) offers a comprehensive means to reflect system complexity such as nonlinearities, stochasticity, and data degraded from multiple sensors.*

The nonlinear stochastic functions $f(x(k))$ and $g(x(k))$ are assumed to have the following first moments for all $x(k)$:

$$\mathbb{E}\left\{ \begin{bmatrix} f(x(k)) \\ g(x(k)) \end{bmatrix} \bigg| x(k) \right\} = 0, \tag{7.4}$$

with the covariance given by

$$\mathbb{E}\left\{ \begin{bmatrix} f(x(k)) \\ g(x(k)) \end{bmatrix} [f^T(x(j)) \quad g^T(x(j))] \bigg| x(k) \right\} = 0, \qquad k \neq j, \tag{7.5}$$

and

$$\mathbb{E}\left\{ \begin{bmatrix} f(x(k)) \\ g(x(k)) \end{bmatrix} [f^{\mathrm{T}}(x(k)) \quad g^{\mathrm{T}}(x(k))] \, \middle| \, x(k) \right\} = \sum_{i=1}^{q} \Pi_i x^{\mathrm{T}}(k)\Gamma_i x(k), \qquad (7.6)$$

where $\Pi_i$ and $\Gamma_i$ ($i = 1, 2, \ldots, q$) are known positive definite matrices with appropriate dimensions.

For system (7.1), consider the following observer-based controller:

$$\hat{x}(k + 1) = A_f \hat{x}(k) + H_f y(k), \qquad (7.7)$$

$$u(k) = K\hat{x}(k), \qquad (7.8)$$

where $A_f$ and $H_f$ (observer parameters) and $K$ (controller parameter) are to be determined.

From (7.1), (7.7), and (7.8), we obtain the following augmented system:

$$\begin{cases} \xi(k + 1) = \bar{A}\xi(k) + \bar{H}h(x(k)) + \bar{D}\omega(k), \\ \quad z(k) = \bar{L}\xi(k) + D_2\omega(k), \end{cases} \qquad (7.9)$$

where

$$\xi(k) = \begin{bmatrix} x(k) \\ \hat{x}(k) \end{bmatrix},$$

$$\bar{A} = \begin{bmatrix} A & B_1 K \\ H_f \Theta C & A_f \end{bmatrix},$$

$$\bar{H} = \begin{bmatrix} I & 0 \\ 0 & H_f \end{bmatrix},$$

$$\bar{D} = \begin{bmatrix} D_1 \\ H_f D_3 \end{bmatrix},$$

$$\bar{L} = \begin{bmatrix} L & B_2 K \end{bmatrix},$$

$$h(x(k)) = \begin{bmatrix} f(x(k)) \\ g(x(k)) \end{bmatrix}.$$

**Definition 7.1.1** *System (7.9) is said to be exponentially mean-square stable if, with* $\omega(k) = 0$, *there exist constants* $\rho \geq 1$ *and* $\tau \in (0, 1)$ *such that*

$$\mathbb{E}\{\|\xi(k)\|^2\} \leq \rho\tau^k \mathbb{E}\{\|\xi(0)\|^2\}, \quad \forall \xi(0) \in \mathbb{R}^{2n}, \quad k \in \mathbb{I}^+, \qquad (7.10)$$

*for all possible degraded measurements.*

We are now in a position to introduce the performance of dissipativity. Let the energy supply function of system (7.1) be defined by

$$G(\omega, z, T) = \langle z, Qz \rangle_T + 2\langle z, S\omega \rangle_T + \langle \omega, R\omega \rangle_T, \quad \forall T \geq 0, \qquad (7.11)$$

where $Q$, $S$, and $R$ are real matrices with $Q$, $R$ symmetric, $T \geq 0$ is an integer, and $\langle a, b \rangle_T \triangleq \sum_{k=0}^{T} a^T(k)b(k)$. Without loss of generality, we assume that $Q < 0$ and denote $\bar{Q} = \sqrt{-Q}$.

**Definition 7.1.2** *[91] Closed-loop system (7.9) is said to be strictly $(Q, R, S)$ dissipative if, for some scalar $\gamma > 0$, the following inequality*

$$G(\omega, z, T) \geq \gamma \langle \omega, \omega \rangle_T, \quad \forall T \geq 0, \tag{7.12}$$

*holds under a zero initial condition.*

If system (7.1) is asymptotically stable, its steady-state covariance is defined as follows:

$$\bar{X} \triangleq \lim_{k \to \infty} \mathbb{E}\{x(k)x^T(k)\}. \tag{7.13}$$

**Assumption 7.1** *The matrices $\Pi_i$ and $\Gamma_i$ in (7.6) have the following form:*

$$\Pi_i = \bar{\pi}_i \bar{\pi}_i^T = \begin{bmatrix} \pi_{1i} \\ \pi_{2i} \end{bmatrix} \begin{bmatrix} \pi_{1i} \\ \pi_{2i} \end{bmatrix}^T, \quad \Gamma_i = \eta_i \eta_i^T, \tag{7.14}$$

*where $\pi_{i1}$, $\pi_{2i}$, and $\eta_i$ $(i = 1, 2, \ldots, i)$ are known vectors with appropriate dimensions.*

This chapter aims to determine the observer parameters $A_f$, $H_f$, and the feedback controller parameter $K$ for the system (7.1) such that, for all possible degraded measurements, the following three objectives are achieved simultaneously:

(R1) Augmented system (7.9) is exponentially mean-square stable.
(R2) Augmented system (7.9) is strictly $(Q, S, R)$ dissipative.
(R3) The steady-state variance for each individual state of system (7.1) satisfies

$$\bar{X}_s \leq \delta_s^2, \quad s = 1, 2, \ldots, n, \tag{7.15}$$

where $\bar{X}_s$ stands for the steady-state variance for the $s$th state and $\delta_s^2$ denotes the pre-specified steady-state variance constraint on the $s$th state.

## 7.2 Stability, Dissipativity, and Variance Analysis

Before giving our preliminary results, let us introduce some useful lemmas. For presentation convenience, we denote

$$\hat{A} = \begin{bmatrix} A & B_1 K \\ H_f \bar{\Theta} C & A_f \end{bmatrix},$$

$$\check{A} = \begin{bmatrix} 0 & 0 \\ H_f (\Theta - \bar{\Theta}) C & 0 \end{bmatrix},$$

$$\tilde{A}_i = \begin{bmatrix} 0 & 0 \\ H_f C_i & 0 \end{bmatrix},$$

$$\bar{\Gamma}_i = \begin{bmatrix} \eta_i \\ 0 \end{bmatrix} \begin{bmatrix} \eta_i \\ 0 \end{bmatrix}^{\mathrm{T}} \triangleq \bar{\eta}_i \bar{\eta}_i^{\mathrm{T}},$$

$$\bar{\Theta} = \mathbb{E}\{\Theta\}.$$

**Lemma 7.2.1** *[145] Let $V(\xi(k)) = \xi^{\mathrm{T}}(k) X \xi(k)$ be a Lyapunov functional where $X > 0$. If there exist real scalars $\lambda$, $\mu > 0$, $\nu > 0$, and $0 < \psi < 1$ such that both*

$$\mu \|\xi(k)\|^2 \le V(\xi(k)) \le \nu \|\xi(k)\|^2 \tag{7.16}$$

*and*

$$\mathbb{E}\{V(\xi(k+1))|\xi(k)\} - V(\xi(k)) \le \lambda - \psi V(\xi(k)) \tag{7.17}$$

*hold, then the process $\xi(k)$ satisfies*

$$\mathbb{E}\{\|\xi(k)\|^2\} \le \frac{\nu}{\mu} \|\xi(0)\|^2 (1 - \psi)^k + \frac{\lambda}{\mu\psi}. \tag{7.18}$$

**Lemma 7.2.2** *Given the parameters $A_f$, $H_f$, and $K$. The following statements are equivalent:*

*1.*

$$\rho\left( \hat{A}^{\mathrm{T}} \otimes \hat{A}^{\mathrm{T}} + \sum_{j=1}^{m} \sigma_j^2 \tilde{A}_j^{\mathrm{T}} \otimes \tilde{A}_j^{\mathrm{T}} + \sum_{i=1}^{q} \mathrm{st}(\bar{\Gamma}_i) \mathrm{st}^{\mathrm{T}}(\bar{H}\Pi_i \bar{H}^{\mathrm{T}}) \right) < 1 \tag{7.19}$$

*or*

$$\rho\left( \hat{A} \otimes \hat{A} + \sum_{j=1}^{m} \sigma_j^2 \tilde{A}_j \otimes \tilde{A}_j + \sum_{i=1}^{q} \mathrm{st}(\bar{H}\Pi_i \bar{H}^{\mathrm{T}}) \mathrm{st}^{\mathrm{T}}(\bar{\Gamma}_i) \right) < 1. \tag{7.20}$$

*2. There exists a positive definite matrix $X > 0$ such that*

$$\hat{A}^{\mathrm{T}} X \hat{A} + \sum_{j=1}^{m} \sigma_j^2 \tilde{A}_j^{\mathrm{T}} X \tilde{A}_j + \sum_{i=1}^{q} \bar{\Gamma}_i \mathrm{tr}[X\bar{H}\Pi_i \bar{H}^{\mathrm{T}}] - X < 0. \tag{7.21}$$

*3. There exists a positive definite matrix $Y > 0$ such that*

$$\hat{A} Y \hat{A}^{\mathrm{T}} + \sum_{j=1}^{m} \sigma_j^2 \tilde{A}_j Y \tilde{A}_j^{\mathrm{T}} + \sum_{i=1}^{q} \bar{H}\Pi_i \bar{H}^{\mathrm{T}} \mathrm{tr}[\bar{\Gamma}_i Y] - Y < 0. \tag{7.22}$$

*4. System (7.9) is exponentially mean-square stable.*

The proof of Lemma 7.2.2 can be carried out along similar lines to that of Lemma 2 of Ref. [65] and is therefore omitted. The main difference between this lemma and Theorem 1 of Ref. [133] is that the state matrix of system (7.9) involves the stochastic variable $\Theta$, which describes probabilistic degraded measurements.

The following theorem gives a sufficient condition for the exponential mean-square stability as well as strictly $(Q, S, R)$ dissipativity of the system (7.9).

**Theorem 7.2.3** *Given the parameters $A_f$, $H_f$, $K$, symmetric matrices $Q$, $R$, and a matrix $S$. The closed-loop system (7.9) is exponentially mean-square stable and strictly $(Q, S, R)$ dissipative if there exists a matrix $X > 0$ such that the following matrix inequality holds:*

$$\Omega \triangleq \begin{bmatrix} \Omega_{11} & \hat{A}^T X \bar{D} - \bar{L}^T Q D_2 - \bar{L}^T S \\ * & \bar{D}^T X \bar{D} - D_2^T Q D_2 - D_2^T S - S^T D_2 - R \end{bmatrix} < 0, \qquad (7.23)$$

*where*

$$\Omega_{11} \triangleq \hat{A}^T X \hat{A} + \sum_{j=1}^{m} \sigma_j^2 \tilde{A}_j^T X \tilde{A}_j + \sum_{i=1}^{q} \bar{\Gamma}_i \text{tr}[X \bar{H} \Pi_i \bar{H}^T] - X - \bar{L}^T Q \bar{L}.$$

*Proof.* First, it follows from (7.23) that

$$\hat{A}^T X \hat{A} + \sum_{j=1}^{m} \sigma_j^2 \tilde{A}_j^T X \tilde{A}_j + \sum_{i=1}^{q} \bar{\Gamma}_i \text{tr}[X \bar{H} \Pi_i \bar{H}^T] - X < \bar{L}^T Q \bar{L} < 0. \qquad (7.24)$$

Therefore, from Lemma 7.2.2, system (7.9) is exponentially mean-square stable. When $\omega(k) \neq 0$, we obtain from (7.23) that

$$\mathbb{E}\{V(\xi(k+1))|\xi(k)\} - V(\xi(k))$$

$$- z^T(k)Qz(k) - 2z^T(k)S\omega(k) - \omega^T(k)R\omega(k)$$

$$= \mathbb{E}\{(\bar{A}\xi(k) + \bar{H}h(x(k)) + \bar{D}\omega(k))^T X (\bar{A}\xi(k) + \bar{H}h(x(k)) + \bar{D}\omega(k))|\xi(k)\}$$

$$- \xi^T(k)X\xi(k) - (\bar{L}\xi(k) + D_2\omega(k))^T Q(\bar{L}\xi(k) + D_2\omega(k)) \qquad (7.25)$$

$$- 2(\bar{L}\xi(k) + D_2\omega(k))^T S\omega(k) - \omega^T(k)R\omega(k)$$

$$= \begin{bmatrix} \xi(k) \\ \omega(k) \end{bmatrix}^T \Omega \begin{bmatrix} \xi(k) \\ \omega(k) \end{bmatrix} < 0.$$

Obviously, there always exists a sufficiently small positive scalar $\gamma > 0$ such that

$$\Omega + \begin{bmatrix} 0 & 0 \\ 0 & \gamma I \end{bmatrix} < 0 \qquad (7.26)$$

and therefore

$$\mathbb{E}\{V(\xi(k+1))|\xi(k)\} - V(\xi(k)) + \gamma\omega^T(k)\omega(k)$$

$$< z^T(k)Qz(k) + 2z^T(k)S\omega(k) + \omega^T(k)R\omega(k). \qquad (7.27)$$

Summing (7) from 0 to $T$ with respect to $k$ on both sides, and noticing that $V(\xi(T + 1)) > 0$ and $V(\xi(0)) = 0$, it can be found that

$$G(\omega, z, T) \geq \gamma \langle \omega, \omega \rangle_T, \tag{7.28}$$

which implies that the system (7.9) is strictly $(Q, S, R)$ dissipative. The proof is complete. □

Now let us proceed to analyze the steady-state covariance of the system (7.9). Define the state covariance of system (7.9) as

$$Y(k) \triangleq \mathbb{E}\{\xi(k)\xi^{\mathrm{T}}(k)\}. \tag{7.29}$$

The evolution of $Y(k)$ can be derived as follows:

$$Y(k + 1) = \hat{A}Y(k)\hat{A}^{\mathrm{T}} + \sum_{j=1}^{m} \sigma_j^2 \tilde{A}_j Y(k) \tilde{A}_j^{\mathrm{T}}$$

$$+ \sum_{i=1}^{q} \bar{H}\Pi_i \bar{H}^{\mathrm{T}} \mathrm{tr}[\bar{\Gamma}_i Y(k)] + \bar{D}W\bar{D}^{\mathrm{T}}. \tag{7.30}$$

Furthermore, define the steady-state covariance as

$$\bar{Y} \triangleq \lim_{k \to \infty} Y(k). \tag{7.31}$$

The following theorem presents a sufficient condition that guarantees the exponential mean-square stability of system (7.9) and, at the same time, gives an upper bound of the steady-state covariance.

**Theorem 7.2.4** *Given the parameters $A_f$, $H_f$, and $K$. If there exists a matrix $Y > 0$ such that*

$$\hat{A}Y\hat{A}^{\mathrm{T}} + \sum_{j=1}^{m} \sigma_j^2 \tilde{A}_j Y \tilde{A}_j^{\mathrm{T}} + \sum_{i=1}^{q} \bar{H}\Pi_i \bar{H}^{\mathrm{T}} \mathrm{tr}(\bar{\Gamma}_i Y) - Y + \bar{D}W\bar{D}^{\mathrm{T}} < 0, \tag{7.32}$$

*then system (7.9) is exponentially mean-square stable. Moreover, the steady-state covariance defined in (7.31) exists and satisfies $\bar{Y} \leq Y$.*

*Proof.* First of all, the matrix inequality (7.32) indicates that

$$\hat{A}Y\hat{A}^{\mathrm{T}} + \sum_{j=1}^{m} \sigma_j^2 \tilde{A}_j Y \tilde{A}_j^{\mathrm{T}} + \sum_{i=1}^{q} \bar{H}\Pi_i \bar{H}^{\mathrm{T}} \mathrm{tr}(\bar{\Gamma}_i Y) - Y < -\bar{D}W\bar{D}^{\mathrm{T}} < 0 \tag{7.33}$$

and therefore it follows from Lemma 7.2.2 that system (7.9) is exponentially mean-square stable.

Rewrite equation (7.30) in the following form:

$$\text{st}(Y(k+1)) = \mathcal{A}\,\text{st}(Y(k)) + \text{st}(\bar{D}W\bar{D}^{\mathrm{T}}), \tag{7.34}$$

where

$$\mathcal{A} \triangleq \hat{A} \otimes \hat{A} + \sum_{j=1}^{m} \sigma_j^2 \tilde{A}_j \otimes \tilde{A}_j + \sum_{i=1}^{q} \text{st}(\bar{H}\Pi_i\bar{H}^{\mathrm{T}})\text{st}^{\mathrm{T}}(\bar{\Gamma}_i).$$

From Lemma 7.2.2, the exponential mean-square stability of system (7.9) ensures that the inequality (7.20) holds, which implies the convergence of the covariance $Y(k)$ to the constant matrix $\bar{Y}$ when $k \to \infty$, that is,

$$-\bar{Y} + \hat{A}\bar{Y}\hat{A}^{\mathrm{T}} + \sum_{j=1}^{m} \sigma_j^2 \tilde{A}_j \bar{Y}\tilde{A}_j^{\mathrm{T}} + \sum_{i=1}^{q} \bar{H}\Pi_i\bar{H}^{\mathrm{T}}\text{tr}(\bar{\Gamma}_i\bar{Y}) + \bar{D}W\bar{D}^{\mathrm{T}} = 0. \tag{7.35}$$

Subtracting (7.35) from (7.32), we obtain

$$-(Y - \bar{Y}) + \hat{A}(Y - \bar{Y})\hat{A}^{\mathrm{T}}$$
$$+ \sum_{j=1}^{m} \sigma_j^2 \tilde{A}_j(Y - \bar{Y})\tilde{A}_j^{\mathrm{T}} + \sum_{i=1}^{q} \bar{H}\Pi_i\bar{H}^{\mathrm{T}}\text{tr}[\bar{\Gamma}_i(Y - \bar{Y})] < 0. \tag{7.36}$$

In the following stage, we need to prove that $\tilde{Y} \triangleq Y - \bar{Y} \geq 0$. For this purpose, let us first prove the fact that if system (7.9) is exponentially mean-square stable and there exists a symmetric matrix $\tilde{X}$ such that

$$\hat{A}^{\mathrm{T}}\tilde{X}\hat{A} + \sum_{j=1}^{m} \sigma_j^2 \tilde{A}_j^{\mathrm{T}}\tilde{X}\tilde{A}_j + \sum_{i=1}^{q} \bar{\Gamma}_i\text{tr}(\tilde{X}\bar{H}\Pi_i\bar{H}^{\mathrm{T}}) - \tilde{X} < 0, \tag{7.37}$$

then $\tilde{X} \geq 0$. In fact, if (7.37) holds, then there always exists a matrix $\Xi > 0$ satisfying

$$\hat{A}^{\mathrm{T}}\tilde{X}\hat{A} + \sum_{j=1}^{m} \sigma_j^2 \tilde{A}_j^{\mathrm{T}}\tilde{X}\tilde{A}_j + \sum_{i=1}^{q} \bar{\Gamma}_i\text{tr}(\tilde{X}\bar{H}\Pi_i\bar{H}^{\mathrm{T}}) - \tilde{X} = -\Xi. \tag{7.38}$$

Using the functional $V(\xi(k)) = \xi^{\mathrm{T}}(k)\tilde{X}\xi(k)$ for (7.9), we obtain

$$\mathbb{E}\{V(\xi(k+1))|\xi(k)\} - V(\xi(k))$$
$$= \xi^{\mathrm{T}}(k)\left[\hat{A}^{\mathrm{T}}\tilde{X}\hat{A} + \sum_{j=1}^{m} \sigma_j^2 \tilde{A}_j^{\mathrm{T}}\tilde{X}\tilde{A}_j + \sum_{i=1}^{q} \bar{\Gamma}_i\text{tr}(\tilde{X}\bar{H}\Pi_i\bar{H}^{\mathrm{T}}) - \tilde{X}\right]\xi(k) \tag{7.39}$$
$$= -\xi^{\mathrm{T}}(k)\Xi\xi^{\mathrm{T}}(k).$$

Taking the sum on both sides of (7.39) with respect to $k$ from 0 to $\infty$ results in

$$\lim_{n\to\infty} \mathbb{E}\{\xi^T(n)\tilde{X}\xi(n)\} - \xi^T(0)\tilde{X}\xi(0) = -\lim_{n\to\infty}\sum_{k=0}^{n}\xi^T(k)\Xi\xi(k). \qquad (7.40)$$

Since system (7.9) is exponentially mean-square stable, we have

$$\lim_{n\to\infty} \mathbb{E}\{\xi^T(n)\tilde{X}\xi(n)\} \leq \|\tilde{X}\| \lim_{n\to\infty}\xi^T(n)\xi(n) = 0. \qquad (7.41)$$

Therefore, for any nonzero initial state $\xi(0)$, it can be deduced from (7.40) that

$$\xi^T(0)\tilde{X}\xi(0) = \lim_{n\to\infty}\sum_{k=0}^{n}\xi^T(k)\Xi\xi(k) \geq 0, \qquad (7.42)$$

which means $\tilde{X} \geq 0$.

Now let us construct an auxiliary system as follows:

$$\bar{\xi}(k+1) = \bar{A}^T\bar{\xi}(k) + \bar{h}(\bar{\xi}(k)), \qquad (7.43)$$

where $\bar{h}(\bar{\xi}(k))$ satisfies

$$\mathbb{E}\{\bar{h}(\bar{\xi}(k))|\bar{\xi}(k)\} = 0,$$

$$\mathbb{E}\{\bar{h}(\bar{\xi}(k))\bar{h}^T(\bar{\xi}(j))|\bar{\xi}(k)\} = 0, \qquad k \neq j,$$

$$\mathbb{E}\{\bar{h}(\bar{\xi}(k))\bar{h}^T(\bar{\xi}(k))|\bar{\xi}(k)\} = \sum_{i=1}^{q}\bar{\Gamma}_i\bar{\xi}^T(k)\bar{H}\Pi_i\bar{H}^T\bar{\xi}(k). \qquad (7.44)$$

It follows from the exponential mean-square stability of system (7.9) and Lemma 7.2.2 that

$$\rho\left(\hat{A}^T \otimes \hat{A}^T + \sum_{j=1}^{m}\sigma_j^2\tilde{A}_j^T \otimes \tilde{A}_j^T + \sum_{i=1}^{q}\mathrm{st}(\bar{\Gamma}_i)\mathrm{st}^T(\bar{H}\Pi_i\bar{H}^T)\right) < 1. \qquad (7.45)$$

Thus, the auxiliary system (7.43) is also exponentially mean-square stable. Then, from the previously proven fact, if there exists a symmetric matrix $\tilde{Y}$ such that

$$(\hat{A}^T)^T\tilde{Y}\hat{A}^T + \sum_{j=1}^{m}\sigma_j^2(\tilde{A}_j^T)^T\tilde{Y}\tilde{A}_j^T + \sum_{i=1}^{q}\bar{H}\Pi_i\bar{H}^T\mathrm{tr}(\bar{\Gamma}_i\tilde{Y}) - \tilde{Y} < 0, \qquad (7.46)$$

it can be concluded that $\tilde{Y} \geq 0$. The proof is complete.                    $\square$

Based on the results we have obtained so far concerning the exponential mean-square stability, the dissipative property, as well as steady-state covariance, we are now ready to cope with the addressed multi-objective controller design problem.

## 7.3   Observer-Based Controller Design

In this section, we will first propose a sufficient condition for the solvability of the addressed problem in terms of the feasibility of certain constrained LMIs. Then, an algorithm is presented via the cone complementarity linearization method to solve the addressed non-convex optimization problem.

### 7.3.1   Solvability of the Multi-Objective Control Problem

To begin with, a corollary is given which combines the exponential mean-square stability, system dissipativity, and steady-state covariance constraints.

**Corollary 7.3.1** *Given the parameters $A_f$, $H_f$, $K$, matrices $Q$, $R$, and $S$ with $Q$ and $R$ being symmetric. Denote $Y_0 \triangleq \mathrm{diag}\{\delta_1^2, \delta_2^2, \ldots, \delta_n^2\}$ where $\delta_s^2$ $(s = 1, 2, \ldots, n)$ are the pre-specified upper bounds on the steady-state variance of each individual state. If there exist a matrix $Y > 0$ and scalars $\alpha_i > 0$, $\beta_i > 0$ satisfying $\alpha_i \beta_i = 1$ $(i = 1, 2, \ldots, q)$ such that*

$$\begin{bmatrix} I & 0 \end{bmatrix} Y \begin{bmatrix} I \\ 0 \end{bmatrix} - Y_0 < 0, \tag{7.47}$$

$$\begin{bmatrix} -\alpha_i^{-1} & \bar{\pi}_i^{\mathrm{T}} \bar{H}^{\mathrm{T}} \\ * & -Y \end{bmatrix} < 0, \tag{7.48}$$

$$\begin{bmatrix} -\beta_i^{-1} & \bar{\eta}_i^{\mathrm{T}} Y \\ * & -Y \end{bmatrix} < 0, \tag{7.49}$$

$$\begin{bmatrix} -Y & \hat{A}Y & \bar{A}\mathcal{Y} & \hat{H} & \bar{D} \\ * & -Y & 0 & 0 & 0 \\ * & * & -\mathcal{Y} & 0 & 0 \\ * & * & * & -\bar{\beta} & 0 \\ * & * & * & * & -W^{-1} \end{bmatrix} < 0, \tag{7.50}$$

$$\begin{bmatrix} -Y & -Y(\bar{L}^{\mathrm{T}}QD_2 - \bar{L}^{\mathrm{T}}S) & Y\hat{A}^{\mathrm{T}} & YA^{\mathrm{T}} & Y\hat{\eta} & Y\bar{L}^{\mathrm{T}}\bar{Q} \\ * & -D_2^{\mathrm{T}}QD_2 - D_2^{\mathrm{T}}S - S^{\mathrm{T}}D_2 - R & \bar{D}^{\mathrm{T}} & 0 & 0 & 0 \\ * & * & -Y & 0 & 0 & 0 \\ * & * & * & -\mathcal{Y} & 0 & 0 \\ * & * & * & * & -\bar{\alpha} & 0 \\ * & * & * & * & * & -I \end{bmatrix} < 0, \tag{7.51}$$

*where*

$$\mathcal{A}^{\mathrm{T}} = \begin{bmatrix} \sigma_1 \tilde{A}_1^{\mathrm{T}} & \sigma_2 \tilde{A}_2^{\mathrm{T}} & \cdots & \sigma_m \tilde{A}_m^{\mathrm{T}} \end{bmatrix},$$

$$\bar{\mathcal{A}} = \begin{bmatrix} \sigma_1 \tilde{A}_1 & \sigma_2 \tilde{A}_2 & \cdots & \sigma_m \tilde{A}_m \end{bmatrix},$$

$$\mathcal{Y} = \begin{bmatrix} Y & 0 & \cdots & 0 \\ 0 & Y & \cdots & 0 \\ \vdots & \vdots & \ddots & \vdots \\ 0 & 0 & \cdots & Y \end{bmatrix},$$

$$\hat{H} = \begin{bmatrix} \bar{H}\bar{\pi}_1 & \bar{H}\bar{\pi}_2 & \cdots & \bar{H}\bar{\pi}_q \end{bmatrix},$$

$$\hat{\eta} = \begin{bmatrix} \bar{\eta}_1 & \bar{\eta}_2 & \cdots & \bar{\eta}_q \end{bmatrix},$$

$$\bar{\alpha} = \begin{bmatrix} \alpha_1 I & 0 & \cdots & 0 \\ 0 & \alpha_2 I & \cdots & 0 \\ \vdots & \vdots & \ddots & \vdots \\ 0 & 0 & \cdots & \alpha_q I \end{bmatrix},$$

$$\bar{\beta} = \begin{bmatrix} \beta_1 I & 0 & \cdots & 0 \\ 0 & \beta_2 I & \cdots & 0 \\ \vdots & \vdots & \ddots & \vdots \\ 0 & 0 & \cdots & \beta_q I \end{bmatrix},$$

*then the system is exponentially mean-square stable and strictly $(Q, S, R)$ dissipative, while its individual steady-state variance is not more than the corresponding pre-specified upper bound.*

*Proof.* Based on the results we have obtained in Theorem 7.2.3 and Theorem 7.2.4, it suffices to prove that inequality (7.49) with (7.50) guarantee that (7.32) holds, and inequality (7.48) with (7.51) imply (7.23).

Firstly, by Lemma 2.2.2, we can see that (7.50) is equivalent to

$$\hat{A} Y \hat{A}^{\mathrm{T}} + \sum_{j=1}^{m} \sigma_j^2 \tilde{A}_j Y \tilde{A}_j^{\mathrm{T}} + \sum_{i=1}^{q} \bar{H} \Pi_i \bar{H}^{\mathrm{T}} \beta_i^{-1} - Y + \bar{D} W \bar{D}^{\mathrm{T}} < 0. \tag{7.52}$$

Then, by the Schur Complement Lemma it is not difficult to see that inequality (7.49) indicates $\bar{\eta}_i^{\mathrm{T}} Y \bar{\eta}_i < \beta_i^{-1}$ or, equivalently, $\mathrm{tr}(\Gamma_i Y) < \beta_i^{-1}$, and therefore

$$\hat{A} Y \hat{A}^{\mathrm{T}} + \sum_{j=1}^{m} \sigma_j^2 \tilde{A}_j Y \tilde{A}_j^{\mathrm{T}} + \sum_{i=1}^{q} \bar{H} \Pi_i \bar{H}^{\mathrm{T}} \mathrm{tr}(\bar{\Gamma}_i Y) - Y + \bar{D} W \bar{D}^{\mathrm{T}} < 0. \tag{7.53}$$

It follows directly from Theorem 7.2.4 that the system (7.9) is exponentially mean-square stable and the steady-state covariance defined by (7.31) exists and satisfies $\bar{Y} < Y$, where $\bar{Y}$ satisfies (7.35). Moreover, from (7.47), we can see that the steady-state covariance of the system (7.1) defined in (7.13) satisfies

$$\bar{X} = \begin{bmatrix} I & 0 \end{bmatrix} \bar{Y} \begin{bmatrix} I \\ 0 \end{bmatrix} < \begin{bmatrix} I & 0 \end{bmatrix} Y \begin{bmatrix} I \\ 0 \end{bmatrix} < Y_0, \tag{7.54}$$

which means that the steady-state covariance constraint is also achieved. Similarly, it is not difficult to prove that the exponential mean-square stability and system dissipativity can be ensured simultaneously by inequality (7.48) together with (7.51). The proof is complete. $\qquad\square$

**Theorem 7.3.2** *Given pre-specified steady-state variance upper bounds $\delta_1^2, \delta_2^2, \ldots, \delta_n^2$, matrices Q, S, and R with Q and R being symmetric, and scalars $\vartheta_j > 0$ and $\zeta_j > 0\,(j = 1, 2, \ldots, m)$. If there exist matrices $M > 0$, $N > 0$, real matrices $\bar{A}_f$, $\bar{H}_f$, $\bar{K}$, and scalars $\alpha_i > 0$, $\beta_i > 0$ $(i = 1, 2, \ldots, q)$ such that*

$$\alpha_i \beta_i = 1 \qquad (i = 1, 2, \ldots, q), \tag{7.55}$$

$$e_s^{\mathrm{T}} M e_s - \delta_s^2 < 0 \qquad (s = 1, 2, \ldots, n), \tag{7.56}$$

$$\begin{bmatrix} -\beta_i & \pi_{1i}^{\mathrm{T}} N + \pi_{2i}^{\mathrm{T}} \bar{H}_f^{\mathrm{T}} & \pi_{1i}^{\mathrm{T}} \\ * & -N & -I \\ * & * & -M \end{bmatrix} < 0, \tag{7.57}$$

$$\begin{bmatrix} -\alpha_i & \eta_i^{\mathrm{T}} & \eta_i^{\mathrm{T}} M \\ * & -N & -I \\ * & * & -M \end{bmatrix} < 0, \tag{7.58}$$

$$\begin{bmatrix} -N & -I & NA + \bar{H}_f \bar{\Theta} C & \bar{A}_f & \hat{C} & \Phi_{16} \\ * & -M & A & AM + B_1 \bar{K} & 0 & \Phi_{26} \\ * & * & -N & -I & 0 & 0 \\ * & * & * & -M & 0 & 0 \\ * & * & * & * & -\bar{y} & \Phi_{56} \\ * & * & * & * & * & \Phi_{66} \end{bmatrix} < 0, \tag{7.59}$$

$$\begin{bmatrix} -N & -I & L^{\mathrm{T}} \bar{S} & A^{\mathrm{T}} N + C^{\mathrm{T}} \bar{\Theta} \bar{H}_f^{\mathrm{T}} & A^{\mathrm{T}} & \bar{C} & \Upsilon_{17} \\ * & -M & \bar{M} \bar{S} & \bar{A}_f^{\mathrm{T}} & MA^{\mathrm{T}} + \bar{K}^{\mathrm{T}} B_1^{\mathrm{T}} & 0 & \Upsilon_{27} \\ * & * & -\bar{R} & D_1^{\mathrm{T}} N + D_3^{\mathrm{T}} \bar{H}_f^{\mathrm{T}} & D_1^{\mathrm{T}} & 0 & 0 \\ * & * & * & -N & -I & 0 & 0 \\ * & * & * & * & -M & 0 & 0 \\ * & * & * & * & * & -\bar{y} & \Upsilon_{67} \\ * & * & * & * & * & * & \Upsilon_{77} \end{bmatrix} < 0, \tag{7.60}$$

*where*

$$e_s = [\underbrace{0 \quad \cdots \quad 0}_{s-1} \quad 1 \quad \underbrace{0 \quad \cdots \quad 0}_{n-s}]^{\mathrm{T}},$$

$$\bar{M} = ML^{\mathrm{T}} + \bar{K}^{\mathrm{T}}B_2^{\mathrm{T}},$$

$$\bar{S} = S - QD_2,$$

$$\bar{R} = D_2^{\mathrm{T}}QD_2 + D_2^{\mathrm{T}}S + S^{\mathrm{T}}D_2 + R,$$

$$\tilde{C} = \begin{bmatrix} \sigma_1 MC_1^{\mathrm{T}} & \sigma_2 MC_2^{\mathrm{T}} & \cdots & \sigma_m MC_m^{\mathrm{T}} \end{bmatrix},$$

$$\hat{C} = \begin{bmatrix} \sigma_1 \bar{H}_f C_1 & 0 & \sigma_2 \bar{H}_f C_2 & 0 & \cdots & \sigma_m \bar{H}_f C_m & 0 \end{bmatrix},$$

$$\tilde{H}_f = \begin{bmatrix} \zeta_1 \bar{H}_f & \zeta_2 \bar{H}_f & \cdots & \zeta_m \bar{H}_f \end{bmatrix},$$

$$\tilde{\eta} = \begin{bmatrix} \eta_1 & \eta_2 & \cdots & \eta_q \end{bmatrix},$$

$$\bar{C} = \begin{bmatrix} \sigma_1 C_1^{\mathrm{T}} \bar{H}_f^{\mathrm{T}} & 0 & \sigma_2 C_2^{\mathrm{T}} \bar{H}_f^{\mathrm{T}} & 0 & \cdots & \sigma_m C_m^{\mathrm{T}} \bar{H}_f^{\mathrm{T}} & 0 \end{bmatrix},$$

$$\hat{\Pi} = \begin{bmatrix} N\pi_{11} + \bar{H}_f \pi_{21} & N\pi_{12} + \bar{H}_f \pi_{22} & \cdots & N\pi_{1q} + \bar{H}_f \pi_{2q} \end{bmatrix},$$

$$\tilde{\Pi} = \begin{bmatrix} \pi_{11} & \pi_{12} & \cdots & \pi_{1q} \end{bmatrix},$$

$$\Phi_{16} = \begin{bmatrix} \tilde{H}_f & 0 & \hat{\Pi} & ND_1 + \bar{H}_f D_3 \end{bmatrix},$$

$$\Phi_{26} = \begin{bmatrix} 0 & 0 & \tilde{\Pi} & D_1 \end{bmatrix},$$

$$\Phi_{56} = \begin{bmatrix} 0 & \mathcal{C} & 0 & 0 \end{bmatrix},$$

$$\Phi_{66} = \begin{bmatrix} -Z & 0 & 0 & 0 \\ 0 & -Z & 0 & 0 \\ 0 & 0 & -\bar{\beta} & 0 \\ 0 & 0 & 0 & -W^{-1} \} \end{bmatrix},$$

$$Z = \begin{bmatrix} \zeta_1 I & 0 & \cdots & 0 \\ 0 & \zeta_2 I & \cdots & 0 \\ \vdots & \vdots & \ddots & \vdots \\ 0 & 0 & \cdots & \zeta_m I \end{bmatrix},$$

$$\Upsilon_{17} = \begin{bmatrix} 0 & 0 & \tilde{\eta} & L^{\mathrm{T}}\bar{Q} \end{bmatrix},$$

$$\Upsilon_{27} = \begin{bmatrix} \tilde{C} & 0 & M\tilde{\eta} & (ML^{\mathrm{T}} + \bar{K}^{\mathrm{T}}B_2^{\mathrm{T}})\bar{Q} \end{bmatrix},$$

$$\Upsilon_{67} = \begin{bmatrix} 0 & -\mathcal{H} & 0 & 0 \end{bmatrix},$$

$$\Upsilon_{77} = \begin{bmatrix} -\bar{Z} & 0 & 0 & 0 \\ 0 & -\bar{Z} & 0 & 0 \\ 0 & 0 & -\bar{Z} & 0 \\ 0 & 0 & 0 & -I \end{bmatrix},$$

$$\bar{Z} = \begin{bmatrix} \vartheta_1 I & 0 & \cdots & 0 \\ 0 & \vartheta_2 I & \cdots & 0 \\ \vdots & \vdots & \ddots & \vdots \\ 0 & 0 & \cdots & \vartheta_m I \end{bmatrix},$$

$$\mathcal{C} = \text{diag}\{0, \sigma_1 M C_1^{\mathrm{T}}, 0, \sigma_2 M C_2^{\mathrm{T}}, 0, \ldots, 0, \sigma_m M C_m^{\mathrm{T}}\},$$

$$\mathcal{H} = \text{diag}\{\vartheta_1 \bar{H}_f, 0, \vartheta_2 \bar{H}_f, 0, \ldots, \vartheta_m \bar{H}_f, 0\},$$

$$\bar{\mathcal{Y}} = \text{diag}\left\{ \begin{bmatrix} N & I \\ I & M \end{bmatrix}, \begin{bmatrix} N & I \\ I & M \end{bmatrix}, \ldots, \begin{bmatrix} N & I \\ I & M \end{bmatrix} \right\},$$

*then system (7.9) is exponentially mean-square stable and strictly $(Q, S, R)$ dissipative and, meanwhile, the individual steady-state variance constraint is also satisfied. Moreover, the desired estimator parameters and feedback controller parameter can be obtained by*

$$K = \bar{K}(U^{\mathrm{T}})^{-1},$$

$$H_f = V^{-1}\bar{H}_f, \tag{7.61}$$

$$A_f = V^{-1}(\bar{A}_f - (NA + VH_f\bar{\Theta}C)M - NB_1KU^{\mathrm{T}})(U^{\mathrm{T}})^{-1},$$

*where the nonsingular matrices $U$ and $V$ satisfy*

$$UV^{\mathrm{T}} = I - MN, \tag{7.62}$$

*which can be determined by the singular value decomposition of $I - MN$.*

Proof. Firstly, under the conditions of this theorem, it is easy to see that

$$\begin{bmatrix} -N & -I \\ -I & -M \end{bmatrix} < 0, \tag{7.63}$$

which, by the Schur Complement Lemma gives that $-N + M^{-1} < 0$, implying the non-singularity of $I - MN$. Therefore, there always exist nonsingular matrices $U$ and $V$ such that (7.62) is true.

Introduce the following construction of $Y$:

$$Y = \begin{bmatrix} M & U \\ U^{\mathrm{T}} & \Xi_1 \end{bmatrix},$$

$$Y^{-1} = \begin{bmatrix} N & V \\ V^{\mathrm{T}} & \Xi_2 \end{bmatrix},$$

$$\Xi_1 = -U^{\mathrm{T}}NV^{-\mathrm{T}},$$

$$\Xi_2 = -V^{\mathrm{T}}MU^{-\mathrm{T}}, \tag{7.64}$$

and define

$$\Psi_1 = \begin{bmatrix} N & I \\ V^{\mathrm{T}} & 0 \end{bmatrix},$$

$$\Psi_2 = \begin{bmatrix} I & M \\ 0 & U^{\mathrm{T}} \end{bmatrix}. \tag{7.65}$$

Then we have

$$Y\Psi_1 = \Psi_2,$$

$$UV^{\mathrm{T}} = I - MN. \tag{7.66}$$

Next, let us prove that the inequality (7.48) is equivalent to (7.57). To start with, performing the congruence transformation to (7.48) on both sides by $\mathrm{diag}\{1, \Psi_1^{\mathrm{T}}\}$, we obtain

$$\begin{bmatrix} 1 & 0 \\ 0 & \Psi_1^{\mathrm{T}} \end{bmatrix} \begin{bmatrix} -\alpha_i^{-1} & \bar{\pi}_i^{\mathrm{T}}\bar{H}^{\mathrm{T}} \\ * & -Y \end{bmatrix} \begin{bmatrix} 1 & 0 \\ 0 & \Psi_1 \end{bmatrix} < 0$$

$$\Longleftrightarrow \begin{bmatrix} -\alpha_i^{-1} & \bar{\pi}_i^{\mathrm{T}}\bar{H}^{\mathrm{T}}\Psi_1 \\ * & -\Psi_1^{\mathrm{T}}Y\Psi_1 \end{bmatrix} < 0 \tag{7.67}$$

$$\Longleftrightarrow \begin{bmatrix} -\beta_i & \pi_{1i}^{\mathrm{T}}N + \pi_{2i}^{\mathrm{T}}\bar{H}_f^{\mathrm{T}} & \pi_{1i}^{\mathrm{T}} \\ * & -N & -I \\ * & * & -M \end{bmatrix} < 0,$$

where $\beta_i \triangleq \alpha_i^{-1}$ ($i = 1, 2, \ldots, q$). Therefore, inequality (7.48) is equivalent to (7.57). Similarly, we can prove that inequality (7.49) holds if and only if inequality (7.58) holds. It is worth pointing out that here we use the equality constraints $\alpha_i\beta_i = 1$ ($i = 1, 2, \ldots, q$) to avoid the presence of the variable $\alpha_i$ and its reciprocal $\alpha_i^{-1}$ in the same set of LMIs.

In the following, we will show that the inequalities (7.50) and (7.51) are implied by inequalities (7.59) and (7.60) respectively. Performing the congruence transformation to (7.50) on both sides by $\mathrm{diag}\{\Psi_1^{\mathrm{T}}, \Psi_1^{\mathrm{T}}, \Psi_1^{\mathrm{T}}, \ldots, \Psi_1^{\mathrm{T}}, I, I\}$ results in

$$
\begin{bmatrix}
-\Psi_1^{\mathrm{T}} Y \Psi_1 & \Psi_1^{\mathrm{T}} \hat{A} Y \Psi_1 & \Psi_1^{\mathrm{T}} \bar{A} \mathscr{Y} \Psi_1 & \Psi_1^{\mathrm{T}} \hat{H} & \Psi_1^{\mathrm{T}} \bar{D} \\
* & -\Psi_1^{\mathrm{T}} Y \Psi_1 & 0 & 0 & 0 \\
* & * & -\bar{\Psi}_1^{\mathrm{T}} \mathscr{Y} \Psi_1 & 0 & 0 \\
* & * & * & -\bar{\beta} & 0 \\
* & * & * & * & -W^{-1}
\end{bmatrix} < 0, \qquad (7.68)
$$

where $\bar{\Psi}_1 = \mathrm{diag}\{\Psi_1, \Psi_1, \ldots, \Psi_1\}$.

For the term $\Psi_1^{\mathrm{T}} \bar{A} \mathscr{Y} \bar{\Psi}_1$ in (7.68), we conduct the following calculation:

$$
\begin{aligned}
& \Psi_1^{\mathrm{T}} \bar{A} \mathscr{Y} \bar{\Psi}_1 \\
=& \Psi_1^{\mathrm{T}} \begin{bmatrix} \sigma_1 \tilde{A}_1 & \sigma_2 \tilde{A}_2 & \cdots & \sigma_m \tilde{A}_m \end{bmatrix} \mathrm{diag}\{Y\Psi_1, Y\Psi_1, \ldots, Y\Psi_1\} \\
=& \begin{bmatrix} \sigma_1 \Psi_1^{\mathrm{T}} \tilde{A}_1 Y \Psi_1 & \sigma_2 \Psi_1^{\mathrm{T}} \tilde{A}_2 Y \Psi_1 & \cdots & \sigma_m \Psi_1^{\mathrm{T}} \tilde{A}_m Y \Psi_1 \end{bmatrix} \\
=& \begin{bmatrix}
\sigma_1 \bar{H}_f C_1 & \sigma_1 \bar{H}_f C_1 M & \sigma_2 \bar{H}_f C_2 & \sigma_2 \bar{H}_f C_2 M \\
0 & 0 & 0 & 0
\end{bmatrix} \qquad (7.69)
\end{aligned}
$$

$$
\begin{bmatrix}
\cdots & \sigma_m \bar{H}_f C_m & \sigma_m \bar{H}_f C_m M \\
\cdots & 0 & 0
\end{bmatrix}.
$$

Notice that the matrix variables $\bar{H}_f$ and $M$ are not linear in the term $\sigma_i \bar{H}_f C_i M$. Here, for arbitrary scalars $\zeta_i > 0$ ($i = 1, 2, \ldots, m$), it is true that

$$
\begin{bmatrix}
0 & \bar{H}_f C_i M \\
M^{\mathrm{T}} C_i^{\mathrm{T}} \bar{H}_f^{\mathrm{T}} & 0
\end{bmatrix} \leq
\begin{bmatrix}
\zeta_i \bar{H}_f \bar{H}_f^{\mathrm{T}} & 0 \\
0 & \zeta_i^{-1} M^{\mathrm{T}} C_i^{\mathrm{T}} C_i M
\end{bmatrix}. \qquad (7.70)
$$

Then, it follows directly from (7.68) with (7.70) that the matrix inequality (7.50) is true if (7.59) is true. Similarly, we could easily prove that the inequality (7.60) implies (7.51). Therefore, according to Corollary 7.3.1, system (7.9) is exponentially mean-square stable and strictly $(Q, R, S)$ dissipative, and the steady-state covariance exists, satisfying $\bar{X} \leq M$ by (7.54). Next, it is obvious that (7.47) is equivalent to

$$
M - Y_0 < 0. \qquad (7.71)
$$

Thus, $\bar{X} \leq M < Y_0$. Now, from the $n$ LMIs in (7.56), we can see that the individual variance of each system states is not more than the pre-specified value. In other words, the design requirements $(R1)$, $(R2)$, and $(R3)$ are simultaneously satisfied. The proof is complete. $\qquad\qquad \square$

## 7.3.2    Computational Algorithm

It is worth mentioning that the obtained conditions in Theorem 7.3.2 are not all strict LMIs, which, as a result, cannot be solved directly by applying Matlab LMI Toolbox. However, with the so-called cone complementarity linearization (CCL) method proposed in Refs [141, 142], we can convert the original non-convex feasibility problem of certain LMIs into some sequential optimization problems subject to LMI constraints. To this end, we introduce a new condition by $\alpha_i\beta_i \geq 1$, which, by the Schur Complement Lemma is equivalent to

$$\begin{bmatrix} -\alpha_i & 1 \\ 1 & -\beta_i \end{bmatrix} \leq 0, \qquad i = 1, 2, \dots, q. \tag{7.72}$$

Then, using the CCL method, we suggest the following minimization problem involving LMI conditions instead of the original non-convex problem formulated in Theorem 7.3.2.

**Problem MCD (Multi-Objective Controller Design)**

$$\min \sum_{i=1}^{q} \alpha_i\beta_i \text{ subject to (7.56) to (7.60) and (7.72)}. \tag{7.73}$$

If the solution of the above minimization problem is $q$, that is, $\min(\sum_{i=1}^{q} \alpha_i\beta_i) = q$, then the condition in Theorem 7.3.2 is solvable. It should be pointed out that this algorithm does not guarantee finding a global optimal solution for the problem above. Nevertheless, the proposed minimization problem is much easier to be solved than the original non-convex feasibility problem.

**Algorithm MCD**

*Step 1.* Find a feasible set $(M^{(0)}, N^{(0)}, \bar{A}_f^{(0)}, \bar{H}_f^{(0)}, \bar{K}^{(0)}, \alpha_i^{(0)}, \beta_i^{(0)})$ satisfying (7.56) to (7.60) and (7.72). Set $d = 0$.
*Step 2.* Solve the following optimization problem:

$$\min \sum_{i=1}^{q} (\alpha_i^{(d)}\beta_i + \alpha_i\beta_i^{(d)})$$

subject to (7.56) to (7.60) and (7.72)

and denote $g^*$ as the optimized value.
*Step 3.* Substitute the obtained variables $(M, N, \bar{A}_f, \bar{H}_f, \bar{K}, \alpha_i, \beta_i)$ into (7.56) to (7.60). If conditions (7.56) to (7.60) are satisfied with

$$|g^* - 2q| < \upsilon,$$

where $\upsilon$ is a sufficiently small positive scalar, then output the feasible solutions $(M, N, \bar{A}_f, \bar{H}_f, \bar{K}, \alpha_i, \beta_i)$ and obtain the desired parameters $A_f$, $H_f$, and $K$ by (2) and (7.62). EXIT.

*Step 4.* If $d > N$, where $N$ is the maximum number of iterations allowed, EXIT.
*Step 5.* Set $d = d + 1$, $(M^{(d)}, N^{(d)}, \bar{A}_f^{(d)}, \bar{H}_f^{(d)}, \bar{K}^{(d)}, \alpha_i^{(d)}, \beta_i^{(d)}) = (M, N, \bar{A}_f, \bar{H}_f, \bar{K}, \alpha_i, \beta_i)$,
and go to Step 2.

## 7.4  Numerical Example

In this section, we present an illustrative example to demonstrate the effectiveness of
the proposed algorithm.

Consider a nonlinear stochastic system with the following parameters:

$$A = \begin{bmatrix} 0.2 & -0.05 \\ -0.1 & 0.08 \end{bmatrix}, \quad B_1 = \begin{bmatrix} 0.03 \\ -0.5 \end{bmatrix}, \quad D_1 = \begin{bmatrix} 0.1 \\ 0.03 \end{bmatrix},$$

$$L = \begin{bmatrix} 0.05 & -0.07 \end{bmatrix}, \quad B_2 = 0.04, \quad D_2 = 0.25,$$

with the measured output equation

$$y(k) = \Theta \begin{bmatrix} -0.4 & 0.3 \\ 0.2 & -0.1 \end{bmatrix} x(k) + g(x(k)) + \begin{bmatrix} 0.02 \\ 0.01 \end{bmatrix} \omega(k).$$

The stochastic nonlinear functions are taken to be

$$f(x(k)) = \begin{bmatrix} 0.2 \\ 0.3 \end{bmatrix} (0.3 \, \text{sign}[x_1(k)]x_1(k)v_1(k) + 0.4 \, \text{sign}[x_2(k)]x_2(k)v_2(k)),$$

$$g(x(k)) = \begin{bmatrix} 0.1 \\ 0.4 \end{bmatrix} (0.3 \, \text{sign}[x_1(k)]x_1(k)v_1(k) + 0.4 \, \text{sign}[x_2(k)]x_2(k)v_2(k)),$$

where $x_i(k)$ is the $i$th component of $x(k)$ and $v_i(k)$ is a zero mean independent Gaussian
white noise process with unity covariance which is also assumed to be independent
from $\omega(k)$. It is easy to check that $f(x(k))$ and $g(x(k))$ satisfy

$$\mathbb{E} \left\{ \begin{bmatrix} f(x(k)) \\ g(x(k)) \end{bmatrix} | x(k) \right\} = 0,$$

$$\mathbb{E} \left\{ \begin{bmatrix} f(x(k)) \\ g(x(k)) \end{bmatrix} \begin{bmatrix} f^T(x(k)) & g^T(x(k)) \end{bmatrix} | x(k) \right\}$$

$$= \begin{bmatrix} 0.2 \\ 0.3 \\ 0.1 \\ 0.4 \end{bmatrix} \begin{bmatrix} 0.2 \\ 0.3 \\ 0.1 \\ 0.4 \end{bmatrix}^T x^T(k) \left( \begin{bmatrix} 0.3 \\ 0 \end{bmatrix} \begin{bmatrix} 0.3 \\ 0 \end{bmatrix}^T + \begin{bmatrix} 0 \\ 0.4 \end{bmatrix} \begin{bmatrix} 0 \\ 0.4 \end{bmatrix}^T \right) x(k).$$

Hence,

$$\pi_{11} = \pi_{12} = \begin{bmatrix} 0.2 \\ 0.3 \end{bmatrix}, \quad \pi_{21} = \pi_{22} = \begin{bmatrix} 0.1 \\ 0.4 \end{bmatrix},$$

$$\eta_1 = \begin{bmatrix} 0.3 \\ 0 \end{bmatrix}, \quad \eta_2 = \begin{bmatrix} 0 \\ 0.4 \end{bmatrix}.$$

In addition, we assume that the probabilistic density functions of $\theta_1$ and $\theta_2$ in $[0, 1]$ are described by

$$p_1(s_1) = \begin{cases} 0.8, & s_1 = 0 \\ 0.1, & s_1 = 0.5 \\ 0.1, & s_1 = 1 \end{cases} \quad \text{and} \quad p_2(s_2) = \begin{cases} 0.7, & s_2 = 0 \\ 0.2, & s_2 = 0.5 \\ 0.1, & s_2 = 1 \end{cases}$$

from which the expectations and variances can be easily calculated as $\bar{\theta}_1 = 0.15$, $\bar{\theta}_2 = 0.2$, $\sigma_1^2 = 0.1025$, and $\sigma_2^2 = 0.11$. Select $Q = -1.2$, $S = 0.8$, and $R = 1.6$. Choose the required steady-state variance constraints as $\delta_1^2 = 0.36$ and $\delta_2^2 = 0.64$.

Applying standard numerical software to solve Problem MCD, we can obtain the observer and feedback controller parameters as follows:

$$A_f = \begin{bmatrix} 0.6944 & 0.4181 \\ 0.9199 & 0.5465 \end{bmatrix},$$

$$H_f = \begin{bmatrix} 0.0155 & 0.0836 \\ 0.0221 & 0.1168 \end{bmatrix},$$

$$K = \begin{bmatrix} 0.3862 & 0.0120 \end{bmatrix},$$

$$\alpha_1 = 0.9586, \quad \alpha_2 = 0.9664,$$

$$\beta_1 = 1.0432, \quad \beta_2 = 1.0348.$$

The time responses of the individual states $x_1(k)$, $x_2(k)$ and their estimates $\hat{x}_1(k)$, $\hat{x}_2(k)$ are shown in Figures 7.1 and 7.2. The evolution of steady-state variance is shown in Figure 7.3 while the time response of the output signal $z(k)$ is shown in Figure 7.4, from which we can see clearly that all the designing requirements have been achieved.

## 7.5   Summary

In this chapter, we have designed an observer-based controller for a class of nonlinear stochastic systems such that, for all possible degraded measurements, the closed-loop system is exponentially mean-square stable, the system dissipativity is achieved, and the steady-state variance of individual state components is not more than the

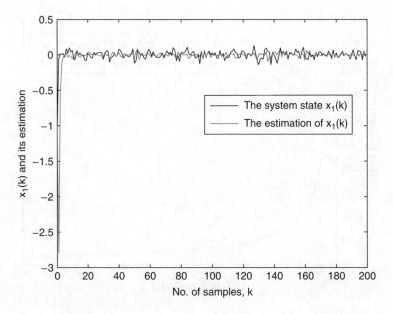

**Figure 7.1**  System state $x_1(k)$ and its estimate $\hat{x}_1(k)$

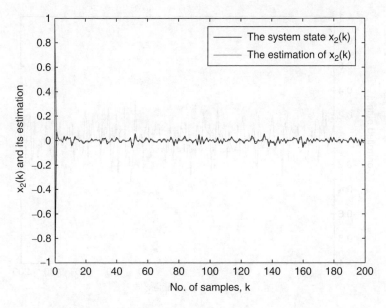

**Figure 7.2**  System state $x_2(k)$ and its estimate $\hat{x}_2(k)$

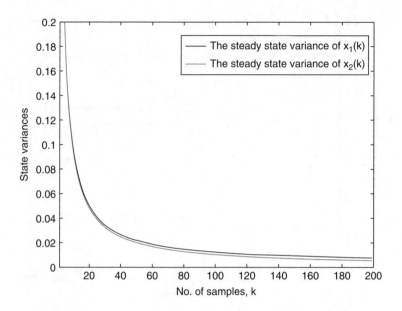

**Figure 7.3**   The individual steady state variance of each state

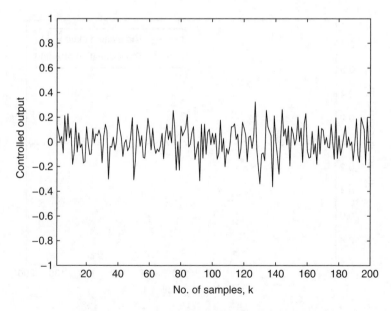

**Figure 7.4**   The controlled output signal $z(k)$

pre-specified values. The nonlinearities considered here are characterized statistically, which can cover several classes of commonly encountered nonlinearities. The solvability of the addressed problem has been expressed as the feasibility of a set of LMIs with equality constraints. An algorithm has been proposed to convert the original non-convex feasibility problem into an optimal minimization problem, which is much easier to solve by standard numerical software. An illustrative example has been presented to demonstrate the effectiveness and applicability of the provided design method.

# 8

# Variance-Constrained $H_\infty$ Control with Multiplicative Noises

Among many practical systems, plants may be modeled by bilinear systems (which are also known as systems with multiplicative noises) since some characteristics of nonlinear systems can be closely approximated by bilinear models rather than linearized models. The bilinear system is a kind of "nearly linear" yet nonlinear system. The related problems of bilinear systems, such as the state analysis and the parameter estimation, are much more difficult to solve than those of the linear systems. Up to now, it has been known that bilinear systems could describe many real processes in the fields of socioeconomics, ecology, agriculture, biology, and industry, etc. In particular, bilinear models are widely used to model nonlinear processes in signal and image processing, as well as communication systems analysis, such as channel equalization, echo cancelation, nonlinear tracking, electroencephalogram (EEG) signal classification, multiplicative disturbance tracking, to name but a few key ones. As a result, control problems have been extensively studied for bilinear systems and many research results have been reported in the literature. However, unfortunately, up to now, the multi-objective control problems that take multiple performance requirements, including variance constraints, $H_\infty$ specifications, and other performances, have received little attention for stochastic systems with multiplicative noises, and remain open and challenging, though similar problems have already been investigated for linear systems.

In view of the above observation, in this chapter the robust variance-constrained $H_\infty$ control problem is considered for uncertain stochastic systems with multiplicative noises. The norm-bounded parametric uncertainties enter into both the system and output matrices. The purpose of the problem is to design a state feedback controller such that, for all admissible parameter uncertainties, (1) the closed-loop system is exponentially mean-square quadratically stable; (2) the individual steady-state variance satisfies given upper bound constraints; and (3) the prescribed noise attenuation

*Variance-Constrained Multi-Objective Stochastic Control and Filtering*, First Edition.
Lifeng Ma, Zidong Wang and Yuming Bo.
© 2015 John Wiley & Sons, Ltd. Published 2015 by John Wiley & Sons, Ltd.

level is guaranteed in an $H_\infty$ sense with respect to the additive noise disturbances. A general framework is established to solve the addressed multi-objective problem by using a linear matrix inequality (LMI) approach, where the required stability, the $H_\infty$ characterization, and variance constraints are all easily enforced. Within such a framework, two additional optimization problems are formulated: one is to optimize the $H_\infty$ performance and the other is to minimize the weighted sum of the system state variances. A numerical example is provided to illustrate the effectiveness of the proposed design algorithm.

The remainder of this chapter is organized as follows: In Section 8.1, the robust variance-constrained $H_\infty$ control problem for stochastic systems with both multiplicative noises and norm-bounded parameter uncertainties is formulated. The conditions for stability, $H_\infty$ performance, and state variance are expressed in terms of LMI in Section 8.2. An LMI algorithm is developed in Section 8.3 for designing robust variance-constrained $H_\infty$ state feedback controllers with both multiplicative noises and deterministic norm-bounded parameter uncertainty. A numerical example is presented in Section 8.4 to show the applicability of the algorithm and some concluding remarks are provided in Section 8.5.

## 8.1    Problem Formulation

Consider the following class of *stochastic* discrete-time systems with both *multiplicative noises* and *deterministic* norm-bounded parameter uncertainties:

$$x(k+1) = (A + H_1 FE + A_s \eta(k))x(k) + B_1 w(k) + B_2 u(k),$$

$$z(k) = (C_1 + H_2 FE)x(k) + D_{11}w(k) + D_{12}u(k), \tag{8.1}$$

where $x(k) \in \mathbb{R}^n$ is the state, $u(k) \in \mathbb{R}^r$ is the control input, and $z(k) \in \mathbb{R}^p$ is the controlled output. The process noise $w(k) \in \mathbb{R}^q$ is a zero mean Gaussian white noise sequence with covariance $R > 0$ and the stochastic multiplicative noise $\eta(k) \in \mathbb{R}$ is also a zero mean Gaussian white noise sequence but with unity covariance. The real matrices $A, A_s, B_1, B_2, C_1, D_{11}, D_{12}, H_1, H_2$, and $E$ are known matrices with appropriate dimensions.

The real matrix $F \in \mathbb{R}^{i \times j}$, which could be time-varying, represents the deterministic norm-bounded parameter uncertainty and satisfies

$$FF^{\mathrm{T}} \leq I. \tag{8.2}$$

The parameter uncertainty $F$ is said to be admissible if it satisfies the condition (8.2).

**Remark 8.1** *The intensity of the multiplicative noise $\eta(k)$, which causes the bilinearities or stochastic uncertainties, can be scaled and absorbed in the matrix $A_s$. Hence, without loss of generality, we could assume that $\eta(k)$ is of unity covariance.*

Applying the state feedback control law

$$u(k) = Kx(k), \tag{8.3}$$

to the system (8.1), we obtain the following closed-loop system:

$$x(k+1) = (A_K + A_s\eta(k))x(k) + B_1 w(k),$$
$$z(k) = C_K x(k) + D_{11} w(k), \tag{8.4}$$

where $K$ is the state feedback gain and

$$A_K \triangleq A + B_2 K + H_1 FE, \tag{8.5}$$
$$C_K \triangleq C_1 + D_{12}K + H_2 FE. \tag{8.6}$$

Before giving our design goal, we introduce the following stability concept for the system (8.4).

**Definition 8.1.1** *The system* (8.4) *is said to be* exponentially mean-square quadratically stable *if, with* $w(k) = 0$, *there exist constants* $\alpha \geq 1$ *and* $\beta \in (0, 1)$ *such that*

$$\mathbb{E}\{\|x(k)\|^2\} \leq \alpha\beta^k \mathbb{E}\{\|x_0\|^2\}, \quad \forall x_0 \in \mathbb{R}^n, \quad k \in \mathbb{I}^+, \tag{8.7}$$

*for all admissible uncertainties.*

The aim in this chapter is to design a state feedback controller of the form (8.3), such that for all admissible deterministic uncertainties, the following three requirements are simultaneously satisfied for the system (8.4):

(R1) The system (8.4) is exponentially mean-square quadratically stable.
(R2) For a given scalar $\gamma > 0$ and all nonzero $w(k)$ and zero initial condition $x_0 = 0$, the controlled output $z(k)$ satisfies

$$\sum_{k=0}^{N} \mathbb{E}\{\|z(k)\|^2\} \leq \gamma^2 \sum_{k=0}^{N} \mathbb{E}\{\|w(k)\|^2\}. \tag{8.8}$$

(R3) The individual steady-state state variances satisfy the following constraints:

$$\text{Var}\{x_i(k)\} \triangleq \lim_{k \to \infty} \mathbb{E}\{x_i(k)x_i^T(k)\} < \sigma_i^2, \tag{8.9}$$

where $x(k) = \begin{bmatrix} x_1(k) & x_2(k) & \dots & x_n(k) \end{bmatrix}^T$ and $\sigma_i^2 > 0$ $(i = 1, 2, \dots, n)$ are given scalars specifying the acceptable variance upper bounds obtained from the engineering requirements.

The problem addressed above is referred to as the robust $H_\infty$ control problem with variance constraints.

## 8.2    Stability, $H_\infty$ Performance, and Variance Analysis

In this section, the multi-objective (stability, $H_\infty$ performance, and variance analysis) will be considered for stochastic discrete-time systems with both multiplicative noises and deterministic norm-bounded parameter uncertainties.

### 8.2.1    Stability

Before deriving the stability conditions, two useful lemmas are given as follows.

**Lemma 8.2.1** *Let $V(x(k)) = x(k)^T P x(k)$ be a Lyapunov functional where $P > 0$. If there exist real scalars $\lambda$, $\mu > 0$, $\nu > 0$, and $0 < \psi < 1$ such that both*

$$\mu \|x(k)\|^2 \le V(x(k)) \le \nu \|x(k)\|^2 \tag{8.10}$$

*and*

$$\mathbb{E}\{V(x(k+1))|x(k)\} - V(x(k)) \le \lambda - \psi V(x(k)) \tag{8.11}$$

*hold, then the process $x(k)$ satisfies*

$$\mathbb{E}\{\|x(k)\|^2\} \le \frac{\nu}{\mu}\|x_0\|^2 (1 - \psi)^k + \frac{\lambda}{\mu\psi}. \tag{8.12}$$

**Lemma 8.2.2** *Consider a system*

$$\xi(k+1) = (M + N\eta(k))\xi(k), \tag{8.13}$$

*where $\eta(k)$ is a zero mean Gaussian white noise sequence and $M$, $N$ are constant matrices with appropriate dimensions. If the system (8.13) is exponentially mean-square stable, i.e., there exist constants $\alpha \ge 1$ and $\beta \in (0, 1)$ such that*

$$\mathbb{E}\{\|\xi(k)\|^2\} \le \alpha\beta^k \mathbb{E}\{\|\xi_0\|^2\}, \quad \forall \xi_0 \in \mathbb{R}^n, \quad k \in \mathbb{I}^+, \tag{8.14}$$

*and there exists a symmetric matrix $Y$ satisfying*

$$MYM^T - Y + NYN^T < 0, \tag{8.15}$$

*then $Y \ge 0$.*

   *Proof.* It follows from (8.15) that

$$MYM^T - Y + NYN^T = -\Theta \tag{8.16}$$

for some $\Theta > 0$. Define a functional $W(\xi(k)) = \xi^T(k)Y\xi(k)$. Applying the super-Martingale property for the system (8.13) yields

$$\mathbb{E}\{W(\xi(k+1))|\xi(k)\} - W(\xi(k)) = \xi^T(k)(MYM^T - Y + NYN^T)\xi(k)$$

$$= -\xi^T(k)\Theta\xi(k). \tag{8.17}$$

Summing (8.17) from 0 to $n$ with respect to $k$, we obtain

$$\mathbb{E}(\xi^{\mathrm{T}}(n)Y\xi(n)) - \xi^{\mathrm{T}}(0)Y\xi(0) = -\sum_{k=0}^{n}\xi^{\mathrm{T}}(k)\Theta\xi(k). \tag{8.18}$$

Let $n \to \infty$ in (8.18). It then follows from the exponential mean-square stability of the system (8.13) and the fact that

$$\lim_{n\to\infty}\mathbb{E}(\xi^{\mathrm{T}}(n)Y\xi(n)) \leq \|Y\| \lim_{n\to\infty}\mathbb{E}(\xi^{\mathrm{T}}(n)\xi(n))$$

that $\lim_{n\to\infty}\mathbb{E}(\xi^{\mathrm{T}}(n)Y\xi(n)) = 0$. Hence, we have from (8.18) that

$$\xi^{\mathrm{T}}(0)Y\xi(0) = \sum_{k=0}^{\infty}\xi^{\mathrm{T}}(k)\Theta\xi(k) \geq 0. \tag{8.19}$$

Since (8.19) holds for any nonzero initial state $\xi(0)$, we arrive at the conclusion that $Y \geq 0$. □

According to Definition 8.1.1, we have the following theorem, which provides the sufficient and necessary conditions for the exponential quadratic stability of the system (8.4).

**Theorem 8.2.3** *Given the feedback gain matrix $K$. The system (8.4) is exponentially mean-square quadratically stable if, for all admissible uncertainties, there exists a positive definite matrix $P$ satisfying*

$$A_K^{\mathrm{T}}PA_K - P + A_s^{\mathrm{T}}PA_s < 0. \tag{8.20}$$

*Proof.* To prove the sufficiency, we define Lyapunov functional $V(x(k)) = x^{\mathrm{T}}(k)Px(k)$, where $P > 0$ is the solution to (8.20). By using the super-Martingale property for the system (8.4) with $w(k) = 0$, we obtain

$$\mathbb{E}\{V(x(k+1))|x(k)\} - V(x(k))$$
$$= x^{\mathrm{T}}(k)A_K^{\mathrm{T}}PA_Kx(k) + x^{\mathrm{T}}(k)\mathbb{E}\{A_s^{\mathrm{T}}PA_s\eta^2(k)\}x(k) - x^{\mathrm{T}}(k)Px(k)$$
$$= x^{\mathrm{T}}(k)(A_K^{\mathrm{T}}PA_K - P + A_s^{\mathrm{T}}PA_s)x(k). \tag{8.21}$$

We know from (8.20) that there must exist a sufficiently small scalar $\alpha$ satisfying $0 < \alpha < \lambda_{\max}(P)$ and

$$A_K^{\mathrm{T}}PA_K - P + A_s^{\mathrm{T}}PA_s < -\alpha I. \tag{8.22}$$

Therefore, it follows that

$$\mathbb{E}\{V(x(k+1))|x(k)\} - V(x(k)) \leq -\alpha x^{\mathrm{T}}(k)x(k) \leq -\frac{\alpha}{\lambda_{\max}(P)}V(x(k)). \tag{8.23}$$

Then, the proof of the sufficiency follows immediately from Lemma 8.2.1. □

**Corollary 8.2.4** *Given the feedback gain matrix K. The system (8.4) is exponentially mean-square quadratically stable if and only if, for all admissible uncertainties, there exists a positive definite matrix Q satisfying*

$$A_K Q A_K^T - Q + A_s Q A_s^T < 0. \tag{8.24}$$

*Proof.* The proof follows easily from Theorem 8.2.3 and the fact that $\rho(\Phi) = \rho(\Phi^T)$, where $\Phi$ is a square matrix and $\rho(\cdot)$ is the spectral radius. □

## 8.2.2 $H_\infty$ Performance

Contrary to the standard $H_\infty$ performance formulation, we shall use the expression (8.8) to describe the $H_\infty$ performance of the stochastic system, where the expectation operator is utilized on both the controlled output and the disturbance input.

We are now in the situation to derive the sufficient conditions for establishing the $H_\infty$-norm performance.

**Theorem 8.2.5** *For a given $\gamma > 0$ and a given feedback gain matrix K, the system (8.4) is exponentially mean-square quadratically stable and achieves the $H_\infty$-norm constraint (8.8) for all nonzero w(k), if there exists a positive definite matrix P satisfying*

$$\begin{bmatrix} A_K^T P A_K - P + A_s^T P A_s + C_K^T C_K & A_K^T P B_1 + C_K^T D_{11} \\ B_1^T P A_K + D_{11}^T C_K & B_1^T P B_1 - \gamma^2 I + D_{11}^T D_{11} \end{bmatrix} < 0, \tag{8.25}$$

*for all admissible uncertainties.*

*Proof.* It is obvious that (8.25) implies (8.20); hence it follows from Theorem 8.2.3 that the system (8.4) is exponentially mean-square quadratically stable.

Next, for any nonzero $w(k)$, it follows from (8.25) that

$$\mathbb{E}\{V(x(k+1))|x(k)\} - V(x(k)) + \mathbb{E}\{z^T(k)z(k)\} - \gamma^2 \mathbb{E}\{w^T(k)w(k)\}$$
$$= x^T(k)(A_K^T P A_K - P + A_s^T P A_s)x(k) + x^T(k)A_K^T P B_1 w(k)$$
$$+ w^T(k)B_1^T P A_K x(k) + w^T(k)B_1^T P B_1 w(k)$$
$$+ x^T(k)C_K^T C_K x(k) + x^T(k)C_K^T D_{11} w(k) + w^T(k)D_{11}^T C_K x(k)$$
$$+ w^T(k)D_{11}^T D_{11} w(k) - \gamma^2 w^T(k)w(k)$$
$$= x^T(k)(A_K^T P A_K - P + A_s^T P A_s + C_K^T C_K)x(k) + x^T(k)(A_K^T P B_1 + C_K^T D_{11})w(k)$$
$$+ w^T(k)(B_1^T P A_K + D_{11}^T C_K)x(k) + w^T(k)(B_1^T P B_1 + D_{11}^T D_{11} - \gamma^2 I)w(k)$$
$$= \begin{bmatrix} x(k) \\ w(k) \end{bmatrix}^T \begin{bmatrix} \bar{A}_K & A_K^T P B_1 + C_K^T D_{11} \\ B_1^T P A_K + D_{11}^T C_K & B_1^T P B_1 - \gamma^2 I + D_{11}^T D_{11} \end{bmatrix} \begin{bmatrix} x(k) \\ w(k) \end{bmatrix} < 0. \tag{8.26}$$

where $\bar{A}_K \triangleq A_K^T P A_K - P + A_s^T P A_s + C_K^T C_K$.

Now, summing (8.26) from 0 to $\infty$ with respect to $k$ leads to

$$\sum_{k=0}^{\infty} [\mathbb{E}\{V(x(k+1))|x(k)\} - V(x(k))$$

$$+ \mathbb{E}\{z^{\mathrm{T}}(k)z(k)\} - \gamma^2 \mathbb{E}\{w^{\mathrm{T}}(k)w(k)\}] < 0 \tag{8.27}$$

or

$$\sum_{k=0}^{\infty} \mathbb{E}\{\|z(k)\|^2\} < \gamma^2 \sum_{k=0}^{\infty} \mathbb{E}\{\|w(k)\|^2\} + V(x_0) - V(x_\infty). \tag{8.28}$$

Since $x_0 = 0$ and the system (8.4) is exponentially mean-square quadratically stable, it is straightforward to see that

$$\sum_{k=0}^{\infty} \mathbb{E}\{\|z(k)\|^2\} < \gamma^2 \sum_{k=0}^{\infty} \mathbb{E}\{\|w(k)\|^2\}, \tag{8.29}$$

which ends the proof. $\qquad\qquad\square$

Note that the inequality (8.25) is not linear on the closed-loop matrix $A_K$. In the interest of establishing an LMI framework for the controller design, we now restate Theorem 8.2.5 in terms of an LMI as follows.

**Theorem 8.2.6** *For a given $\gamma > 0$ and a given feedback gain matrix $K$, the system (8.4) is exponentially mean-square quadratically stable and achieves the $H_\infty$-norm constraint (8.8) for all nonzero $w(k)$, if there exists a positive definite matrix $Q$ satisfying*

$$\begin{bmatrix} -Q & A_K Q & 0 & 0 & B_1 \\ QA_K^{\mathrm{T}} & -Q & QA_s^{\mathrm{T}} & QC_K^{\mathrm{T}} & 0 \\ 0 & A_s Q & -Q & 0 & 0 \\ 0 & C_K Q & 0 & -I & D_{11} \\ B_1^{\mathrm{T}} & 0 & 0 & D_{11}^{\mathrm{T}} & -\gamma^2 I \end{bmatrix} < 0. \tag{8.30}$$

*Proof.* Using the Schur Complement Lemma twice, we can see that (8.25) is equivalent to

$$\begin{bmatrix} -P + A_s^{\mathrm{T}} P A_s & 0 & A_K^{\mathrm{T}} & C_K^{\mathrm{T}} \\ 0 & -\gamma^2 I & B_1^{\mathrm{T}} & D_{11}^{\mathrm{T}} \\ A_K & B_1 & -P^{-1} & 0 \\ C_K & D_{11} & 0 & -I \end{bmatrix} < 0 \tag{8.31}$$

or

$$\begin{bmatrix} -P & 0 & A_K^T & C_K^T & A_s^T \\ 0 & -\gamma^2 I & B_1^T & D_{11}^T & 0 \\ A_K & B_1 & -P^{-1} & 0 & 0 \\ C_K & D_{11} & 0 & -I & 0 \\ A_s & 0 & 0 & 0 & -P^{-1} \end{bmatrix} < 0. \tag{8.32}$$

Performing *twice* the congruence transformation to (8.32) by

$$\begin{bmatrix} I & 0 & 0 & 0 & 0 \\ 0 & 0 & 0 & 0 & I \\ 0 & I & 0 & 0 & 0 \\ 0 & 0 & 0 & I & 0 \\ 0 & 0 & I & 0 & 0 \end{bmatrix} \quad \text{and} \quad \begin{bmatrix} 0 & I & 0 & 0 & 0 \\ I & 0 & 0 & 0 & 0 \\ 0 & 0 & I & 0 & 0 \\ 0 & 0 & 0 & I & 0 \\ 0 & 0 & 0 & 0 & I \end{bmatrix}, \tag{8.33}$$

we can see that (8.32) is equivalent to

$$\begin{bmatrix} -P^{-1} & A_K & 0 & 0 & B_1 \\ A_K^T & -P & A_s^T & C_K^T & 0 \\ 0 & A_s & -P^{-1} & 0 & 0 \\ 0 & C_K & 0 & -I & D_{11} \\ B_1^T & 0 & 0 & D_{11}^T & -\gamma^2 I \end{bmatrix} < 0. \tag{8.34}$$

Let $P = Q^{-1}$ in (8.34) and then again applying the congruence transformation by diag$\{I, \ Q, \ I, \ I, \ I\}$, we arrive at (8.30), and the proof is complete. $\qquad\square$

### 8.2.3    Variance Analysis

Define the steady-state covariance by

$$\hat{Q} \triangleq \lim_{k \to \infty} \mathbb{E}\{x(k)x(k)^T\}$$

$$= \lim_{k \to \infty} \mathbb{E}\{ \begin{bmatrix} x_1(k) & x_2(k) & \dots & x_n(k) \end{bmatrix} \begin{bmatrix} x_1(k) & x_2(k) & \dots & x_n(k) \end{bmatrix}^T \}. \tag{8.35}$$

Obviously, if the system (8.4) is exponentially mean-square quadratically stable, then in the steady state, $\hat{Q}$ exists and satisfies the following equation:

$$A_K \hat{Q} A_K^T - \hat{Q} + A_s \hat{Q} A_s^T + B_1 R B_1^T = 0. \tag{8.36}$$

**Theorem 8.2.7** *If there exists a positive definite matrix $Q$ satisfying*

$$
\begin{bmatrix}
-Q & A_K Q & A_s Q & B_1 \\
QA_K^T & -Q & 0 & 0 \\
QA_s^T & 0 & -Q & 0 \\
B_1^T & 0 & 0 & -R^{-1}
\end{bmatrix} < 0,
\tag{8.37}
$$

*then the system (8.4) is exponentially mean-square quadratically stable and $\hat{Q} \leq Q$.*

*Proof.* We first prove that (8.37) is equivalent to

$$
A_K Q A_K^T - Q + A_s Q A_s^T + B_1 R B_1^T < 0.
\tag{8.38}
$$

By using the Schur Complement Lemma to (8.38), we have

$$
\begin{bmatrix}
-Q + A_s Q A_s^T + B_1 R B_1^T & A_K \\
A_K^T & -Q^{-1}
\end{bmatrix} < 0
\tag{8.39}
$$

$$
\iff
\begin{bmatrix}
-Q + B_1 R B_1^T & A_K & A_s \\
A_K^T & -Q^{-1} & 0 \\
A_s^T & 0 & -Q^{-1}
\end{bmatrix} < 0
\tag{8.40}
$$

$$
\iff
\begin{bmatrix}
-Q & A_K & A_s & B_1 \\
A_K^T & -Q^{-1} & 0 & 0 \\
A_s^T & 0 & -Q^{-1} & 0 \\
B_1^T & 0 & 0 & -R^{-1}
\end{bmatrix} < 0.
\tag{8.41}
$$

Performing the congruence transformation to (8.41) by diag$\{I, Q, Q, I\}$ yields (8.37). Hence, there exists a matrix $Q > 0$ satisfying (8.37) if and only if there exists a matrix $Q > 0$ satisfying (8.38).

Next, it follows directly from (8.38) and Theorem 8.2.3 that the system (8.4) is exponentially mean-square quadratically stable. Hence, $\hat{Q}$ exists and meets (8.36).

Subtracting (8.36) from (8.38) gives

$$
A_K(Q - \hat{Q})A_K^T - (Q - \hat{Q}) + A_s(Q - \hat{Q})A_s^T < 0,
\tag{8.42}
$$

which indicates from Lemma 8.2.2 that $Q - \hat{Q} \geq 0$. The proof is now completed. $\square$

The results provided in the above theorem will be essential for designing the controllers, which guarantee the stability, $H_\infty$ performance, and variance constraints for the uncertain stochastic systems with multiplicative noises in the next section.

## 8.3   Robust State Feedback Controller Design

In this section, we will present the solution to the robust $H_\infty$ state feedback controller design problem with variance constraints for the *stochastic* discrete-time systems with both multiplicative noises and *deterministic* norm-bounded parameter uncertainty. That is, we will design the controller that achieves the requirements (R1), (R2), and (R3) described in Section 8.1.

The following theorem provides an LMI approach to the addressed multi-objective (stability, $H_\infty$ performance, and variance constraints) design problem for the uncertain stochastic discrete-time systems with multiplicative noises.

**Theorem 8.3.1** *Given $\gamma > 0$ and $\sigma_i^2 > 0$ $(i = 1, 2, \dots, n)$. If there exist a positive definite matrix $Q > 0$, a real matrix $G$, and positive scalars $\varepsilon_1$ and $\varepsilon_2$ such that the following set of linear matrix inequalities (LMIs),*

$$
\begin{bmatrix}
-Q & AQ + B_2 G & 0 & 0 & B_1 & \varepsilon_1 H_1 & 0 \\
QA^T + G^T B_2^T & -Q & QA_s^T & QC_1^T + G^T D_{12}^T & 0 & 0 & QE^T \\
0 & A_s Q & -Q & 0 & 0 & 0 & 0 \\
0 & C_1 Q + D_{12} G & 0 & -I & D_{11} & \varepsilon_1 H_2 & 0 \\
B_1^T & 0 & 0 & D_{11}^T & -\gamma^2 I & 0 & 0 \\
\varepsilon_1 H_1^T & 0 & 0 & \varepsilon_1 H_2^T & 0 & -\varepsilon_1 I & 0 \\
0 & EQ & 0 & 0 & 0 & 0 & -\varepsilon_1 I
\end{bmatrix} < 0,
$$
$$\tag{8.43}$$

$$
\begin{bmatrix}
-Q & AQ + B_2 G & A_s Q & B_1 & \varepsilon_2 H_1 & 0 \\
QA^T + G^T B_2^T & -Q & 0 & 0 & 0 & QE^T \\
QA_s^T & 0 & -Q & 0 & 0 & 0 \\
B_1^T & 0 & 0 & -R^{-1} & 0 & 0 \\
\varepsilon_2 H_1^T & 0 & 0 & 0 & -\varepsilon_2 I & 0 \\
0 & EQ & 0 & 0 & 0 & -\varepsilon_2 I
\end{bmatrix} < 0,
$$
$$\tag{8.44}$$

$$
\begin{bmatrix} 1 & 0 & 0 & \cdots & 0 \end{bmatrix} Q \begin{bmatrix} 1 & 0 & 0 & \cdots & 0 \end{bmatrix}^T < \sigma_1^2,
$$
$$\tag{8.45}$$

$$
\begin{bmatrix} 0 & 1 & 0 & \cdots & 0 \end{bmatrix} Q \begin{bmatrix} 0 & 1 & 0 & \cdots & 0 \end{bmatrix}^T < \sigma_2^2,
$$
$$\tag{8.46}$$

$$
\vdots
$$

$$
\begin{bmatrix} 0 & 0 & \cdots & 0 & 1 \end{bmatrix} Q \begin{bmatrix} 0 & 0 & \cdots & 0 & 1 \end{bmatrix}^T < \sigma_n^2,
$$
$$\tag{8.47}$$

*are feasible, then there exists a state feedback controller of the form (8.3) such that three requirements (R1), (R2), and (R3) are satisfied for all admissible deterministic uncertainties. Moreover, the desired controller (8.3) can be determined by*

$$
K = GQ^{-1}.
$$
$$\tag{8.48}$$

*Proof.* We first prove that (8.30) holds if and only if (8.43) holds and (8.37) is true if and only if (8.44 ) is true. To do this, we rewrite in the following form:

$$
\begin{bmatrix}
-Q & (A+B_2K)Q & 0 & 0 & B_1 \\
Q(A+B_2K)^{\mathrm{T}} & -Q & QA_s^{\mathrm{T}} & Q(C_1+D_{12}K)^{\mathrm{T}} & 0 \\
0 & A_sQ & -Q & 0 & 0 \\
0 & (C_1+D_{12}K)Q & 0 & -I & D_{11} \\
B_1^{\mathrm{T}} & 0 & 0 & D_{11}^{\mathrm{T}} & -\gamma^2 I
\end{bmatrix}
$$

$$
+ \begin{bmatrix} H_1 \\ 0 \\ 0 \\ H_2 \\ 0 \end{bmatrix} F \begin{bmatrix} 0 & EQ & 0 & 0 & 0 \end{bmatrix} + \begin{bmatrix} 0 & EQ & 0 & 0 & 0 \end{bmatrix}^{\mathrm{T}} F^{\mathrm{T}} \begin{bmatrix} H_1 \\ 0 \\ 0 \\ H_2 \\ 0 \end{bmatrix}^{\mathrm{T}} < 0. \tag{8.49}
$$

In order to cope with the uncertainty factor $F$, we apply Lemma 2.3.1 to (8.49) and then have the conclusion that (8.49) holds if and only if there exists a positive scalar $\varepsilon_1$ such that the following LMI holds:

$$
\begin{bmatrix}
-Q & (A+B_2K)Q & 0 & 0 & B_1 & \varepsilon_1 H_1 & 0 \\
Q(A+B_2K)^{\mathrm{T}} & -Q & QA_s^{\mathrm{T}} & Q(C_1+D_{12}K)^{\mathrm{T}} & 0 & 0 & QE^{\mathrm{T}} \\
0 & A_sQ & -Q & 0 & 0 & 0 & 0 \\
0 & (C_1+D_{12}K)Q & 0 & -I & D_{11} & \varepsilon_1 H_2 & 0 \\
B_1^{\mathrm{T}} & 0 & 0 & D_{11}^{\mathrm{T}} & -\gamma^2 I & 0 & 0 \\
\varepsilon_1 H_1^{\mathrm{T}} & 0 & 0 & \varepsilon_2 H_1^{\mathrm{T}} & 0 & -\varepsilon_1 I & 0 \\
0 & EQ & 0 & 0 & 0 & 0 & -\varepsilon_1 I
\end{bmatrix} < 0.
$$

$$\tag{8.50}$$

Similarly, we rewrite (8.37) in the following form:

$$
\begin{bmatrix}
-Q & (A+B_2K)Q & A_sQ & B_1 \\
Q(A+B_2K)^{\mathrm{T}} & -Q & 0 & 0 \\
QA_s^{\mathrm{T}} & 0 & -Q & 0 \\
B_1^{\mathrm{T}} & 0 & 0 & -R^{-1}
\end{bmatrix} +
$$

$$
\begin{bmatrix} H_1 \\ 0 \\ 0 \\ 0 \end{bmatrix} F \begin{bmatrix} 0 & EQ & 0 & 0 \end{bmatrix} + \begin{bmatrix} 0 & EQ & 0 & 0 \end{bmatrix}^{\mathrm{T}} F^{\mathrm{T}} \begin{bmatrix} H_1 \\ 0 \\ 0 \\ 0 \end{bmatrix}^{\mathrm{T}} < 0, \tag{8.51}
$$

and apply Lemma 2.3.1 again to (8.51). We know that (8.51) holds if and only if there exists a positive scalar $\varepsilon_2$ such that the following LMI holds:

$$\begin{bmatrix} -Q & (A+B_2K)Q & A_sQ & B_1 & \varepsilon_2 H_1 & 0 \\ Q(A+B_2K)^{\mathrm{T}} & -Q & 0 & 0 & 0 & QE^{\mathrm{T}} \\ QA_s^{\mathrm{T}} & 0 & -Q & 0 & 0 & 0 \\ B_1^{\mathrm{T}} & 0 & 0 & -R^{-1} & 0 & 0 \\ \varepsilon_2 H_1^{\mathrm{T}} & 0 & 0 & 0 & -\varepsilon_2 I & 0 \\ 0 & EQ & 0 & 0 & 0 & -\varepsilon_2 I \end{bmatrix} < 0. \qquad (8.52)$$

Let

$$G = KQ. \qquad (8.53)$$

It is straightforward to see that (8.50) is identical to (8.43) and (8.52 ) is identical to (8.44).

To this end, it follows immediately from Theorem 8.2.6 and Theorem 8.2.7 that, with the feedback gain matrix $K$ given in (8.53) (or (8.48)), the closed-loop system (8.4) is exponentially mean-square quadratically stable, the $H_\infty$-norm constraint (8.8) is achieved for all nonzero $w(k)$, and the steady-state covariance $\hat{Q}$ exists and satisfies $\hat{Q} \le Q$. In other words, the requirements $(R1)$ and $(R2)$ are met. Next, considering the definitions (8.9) and (8.35), we can obtain

$$\mathrm{Var}\{x_i(k)\} = \begin{bmatrix} 0 & \cdots & 0 & 1 & 0 & \cdots & 0 \end{bmatrix} \hat{Q} \begin{bmatrix} 0 & \cdots & 0 & 1 & 0 & \cdots & 0 \end{bmatrix}^{\mathrm{T}}$$

$$\le \begin{bmatrix} 0 & \cdots & 0 & 1 & 0 & \cdots & 0 \end{bmatrix} Q \begin{bmatrix} 0 & \cdots & 0 & 1 & 0 & \cdots & 0 \end{bmatrix}^{\mathrm{T}}. \qquad (8.54)$$

Therefore, the $n$ LMIs given in (8.45) to (8.47) indicates that the requirement $(R3)$ is also met. This completes the proof. □

**Remark 8.2** *The robust $H_\infty$ controller with variance constraints can be obtained by solving the $n+2$ LMIs described in (8.43) to (8.47) in Theorem 8.3.1. Such a set of LMIs can be solved efficiently via the interior point method. Note that the LMIs (8.43) to (8.47) are affine in the scalar positive parameters $\varepsilon_1$, $\varepsilon_2$, the positive definite matrix $Q$, and a real matrix $G$. Hence, they can be defined as LMI variables in order to increase the solvability while reducing the conservatism with respect to the parameter uncertainties.*

Up to now, by means of an LMI approach, we have proposed the controller design procedure, which guarantees the simultaneous satisfaction of the requirements $(R1)$, $(R2)$, and $(R3)$. In order to show the flexibility of the proposed LMI framework, we now discuss the following two optimization problems:

$(P1)$ *The optimal variance-constrained $H_\infty$ control problem for uncertain stochastic systems with multiplicative noises*:

$$\min_{Q>0,\ G,\ \varepsilon_1>0,\ \varepsilon_2>0} \gamma \text{ subject to (8.43) to (8.47)} \qquad (8.55)$$

$$\text{for given } \sigma_1^2,\ \sigma_2^2,\ \ldots,\sigma_n^2. \qquad (8.56)$$

(P2) *The minimum weighted variance $H_\infty$ control problem for uncertain stochastic systems with multiplicative noises:*

$$\min_{Q>0,\ G,\ \varepsilon_1>0,\ \varepsilon_2>0} \sum_{i=1}^{n} \alpha_i \sigma_i^2 \text{ subject to (8.43) to (8.47)} \qquad (8.57)$$

$$\text{for given } \gamma, \qquad (8.58)$$

where $\alpha_i$ $(i = 1, 2, \ldots, n)$ are given weighting coefficients for variances and satisfy $\sum_{i=1}^{n} \alpha_i = 1$.

**Remark 8.3** *In many engineering applications, the performance constraints on the steady-state variances are often specified a priori. That is, the upper bounds $\sigma_1^2$, $\sigma_2^2$, $\ldots$, $\sigma_n^2$ can be prescribed. Hence, in addition to the individual variance constraints, the problem (P1) will help exploit the design freedom to meet the optimal $H_\infty$ performance. This is certainly attractive because the addressed multi-objective problem can be solved while a local optimal performance can also be achieved, and the computation is efficient by using the Matlab LMI Toolbox.*

**Remark 8.4** *In the problem (P2), the variances are weighted against their importance in the real engineering systems and then the feedback gain is sought so as to minimize the weighted sum of the variance. We could, of course, optimize the variances of individual system states by setting the weighting coefficients of certain variances to zeros. Therefore, the problem (P2) is flexible in terms of both the engineering requirements and the computational efficiency.*

## 8.4 Numerical Example

Consider an uncertain stochastic discrete-time system with multiplicative noises described by (8.1) with the model parameters given as follows:

$$A = \begin{bmatrix} -0.1 & 0.3 & -0.2 \\ 0 & -0.25 & 0.1 \\ 0.1 & 0 & 0.5 \end{bmatrix}, \quad B_1 = \begin{bmatrix} 0.3 \\ 0 \\ 0.1 \end{bmatrix},$$

$$B_2 = \begin{bmatrix} -1 \\ 2 \\ 1 \end{bmatrix}, \quad C_1 = \begin{bmatrix} 1 & -1 & 2 \end{bmatrix}, \quad D_{11} = 1,$$

$$D_{12} = 2, \quad A_s = \begin{bmatrix} 0.2 & 0 & 0 \\ 0 & 0.1 & 0 \\ 0 & 0 & 0.2 \end{bmatrix},$$

$$H_1 = \begin{bmatrix} 0.3 \\ 0.2 \\ 0 \end{bmatrix}, \quad H_2 = 0; \quad E = \begin{bmatrix} 1 & 0 & 0 \end{bmatrix}, \quad R = 1.$$

Now, let us examine the following three cases.

***Case 1***: $\gamma^2 = 1.8$, $\sigma_1^2 = 0.5$, $\sigma_2^2 = 0.5$, $\sigma_3^2 = 0.2$.

This case is exactly concerned with the addressed robust $H_\infty$ control problem with specified variance constraints and hence can be tackled by using Theorem 8.3.1 with $n = 3$. By employing the Matlab LMI Toolbox, the solution is given by

$$Q = \begin{bmatrix} 0.2699 & -0.1592 & -0.0710 \\ -0.1592 & 0.4995 & 0.1866 \\ -0.0710 & 0.1866 & 0.1376 \end{bmatrix},$$

$$G = \begin{bmatrix} -0.1348 & 0.0911 & 0.0215 \end{bmatrix},$$

$$\varepsilon_1 = 0.5729, \quad \varepsilon_2 = 0.5601,$$

$$K = \begin{bmatrix} -0.4968 & 0.1243 & -0.2683 \end{bmatrix}.$$

***Case 2***: $\sigma_1^2 = 0.5$, $\sigma_2^2 = 0.5$, $\sigma_3^2 = 0.2$.

In this case, we wish to design the controller that minimizes the $H_\infty$ performance under the variance constraints specified above. That is, we want to solve the problem $(P1)$. Solving the optimization problem (8.56) using the LMI Toolbox yields the optimal value $\gamma_{\mathrm{opt}} = 1.6583$ and

$$Q = \begin{bmatrix} 0.4624 & -0.1193 & -0.1075 \\ -0.1193 & 0.4982 & 0.1597 \\ -0.1075 & 0.1597 & 0.1552 \end{bmatrix},$$

$$G = \begin{bmatrix} -0.1745 & 0.1041 & 0.0223 \end{bmatrix},$$

$$\varepsilon_1 = 1.6594, \quad \varepsilon_2 = 1.2316,$$

$$K = \begin{bmatrix} -0.4047 & 0.2325 & -0.3757 \end{bmatrix}.$$

***Case 3***: $\gamma^2 = 1.8$, $\alpha_1 = 0.3$; $\alpha_2 = 0.4$; $\alpha_3 = 0.3$.

We now deal with the problem $(P2)$. Solving the optimization problem (8.58), we obtain the minimum individual variance values $\sigma_{1\mathrm{min}}^2 = 0.3020$, $\sigma_{2\mathrm{min}}^2 = 0.3410$, $\sigma_{3\mathrm{min}}^2 = 0.0636$, and

$$Q = \begin{bmatrix} 0.3013 & -0.0830 & -0.0232 \\ -0.0830 & 0.3405 & 0.1059 \\ -0.0232 & 0.1059 & 0.0630 \end{bmatrix},$$

$$G = \begin{bmatrix} -0.1215 & 0.0667 & 0.0168 \end{bmatrix},$$

$$\varepsilon_1 = 1.1133, \quad \varepsilon_2 = 0.9002,$$

$$K = \begin{bmatrix} -0.3734 & 0.1360 & -0.0997 \end{bmatrix}.$$

**Remark 8.5** *Within the LMI framework developed in this chapter, we can show that there is some trade off that can be used for satisfying specific performance requirements. For example, the $H_\infty$ performance will be improved if the variance* *constraints become more relaxed (larger). Also, if the value of the $H_\infty$ performance constraint is allowed to be increased, then the steady-state variances can be further reduced. Hence, the proposed approach allows much flexibility in making a compromise between the variances and the $H_\infty$ performance, while the essential multiple objectives can all be achieved simultaneously.*

## 8.5 Summary

In this chapter, a robust $H_\infty$ controller with variance constraints has been designed for a class of stochastic systems with both multiplicative noises and norm-bounded parameter uncertainties. A general framework for solving this problem is established using an LMI approach in conjunction with stability, $H_\infty$ optimization characterization, and variance constraints. Two types of the optimization problems have been proposed by optimizing either the $H_\infty$ performance or the system state variances. Sufficient conditions have been derived in terms of a set of feasible LMIs. We point out that our method can be extended to the output feedback case, and different representations of uncertainties can also be considered. These are possibly the topics of our future research.

# 9

# Robust $H_\infty$ Control with Variance Constraints: the Finite-Horizon Case

In the real world, there are virtually no strict time-invariant systems since the working circumstances, operating points, or equipment deterioration are of inherently time-varying behaviors. Therefore, the *time-varying* stochastic systems have started to receive initial yet scattered attention in recent years. Unfortunately, despite the importance of the time-varying nature in system modeling, the covariance control problem for time-varying nonlinear systems has been largely overlooked, not to mention the simultaneous consideration of the $H_\infty$ constraints.

Recognizing the great importance of the time-varying nature of real-time models, this chapter is concerned with the robust $H_\infty$ control problem for a class of uncertain nonlinear discrete time-varying stochastic systems. All the system parameters are time-varying and the uncertainty enters into the state matrix. The nonlinearities under consideration are described by statistical means, which can cover several classes of well-studied nonlinearities. The purpose of the problem addressed is to design a dynamic output feedback controller such that the $H_\infty$ disturbance rejection attenuation level is achieved in the finite horizon while the state covariance is not more than an individual upper bound at each time point. A novel algorithm is developed to deal with the addressed problem by means of recursive linear matrix inequalities (RLMIs). It is shown that the robust $H_\infty$ control problem is solvable if a series of RLMIs is feasible. An illustrative simulation example is given to show the applicability and effectiveness of the proposed algorithm.

The contribution of this chapter lies in the new problem formulated and new technique developed. Specifically, the contributions are twofold: (1) the robust $H_\infty$ control

*Variance-Constrained Multi-Objective Stochastic Control and Filtering*, First Edition.
Lifeng Ma, Zidong Wang and Yuming Bo.
© 2015 John Wiley & Sons, Ltd. Published 2015 by John Wiley & Sons, Ltd.

problem is considered, for the first time, for a class of *time-varying* systems with stochastic nonlinearities and (2) a novel RLMI approach is developed to handle the problem addressed, which is then demonstrated via a numerical example.

The rest of this chapter is set out as follows: Section 9.1 formulates the robust $H_\infty$ dynamic output feedback controller design problem for the uncertain discrete time-varying nonlinear stochastic systems with a state covariance constraint. In Section 9.2, the $H_\infty$ noise attenuation level and state covariance performances of the closed-loop system are analyzed separately, and a sufficient condition is then presented for the addressed controller design problem via the RLMI method. The controller design technique is given in Section 9.3 by means of a series of RLMIs with properly chosen initial conditions. In Section 9.4, an illustrative numerical example is provided to show the effectiveness and usefulness of the proposed approach. Our summary is drawn in Section 9.5.

## 9.1 Problem Formulation

Consider the following uncertain discrete time-varying nonlinear stochastic system defined on $k \in [0, N]$:

$$\begin{cases} x(k+1) = (A(k) + \Delta A(k)x(k) + B(k)u(k) + f(x(k), k) + D_1(k)\omega(k), \\ \quad y(k) = C(k)x(k) + g(x(k), k) + D_2(k)\omega(k), \end{cases} \tag{9.1}$$

where $x(k) \in \mathbb{R}^n$ is the state, $y(k) \in \mathbb{R}^r$ is the output, $u(k) \in \mathbb{R}^m$ is the control input, $\omega(k) \in \mathbb{R}^p$ is a zero mean Gaussian white noise sequence with covariance $W(k) > 0$, and $A(k)$, $B(k)$, $C(k)$, $D_1(k)$, and $D_2(k)$ are known real time-varying matrices with appropriate dimensions. $\Delta A(k)$ is a real-valued time-varying matrix that represents parametric uncertainty, being of the following form:

$$\Delta A(k) = H(k)F(k)E(k), \quad F(k)F^{\mathrm{T}}(k) \leq I, \tag{9.2}$$

where $H(k)$ and $E(k)$ are known time-varying matrices with appropriate dimensions. The uncertainty in $\Delta A(k)$ is said to be admissible if (9.2) holds.

The nonlinear stochastic functions $f(x(k), k)$ and $g(x(k), k)$ are assumed to have the following first moments for all $x(k)$ and $k$:

$$\mathbb{E}\left\{ \begin{bmatrix} f(x(k), k) \\ g(x(k), k) \end{bmatrix} \middle| x(k) \right\} = 0, \tag{9.3}$$

with the covariance given by

$$\mathbb{E}\left\{ \begin{bmatrix} f(x(k), k) \\ g(x(k), k) \end{bmatrix} \begin{bmatrix} f^{\mathrm{T}}(x(j), j) & g^{\mathrm{T}}(x(j), j) \end{bmatrix} \middle| x(k) \right\} = 0, \quad k \neq j, \tag{9.4}$$

and

$$\mathbb{E}\left\{\begin{bmatrix} f(x(k),k) \\ g(x(k),k) \end{bmatrix} \begin{bmatrix} f^T(x(k),k) & g^T(x(k),k) \end{bmatrix} \middle| x(k) \right\}$$

$$= \sum_{i=1}^q \begin{bmatrix} \Omega_{11}^i & \Omega_{12}^i \\ (\Omega_{12}^i)^T & \Omega_{22}^i \end{bmatrix} x^T(k)\Gamma_i x(k),$$

(9.5)

where $\Omega_{jl}^i$ and $\Gamma_i$ $(j, l = 1, 2; i = 1, 2, \dots, q)$ are known matrices.

Applying the following full-order dynamic output feedback control law

$$\begin{cases} x_g(k+1) = A_g(k)x_g(k) + B_g(k)y(k), \\ u(k) = C_g(k)x_g(k) \end{cases}$$

(9.6)

to the system (9.1), we obtain the following closed-loop system:

$$\begin{cases} z(k+1) = \hat{A}(k)z(k) + \hat{G}(k)h(x(k),k) + \hat{D}(k)\omega(k), \\ y_k = \hat{C}(k)z(k) + g(x(k),k) + D_2(k)\omega(k), \end{cases}$$

(9.7)

where

$$z(k) = \begin{bmatrix} x(k) \\ x_g(k) \end{bmatrix},$$

$$h(x(k),k) = \begin{bmatrix} f(x(k),k) \\ g(x(k),k) \end{bmatrix},$$

$$\hat{A}(k) = \begin{bmatrix} A(k) + \Delta A(k) & B(k)C_g(k) \\ B_g(k)C(k) & A_g(k) \end{bmatrix},$$

$$\hat{G}(k) = \begin{bmatrix} I & 0 \\ 0 & B_g(k) \end{bmatrix},$$

$$\hat{D}(k) = \begin{bmatrix} D_1(k) \\ B_g(k)D_2(k) \end{bmatrix},$$

$$\hat{C}(k) = \begin{bmatrix} C(k) & 0 \end{bmatrix}.$$

In this chapter, it is our objective to design a finite-horizon dynamic output feedback controller of the form (9.6) such that the following two requirements are satisfied simultaneously:

(R1) For the given scalar $\gamma > 0$, matrix $S > 0$, and the initial state $x(0)$, the $H_\infty$ performance index

$$J \triangleq \mathbb{E}\left\{ \|y(k)\|_{[0,N-1]}^2 - \gamma^2\|\omega(k)\|_{[0,N-1]}^2 \right\} - \gamma^2 x^T(0)Sx(0) < 0$$

(9.8)

is achieved for all admissible parameter uncertainty and all stochastic nonlinearities.

(R2) For a sequence of specified definite matrices $\{\Theta(k)\}_{0<k\leq N}$, at each sampling
instant $k$, the system satisfies

$$X(k) \triangleq \mathbb{E}\{x(k)x^{\mathrm{T}}(k)\} \leq \Theta(k), \qquad \forall k. \tag{9.9}$$

The control problem addressed above is referred to as the robust $H_\infty$ output feedback control problem for uncertain nonlinear discrete time-varying stochastic systems
with a covariance constraint.

## 9.2  Performance Analysis

In this section, in terms of two matrix inequalities, the $H_\infty$ and covariance performances will first be analyzed separately for the closed-loop system (9.7). Then, a
theorem that combines the two performance indices in a unified framework is presented via the RLMI algorithm.

### 9.2.1  $H_\infty$ Performance

In this subsection, a sufficient condition is given for the closed-loop system (9.7)
to satisfy the prescribed $H_\infty$ noise attenuation level. For notational convenience, we
denote

$$\hat{\Gamma}_i = \begin{bmatrix} \Gamma_i & 0 \\ 0 & 0 \end{bmatrix}, \qquad \hat{\Omega}_i = \begin{bmatrix} \Omega^i_{11} & \Omega^i_{12} \\ (\Omega^i_{12})^{\mathrm{T}} & \Omega^i_{22} \end{bmatrix}, \qquad \hat{S} = \begin{bmatrix} S & 0 \\ 0 & 0 \end{bmatrix}.$$

**Theorem 9.2.1** *Consider the system (9.7). Let the controller feedback gain matrices*
$A_g(k)$, $B_g(k)$, *and* $C_g(k)$ *be given. For a positive scalar* $\gamma > 0$ *and a positive definite*
*matrix* $S > 0$, *the* $H_\infty$ *performance index requirement defined in (9.8) is achieved*
*for all nonzero* $\omega(k)$ *if, with the initial condition* $Q(0) \leq \gamma^2\hat{S}$, *there exist a sequence*
*of positive definite matrices* $\{Q(k)\}_{1\leq k\leq N+1}$ *satisfying the following recursive matrix*
*inequalities:*

$$\Upsilon \triangleq \begin{bmatrix} \Upsilon_{11} & \Upsilon_{12} \\ \Upsilon_{12}^{\mathrm{T}} & \Upsilon_{22} \end{bmatrix} < 0, \tag{9.10}$$

*where*

$$\Upsilon_{11} = \hat{A}^{\mathrm{T}}(k)Q(k+1)\hat{A}(k) - Q(k) + \hat{C}^{\mathrm{T}}(k)\hat{C}(k) + \sum_{i=1}^{q} \hat{\Gamma}_i \cdot \mathrm{tr}[\Omega^i_{22}]$$

$$+ \sum_{i=1}^{q} \hat{\Gamma}_i \mathrm{tr}[\hat{G}^{\mathrm{T}}(k)Q(k+1)\hat{G}(k)\hat{\Omega}_i],$$

$$\Upsilon_{12} = \hat{A}^{\mathrm{T}}(k)Q(k+1)\hat{D}(k) + \hat{C}^{\mathrm{T}}(k)D_2(k),$$

$$\Upsilon_{22} = -\gamma^2 I + \hat{D}^{\mathrm{T}}(k)Q(k+1)\hat{D}(k) + D_2^{\mathrm{T}}(k)D_2(k).$$

*Proof.* Define

$$J(k) \triangleq z^{\mathrm{T}}(k+1)Q(k+1)z(k+1) - z^{\mathrm{T}}(k)Q(k)z(k). \qquad (9.11)$$

Substituting (9.7) into $J(k)$ leads to

$$\begin{aligned}
\mathbb{E}\{J(k)\} &= \mathbb{E}\{[\hat{A}(k)z(k) + \hat{G}(k)h(x(k),k) + \hat{D}(k)\omega(k)]^{\mathrm{T}}Q(k+1) \\
&\quad \times [\hat{A}(k)z(k) + \hat{G}(k)h(x(k),k) + \hat{D}(k)\omega(k)] - z^{\mathrm{T}}(k)Q(k)z(k)\} \\
&= \mathbb{E}\{z^{\mathrm{T}}(k)\hat{A}^{\mathrm{T}}(k)Q(k+1)\hat{A}(k)z(k) \\
&\quad + \omega^{\mathrm{T}}(k)\hat{D}^{\mathrm{T}}(k)Q(k+1)\hat{D}(k)\omega(k) \\
&\quad + z^{\mathrm{T}}(k)\hat{A}^{\mathrm{T}}(k)Q(k+1)\hat{D}(k)\omega(k) \\
&\quad + \omega^{\mathrm{T}}(k)\hat{D}^{\mathrm{T}}(k)Q(k+1)\hat{A}(k)z(k) - z^{\mathrm{T}}(k)Q(k)z(k) \\
&\quad + (\hat{G}(k)h(x(k),k))^{\mathrm{T}}Q(k+1)(\hat{G}(k)h(x(k),k))\}.
\end{aligned} \qquad (9.12)$$

Taking (9) into consideration, we have

$$\begin{aligned}
&\mathbb{E}\{(\hat{G}(k)h(x(k),k))^{\mathrm{T}}Q(k+1)(\hat{G}(k)h(x(k),k))\} \\
&= x^{\mathrm{T}}(k)\sum_{i=1}^{q}\Gamma_i \mathrm{tr}[\hat{G}^{\mathrm{T}}(k)Q(k+1)\hat{G}(k)\hat{\Omega}_i]x(k) \\
&= z^{\mathrm{T}}(k)\sum_{i=1}^{q}\hat{\Gamma}_i \mathrm{tr}[\hat{G}^{\mathrm{T}}(k)Q(k+1)\hat{G}(k)\hat{\Omega}_i]z(k),
\end{aligned} \qquad (9.13)$$

and therefore

$$\begin{aligned}
&\mathbb{E}\{J(k)\} \\
&= \mathbb{E}\{z^{\mathrm{T}}(k)(\hat{A}^{\mathrm{T}}(k)Q(k+1)\hat{A}(k) - Q(k) \\
&\quad + \sum_{i=1}^{q}\hat{\Gamma}_i \cdot \mathrm{tr}[\hat{G}^{\mathrm{T}}(k)Q(k+1)\hat{G}(k)\hat{\Omega}_i]z(k) \\
&\quad + \omega^{\mathrm{T}}(k)\hat{D}^{\mathrm{T}}(k)Q(k+1)\hat{D}(k)\omega(k) \\
&\quad + z^{\mathrm{T}}(k)\hat{A}^{\mathrm{T}}(k)Q(k+1)\hat{D}(k)\omega(k) \\
&\quad + \omega^{\mathrm{T}}(k)\hat{D}^{\mathrm{T}}(k)Q(k+1)\hat{A}(k)z(k)\}.
\end{aligned} \qquad (9.14)$$

Adding the zero term $y^{\mathrm{T}}(k)y(k) - \gamma^2\omega^{\mathrm{T}}(k)\omega(k) - y^{\mathrm{T}}(k)y(k) + \gamma^2\omega^{\mathrm{T}}(k)\omega(k)$ to $\mathbb{E}\{J(k)\}$ results in

$$\mathbb{E}\{J(k)\} = \mathbb{E}\left\{ [z^{\mathrm{T}}(k) \quad \omega^{\mathrm{T}}(k)] \Upsilon \begin{bmatrix} z(k) \\ \omega(k) \end{bmatrix} - y^{\mathrm{T}}(k)y(k) + \gamma^2\omega^{\mathrm{T}}(k)\omega(k) \right\}. \qquad (9.15)$$

Summing (9.15) on both sides from 0 to $N - 1$ with respect to $k$, we obtain

$$\sum_{k=0}^{N-1} \mathbb{E}\{J(k)\} = \mathbb{E}\{z^{\mathrm{T}}(N)Q(N)z(N)\} - z^{\mathrm{T}}(0)Q(0)z(0)$$

$$= \mathbb{E}\left\{ \sum_{k=0}^{N-1} \begin{bmatrix} z^{\mathrm{T}}(k) & \omega^{\mathrm{T}}(k) \end{bmatrix} \Upsilon \begin{bmatrix} z(k) \\ \omega(k) \end{bmatrix} \right\} \qquad (9.16)$$

$$- \mathbb{E}\left\{ \sum_{k=0}^{N-1} ( y^{\mathrm{T}}(k)y(k) - \gamma^2\omega^{\mathrm{T}}(k)\omega(k) \right\} .$$

Hence, the $H_\infty$ performance index defined in (9.8) is given by

$$J = \mathbb{E}\left\{ \sum_{k=0}^{N-1} \begin{bmatrix} z^{\mathrm{T}}(k) & \omega^{\mathrm{T}}(k) \end{bmatrix} \Upsilon \begin{bmatrix} z(k) \\ \omega(k) \end{bmatrix} \right\}$$
$$- \mathbb{E}\{z^{\mathrm{T}}(N)Q(N)z(N)\} + z^{\mathrm{T}}(0)(-\gamma^2\hat{S} + Q(0)z(0). \qquad (9.17)$$

Noting that $\Upsilon < 0$, $Q(N) > 0$ and the initial condition $Q(0) \leq \gamma^2\hat{S}$, we know that $J < 0$, which completes the proof. $\qquad\qquad\qquad\qquad\qquad\qquad\qquad\qquad\qquad\qquad\qquad\qquad$ □

### 9.2.2   Variance Analysis

In this subsection, we will tackle the state covariance of the closed-loop system (9.7). First define the following performance index for system (9.7) by

$$Z(k) \triangleq \mathbb{E}\{z(k)z^{\mathrm{T}}(k)\}. \qquad (9.18)$$

It is easy to see that

$$X(k) = \begin{bmatrix} I & 0 \end{bmatrix} Z(k) \begin{bmatrix} I \\ 0 \end{bmatrix} . \qquad (9.19)$$

Defining a function as

$$\mathcal{F}(Y(k)) \triangleq \hat{A}(k)Y(k)\,\hat{A}^{\mathrm{T}}(k) + \hat{D}(k)W(k)\hat{D}^{\mathrm{T}}(k)$$
$$+ \sum_{i=1}^{q} \hat{G}(k)\hat{\Omega}_i\hat{G}^{\mathrm{T}}(k)\mathrm{tr}[\hat{\Gamma}_i Y(k)], \qquad (9.20)$$

we now present the following theorem, which gives an upper bound of $Z(k)$.

**Theorem 9.2.2** *Consider the system (9.7). Given are the controller feedback gains $A_g(k)$, $B_g(k)$, and $C_g(k)$. If there exist a sequence of positive definite matrices $\{P(k)\}_{1 \leq k \leq N+1}$ satisfying the following matrix inequality:*

$$P(k + 1) \geq \mathcal{F}(P(k)) \qquad (9.21)$$

*with the initial condition $P(0) = Z(0)$, then $P(k) \geq Z(k)$, $\forall k \in \{1, 2, \ldots, N + 1\}$.*

*Proof.* The Lyapunov-type equation that governs the evolution of $Z_k$ of closed-loop systems (9.7) is given by

$$
Z(k+1)
$$
$$
= \hat{A}(k)Z(k)\hat{A}^T(k) + \hat{D}(k)W(k)\hat{D}^T(k)
$$
$$
+ \mathbb{E}\{\hat{G}(k)h(x(k),k)h^T(x(k),k)\hat{G}^T(k)\}
$$
$$
= \hat{A}(k)Z(k)\hat{A}^T(k) + \hat{D}(k)W(k)\hat{D}^T(k) \qquad (9.22)
$$
$$
+ \sum_{i=1}^{q} \hat{G}(k)\hat{\Omega}_i\hat{G}^T(k) \cdot \text{tr}[\hat{\Gamma}_i Z(k)]
$$
$$
= \mathcal{F}(Z(k)).
$$

The following proof is done by induction. Obviously, $P(0) \geq Z(0)$. Suppose $P(k) \geq Z(k)$; then

$$
P(k+1) \geq \mathcal{F}(P(k)) \geq \mathcal{F}(Z(k)) = Z(k+1), \qquad (9.23)
$$

and the proof is complete. □

In the following stage, to conclude the above analysis, we present a theorem that tends to take both the $H_\infty$ performance index and covariance constraint into consideration in a unified framework via the RLMI method.

**Theorem 9.2.3** *Consider the system (9.7). Given the controller feedback gain $A_g(k)$, $B_g(k)$, and $C_g(k)$, a positive scalar $\gamma > 0$ and a positive definite matrix $S > 0$. If there exist families of positive definite matrices $\{Q(k)\}_{1 \leq k \leq N+1}$, $\{P(k)\}_{1 \leq k \leq N+1}$, and $\{\eta_i(k)\}_{0 \leq k \leq N}$ $(i = 1, 2, \dots, q)$ satisfying the following recursive matrix inequalities:*

$$
\begin{bmatrix} -\eta_i(k) & \pi_i^T \hat{G}^T(k) \\ * & -Q^{-1}(k+1) \end{bmatrix} < 0, \qquad (9.24)
$$

$$
\begin{bmatrix} \hat{\Lambda} & \hat{C}^T(k)D_2(k) & \hat{A}^T(k) \\ * & -\gamma^2 I + D_2^T(k)D_2(k) & \hat{D}^T(k) \\ * & * & -Q^{-1}(k+1) \end{bmatrix} < 0, \qquad (9.25)
$$

$$
\begin{bmatrix} -P(k+1) & \hat{A}(k)P(k) & \hat{\Phi}_{13} & \hat{D}(k) \\ * & -P(k) & 0 & 0 \\ * & * & \hat{\Phi}_{33} & 0 \\ * & * & * & -W^{-1}(k) \end{bmatrix} < 0, \qquad (9.26)
$$

*with the initial condition*

$$
\begin{cases} Q(0) \leq \gamma^2 \hat{S}, \\ P(0) = Z(0), \end{cases} \qquad (9.27)
$$

*where*

$$\hat{\Lambda} = -Q(k) + \hat{C}^{\mathrm{T}}(k)\hat{C}(k) + \sum_{i=1}^{q} \hat{\Gamma}_i(\eta_i(k) + \mathrm{tr}[\Omega_{22}^i]),$$

$$\hat{\Phi}_{13} = \begin{bmatrix} \hat{G}(k)\pi_1 & \hat{G}(k)\pi_2 & \cdots & \hat{G}(k)\pi_q \end{bmatrix},$$

$$\hat{\Phi}_{33} = \begin{bmatrix} -\rho_1 I & 0 & \cdots & 0 \\ 0 & -\rho_2 I & \cdots & 0 \\ \vdots & \vdots & \ddots & \vdots \\ 0 & 0 & \cdots & -\rho_q I \end{bmatrix},$$

$$\rho_i = (\mathrm{tr}[\hat{\Gamma}_i P(k)])^{-1}, \qquad i = 1, 2, \ldots, q,$$

*then, for the closed-loop system (9.7), $J(k) < 0$ and $Z(k) \leq P(k)$ $(0 \leq k \leq N + 1)$.*

*Proof.* Based on the analysis of the $H_\infty$ performance and system state covariance in Subsection 9.2.1 and Subsection 9.2.2, we just need to show that, under initial conditions (9.27), inequalities (9.24) and (9.25) imply (9.10), and inequality (9.26) is equivalent to (9.21).

To proceed, without loss of any generality, we assume that $\hat{\Omega}_i$ can be represented by

$$\hat{\Omega}_i = \pi_i \pi_i^{\mathrm{T}} = \begin{bmatrix} \pi_{1i} \\ \pi_{2i} \end{bmatrix} \begin{bmatrix} \pi_{1i} \\ \pi_{2i} \end{bmatrix}^{\mathrm{T}},$$

where $\pi_i = \begin{bmatrix} \pi_{1i}^{\mathrm{T}} & \pi_{2i}^{\mathrm{T}} \end{bmatrix}^{\mathrm{T}}$ $(i = 1, 2, \ldots, q)$ are column vectors of appropriate dimensions.

By the Schur Complement Lemma, (9.24) is equivalent to

$$\pi_i^{\mathrm{T}} \hat{G}^{\mathrm{T}}(k)Q(k + 1)\hat{G}(k)\pi_i < \eta_i(k), \tag{9.28}$$

which, by the property of matrix trace, can be rewritten as

$$\mathrm{tr}[\hat{G}^{\mathrm{T}}(k)Q(k + 1)\hat{G}(k)\hat{\Omega}_i] < \eta_i(k). \tag{9.29}$$

Using the Schur Complement Lemma again, (9.25) is equivalent to

$$\hat{\Upsilon} \triangleq \begin{bmatrix} \hat{\Upsilon}_{11} & \Upsilon_{12} \\ \Upsilon_{12}^{\mathrm{T}} & \Upsilon_{22} \end{bmatrix} < 0, \tag{9.30}$$

where $\Upsilon_{12}$ and $\Upsilon_{22}$ are defined in Theorem 9.2.1 and

$$\hat{\Upsilon}_{11} = \hat{A}^{\mathrm{T}}(k)Q(k + 1)\hat{A}(k) - Q(k) + \hat{C}^{\mathrm{T}}(k)\hat{C}(k) + \sum_{i=1}^{q} \hat{\Gamma}_i(\eta_i(k) + \mathrm{tr}[\Omega_{22}^i]).$$

Hence, it is easy to see that (9.10) can be implied by (9.24) and (9.25) under the same initial condition.

Similarly, employing the Schur Complement Lemma, we can easily verify that (9.26) is equivalent to (9.21). Thus, according to Theorem 9.2.1 and Theorem 9.2.2, the $H_\infty$ index defined in (9.8) satisfies $J(k) < 0$ and, at the same time, the closed-loop system (9.7) achieves $Z(k) \leq P(k)$, $\forall k \in \{0, 1, \ldots, N+1\}$. The proof is complete. □

## 9.3 Robust Finite-Horizon Controller Design

According to Theorem 9.2.3, an algorithm is proposed in this section to solve the addressed dynamic output feedback control problem for the uncertain discrete time-varying nonlinear stochastic system (9.1). It will be shown that the controller gains can be obtained by solving a certain set of RLMIs. In other words, at each sampling instant $k$ $(k > 0)$, a set of LMIs will be solved to obtain the desired controller gains and, at the same time, certain key parameters are obtained that are needed in solving the LMIs for the $(k + 1)$th instant.

The following theorem provides the controller design procedure for system (9.1).

**Theorem 9.3.1** *For a given disturbance attenuation level $\gamma > 0$, a positive definite matrix $S > 0$, and a sequence of pre-specified upper bounds $\{\Theta(k)\}_{0 \leq k \leq N+1}$, if there exist families of positive definite matrices $\{\mathcal{M}(k)\}_{1 \leq k \leq N+1}$, $\{\mathcal{N}(k)\}_{1 \leq k \leq N+1}$, $\{P_1(k)\}_{1 \leq k \leq N+1}$, $\{P_2(k)\}_{1 \leq k \leq N+1}$, $\{\varepsilon_1(k)\}_{0 \leq k \leq N}$, $\{\varepsilon_2(k)\}_{0 \leq k \leq N}$, and $\{\eta_i(k)\}_{0 \leq k \leq N}$ $(i = 1, 2, \ldots, q)$ and families of real-valued matrices $\{A_g(k)\}_{0 \leq k \leq N}$, $\{B_g(k)\}_{0 \leq k \leq N}$, and $\{C_g(k)\}_{0 \leq k \leq N}$, under initial conditions*

$$\begin{cases} M(0) \leq \gamma^2 S, \\ N(0) = 0, \\ X(0) = P_1(0) \leq \Theta(0), \\ P_2(0) = 0, \end{cases} \tag{9.31}$$

*such that the following recursive LMIs,*

$$\begin{bmatrix} -\eta_i(k) & \pi_{1i}^T & \pi_{2i}^T B_g^T(k) \\ * & -\mathcal{M}(k+1) & 0 \\ * & * & -\mathcal{N}(k+1) \end{bmatrix} < 0, \tag{9.32}$$

$$\Lambda \triangleq \begin{bmatrix} \Lambda_{11} & \Lambda_{21}^T \\ * & \Lambda_{22} \end{bmatrix} < 0, \tag{9.33}$$

$$\Phi \triangleq \begin{bmatrix} \Phi_{11} & \Phi_{12} \\ * & \Phi_{22} \end{bmatrix} < 0, \tag{9.34}$$

$$P_1(k+1) - \Theta(k+1) \leq 0, \tag{9.35}$$

*are satisfied with the parameter updated by*

$$\begin{cases} M(k+1) = \mathcal{M}^{-1}(k+1), \\ N(k+1) = \mathcal{N}^{-1}(k+1), \end{cases} \tag{9.36}$$

*where*

$$\Lambda_{11} = \begin{bmatrix} \tilde{\Lambda}_{11} & 0 & C^{\mathrm{T}}(k)D_2(k) \\ * & -N(k) & 0 \\ * & * & -\gamma^2 I + D_2^{\mathrm{T}}(k)D_2(k) \end{bmatrix},$$

$$\tilde{\Lambda}_{11} = -M(k) + C^{\mathrm{T}}(k)C(k) + \varepsilon_1(k)E^{\mathrm{T}}(k)E(k) + \sum_{i=1}^{q} \Gamma_i(\eta_i(k) + \mathrm{tr}[\Omega_{22}^i]),$$

$$\Lambda_{21} = \begin{bmatrix} A(k) & B(k)C_g(k) & D_1(k) \\ B_g(k)C(k) & A_g(k) & B_g(k)D_2(k) \\ 0 & 0 & 0 \end{bmatrix},$$

$$\Lambda_{22} = \begin{bmatrix} -\mathcal{M}(k+1) & 0 & H(k) \\ * & -\mathcal{N}(k+1) & 0 \\ * & * & -\varepsilon_1(k)I \end{bmatrix},$$

$$\Phi_{11} = \begin{bmatrix} -P_1(k+1) + \varepsilon_2(k)H(k)H^{\mathrm{T}}(k) & 0 & A(k)P_1(k) \\ * & -P_2(k+1) & B_g(k)C(k)P_1(k) \\ * & * & -P_1(k) \end{bmatrix},$$

$$\Phi_{12} = \begin{bmatrix} B(k)C_g(k)P_2(k) & \Phi_{121} & D_1(k) & 0 \\ A_g(k)P_2(k) & \Phi_{122} & B_g(k)D_2(k) & 0 \\ 0 & 0 & 0 & P_1(k)E^{\mathrm{T}}(k) \end{bmatrix},$$

$$\Phi_{121} = \begin{bmatrix} \pi_{11} & \pi_{12} & \cdots & \pi_{1q} \end{bmatrix},$$

$$\Phi_{122} = \begin{bmatrix} B_g(k)\pi_{21} & B_g(k)\pi_{22} & \cdots & B_g(k)\pi_{2q} \end{bmatrix},$$

$$\Phi_{22} = \mathrm{diag}\{-P_2(k), -\rho_1 I, -\rho_2 I, \ldots, -\rho_q I, -W^{-1}(k), -\varepsilon_2(k)I\},$$

$$\rho_i = (\mathrm{tr}[\Gamma_i P_1(k)])^{-1}, \qquad i = 1, 2, \ldots, q,$$

*then the addressed robust $H_\infty$ finite-horizon controller design problem is solved for the stochastic nonlinear system (9.1), i.e., the design requirements (R1) and (R2) can be achieved simultaneously for the time-varying nonlinear stochastic systems. Moreover, the controller gains $A_g(k)$, $B_g(k)$, and $C_g(k)$ at the sampling instant $k$ ($0 \le k \le N$) can be obtained by solving the corresponding set of LMIs at time $k$.*

*Proof.* The proof is based on Theorem 9.2.3. First, supposing the variables $Q(k)$ and $P(k)$ can be decomposed as follows:

$$Q(k) = \begin{bmatrix} M(k) & 0 \\ 0 & N(k) \end{bmatrix},$$

$$Q^{-1}(k) = \begin{bmatrix} \mathcal{M}(k) & 0 \\ 0 & \mathcal{N}(k) \end{bmatrix}, \tag{9.37}$$

$$P(k) = \begin{bmatrix} P_1(k) & 0 \\ 0 & P_2(k) \end{bmatrix},$$

it is easy to see that (9.24) and (9.32) are equivalent to each other.

In order to eliminate the parameter uncertainty $\Delta A(k)$ in (9.25), we rewrite it in the following form:

$$
\begin{bmatrix}
\hat{\Lambda} & \hat{C}^{\mathrm{T}}(k)D_2(k) & \bar{A}^{\mathrm{T}}(k) \\
* & -\gamma^2 I + D_2^{\mathrm{T}}(k)D_2(k) & \hat{D}^{\mathrm{T}}(k) \\
* & * & -Q^{-1}(k+1)
\end{bmatrix}
$$
$$
+\begin{bmatrix} 0 \\ 0 \\ 0 \\ H(k) \\ 0 \end{bmatrix} F(k) \begin{bmatrix} E^{\mathrm{T}} \\ 0 \\ 0 \\ 0 \\ 0 \end{bmatrix}^{\mathrm{T}} + \begin{bmatrix} E^{\mathrm{T}}(k) \\ 0 \\ 0 \\ 0 \\ 0 \end{bmatrix} F(k) \begin{bmatrix} 0 \\ 0 \\ 0 \\ H(k) \\ 0 \end{bmatrix}^{\mathrm{T}} < 0
$$

(9.38)

where

$$
\bar{A}(k) = \begin{bmatrix} A(k) & B(k)C_g(k) \\ B_g(k)C(k) & A_g(k) \end{bmatrix}.
$$

Then, by Lemma 2.3.1, we obtain that matrix inequality (9.38) holds if and only if there exists a sequence of positive scalar $\varepsilon_1(k)$ such that

$$
\begin{bmatrix}
\hat{\Lambda} & \hat{C}^{\mathrm{T}}(k)D_2(k) & \bar{A}^{\mathrm{T}}(k) \\
* & -\gamma^2 I + D_2^{\mathrm{T}}(k)D_2(k) & \hat{D}^{\mathrm{T}}(k) \\
* & * & -Q^{-1}(k+1)
\end{bmatrix}
$$
$$
+\varepsilon^{-1}(k) \begin{bmatrix} 0 \\ 0 \\ 0 \\ H(k) \\ 0 \end{bmatrix} \begin{bmatrix} 0 & 0 & 0 & H^{\mathrm{T}}(k) & 0 \end{bmatrix}
$$

(9.39)

$$
+\varepsilon(k) \begin{bmatrix} E^{\mathrm{T}}(k) \\ 0 \\ 0 \\ 0 \end{bmatrix} \begin{bmatrix} E(k) & 0 & 0 & 0 & 0 \end{bmatrix} < 0
$$

which means that (9.25) is equivalent to (9.33).

Next, in order to prove (9.26) is equivalent to (9.34), we rewrite (9.26) into the following form:

$$
\begin{bmatrix}
-P(k+1) & \tilde{A}(k)P(k) & \hat{\Phi}_{13} & \hat{D}(k) \\
* & -P(k) & 0 & 0 \\
* & * & \hat{\Phi}_{33} & 0 \\
* & * & * & -W^{-1}(k)
\end{bmatrix}
$$

$$
+
\begin{bmatrix}
H(k) \\ 0 \\ 0 \\ 0 \\ 0 \\ \vdots \\ 0 \\ 0 \\ 0
\end{bmatrix}
F(k)
\begin{bmatrix}
0 \\ 0 \\ P_1(k)E^{\mathrm{T}}(k) \\ 0 \\ 0 \\ \vdots \\ 0 \\ 0 \\ 0
\end{bmatrix}^{\mathrm{T}}
+
\begin{bmatrix}
0 \\ 0 \\ P_1(k)E^{\mathrm{T}}(k) \\ 0 \\ 0 \\ \vdots \\ 0 \\ 0 \\ 0
\end{bmatrix}
F(k)
\begin{bmatrix}
H^{\mathrm{T}}(k) \\ 0 \\ 0 \\ 0 \\ 0 \\ \vdots \\ 0 \\ 0 \\ 0
\end{bmatrix}^{\mathrm{T}}
< 0
\qquad (9.40)
$$

which, by Lemma 2.3.1, holds if and only if there exists a sequence of positive constants $\varepsilon_2(k)$ satisfying

$$
\begin{bmatrix}
-P(k+1) & \tilde{A}(k)P(k) & \hat{\Phi}_{13} & \hat{D}(k) \\
* & -P(k) & 0 & 0 \\
* & * & \hat{\Phi}_{33} & 0 \\
* & * & * & -W^{-1}(k)
\end{bmatrix}
$$

$$
+\varepsilon_2(k)
\begin{bmatrix}
H(k) \\ 0 \\ 0 \\ 0 \\ 0 \\ \vdots \\ 0 \\ 0 \\ 0
\end{bmatrix}
\begin{bmatrix}
H(k) \\ 0 \\ 0 \\ 0 \\ 0 \\ \vdots \\ 0 \\ 0 \\ 0
\end{bmatrix}^{\mathrm{T}}
+\varepsilon_2^{-1}(k)
\begin{bmatrix}
0 \\ 0 \\ P_1(k)E^{\mathrm{T}}(k) \\ 0 \\ 0 \\ \vdots \\ 0 \\ 0 \\ 0
\end{bmatrix}
\begin{bmatrix}
0 \\ 0 \\ P_1(k)E^{\mathrm{T}}(k) \\ 0 \\ 0 \\ \vdots \\ 0 \\ 0 \\ 0
\end{bmatrix}^{\mathrm{T}}
< 0
\qquad (9.41)
$$

where

$$
\tilde{A}(k) =
\begin{bmatrix}
A(k)P_1(k) & B(k)C_g(k)P_2(k) \\
B_g(k)C(k)P_1(k) & A_g(k)P_2(k)
\end{bmatrix}.
$$

Then, after some tedious but straightforward calculations, we arrived at the LMI (9.34) Therefore, according to Theorem 9.2.3, we have $J(k) < 0$ and $Z(k) \le P(k)$. From (9.35), it is obvious that $X(k) \le P_1(k) < \Theta(k)$, $\forall k \in \{0, 1, \ldots, N\}$. It can now be concluded that the requirements $(R1)$ and $(R2)$ are simultaneously satisfied. The proof is complete.     □

In the light of Theorem 9.3.1, we can summarize the robust controller design $(RCD)$ algorithm as follows.

*Algorithm RCD*

*Step 1.* Given the $H_\infty$ performance index $\gamma$, the positive definite matrix $S$ and the state initial condition $x(0)$. Set initial values for $\{M(0), N(0), P_1(0), P_2(0)\}$ that satisfy the condition (9.31) and set $k = 0$.

*Step 2.* Obtain the values of matrices $\{\mathcal{M}(k+1), \mathcal{N}(k+1), P_1(k+1), P_2(k+1)\}$ and the desired controller parameters $\{A_g(k), B_g(k), C_g(k)\}$ for the sampling instant $k$ by solving the LMIs (9.32) to (9.35).

*Step 3.* Set $k = k + 1$ and obtain $\{M(k+1), N(k+1)\}$ by the parameter update formula (9.36).

*Step 4.* If $k < N$, then go to Step 2, else go to Step 5.

*Step 5.* Stop.

**Remark 9.1** *It is easy to see that the finite filtering problem for certain nonlinear stochastic time-varying systems can be treated as a special case of the dynamic output feedback control problem studied in this chapter, which can be readily solved by means of the RCD algorithm proposed above with proper modifications.*

## 9.4 Numerical Example

In this section, we present an illustrative example to demonstrate the effectiveness of the proposed algorithms.

Consider a nonlinear stochastic system with the following parameters:

$$A(k) = \begin{bmatrix} 0.425 & -0.493 \\ 0.17 + 0.17\sin(2k) & 0.255 \end{bmatrix},$$

$$B(k) = \begin{bmatrix} -0.68 \\ 0.52 + 0.425\cos(k) \end{bmatrix},$$

$$D_1(k) = \begin{bmatrix} 0.085\sin(0.2k) \\ -0.17 \end{bmatrix},$$

$$C(k) = \begin{bmatrix} -0.13 + 0.255\sin(1.5k) & -0.085 \end{bmatrix},$$

$$D_2(k) = \begin{bmatrix} 0.02 \\ 0.01 \\ 0 \end{bmatrix},$$

$$H(k) = \begin{bmatrix} 0.2 \\ 0.1 \end{bmatrix},$$

$$F = \sin(0.6k),$$

$$E(k) = \begin{bmatrix} 0.1 & 0 \end{bmatrix}.$$

Select the state initial value as $x(0) = \begin{bmatrix} 0.6 & -0.5 \end{bmatrix}$ and $S = \text{diag}\{0.5, 1\}$. Suppose the covariance of $\omega(k)$ is 4.

The nonlinear functions $f(x(k), k)$ and $g(x(k), k)$ are taken as follows:

$$f(x(k), k) = \begin{bmatrix} 0.4 \\ 0.6 \end{bmatrix} (0.3x_1(k)\xi_1(k) + 0.4x_2(k)\xi_2(k)),$$

$$g(x(k), k) = 0.25(0.3x_1(k)\xi_1(k) + 0.4x_2(k)\xi_2(k)),$$

where $x_i(k)$ $(i = 1, 2)$ is the $i$th element of $x(k)$ and $\xi_i(k)$ $(i = 1, 2)$ are zero mean uncorrelated Gaussian white noise processes with unity covariances. We also assume that $\xi_i(k)$ is uncorrelated with $\omega_k$. It can be easily checked that the above class of stochastic nonlinearities satisfies

$$\mathbb{E}\left\{ \begin{bmatrix} f(x(k), k) \\ h(x(k), k) \end{bmatrix} \middle| x(k) \right\} = 0,$$

$$\mathbb{E}\left\{ \begin{bmatrix} f(x(k), k) \\ g(x(k), k) \end{bmatrix} \begin{bmatrix} f^{\mathrm{T}}(x(k), k) & g^{\mathrm{T}}(x(k), k) \end{bmatrix} \middle| x(k) \right\}$$

$$= \begin{bmatrix} 0.4 \\ 0.6 \\ 0.25 \end{bmatrix} \begin{bmatrix} 0.4 \\ 0.6 \\ 0.25 \end{bmatrix}^{\mathrm{T}} x^{\mathrm{T}}(k) \begin{bmatrix} 0.09 & 0 \\ 0 & 0.16 \end{bmatrix} x(k).$$

Set the pre-specified performance indices by $\gamma = 1.2$ and $\{\Theta(k)\}_{1 \leq k \leq N} = \text{diag}\{1, 0.8\}$, and choose the parameter initial values satisfying (9.31). By employing the Matlab LMI Toolbox, we can check the solvability of the addressed problem with the given initial conditions and pre-specified performance indices. Some of the obtained dynamic output feedback controller gains are shown in Table 9.1. The simulation results are shown in Figures. 9.1 to 9.7, which confirm that the desired finite-horizon performance is well achieved and the proposed RCD algorithm is indeed effective.

**Table 9.1** The obtained controller parameters

|  | $A_g$ | | $B_g$ | | $C_g$ | |
|---|---|---|---|---|---|---|
| $k = 0$ | $\begin{bmatrix} 0.1985 & -0.0001 \\ 0.0008 & 0.0022 \end{bmatrix}$ | | $\begin{bmatrix} -0.3791 \\ -0.0328 \end{bmatrix}$ | | $\begin{bmatrix} -0.1100 & 0.3401 \end{bmatrix}$ | |
| $k = 1$ | $\begin{bmatrix} 0.1988 & -0.0006 \\ 0.0019 & 0.0007 \end{bmatrix}$ | | $\begin{bmatrix} -0.5401 \\ 0.1048 \end{bmatrix}$ | | $\begin{bmatrix} -0.0617 & 0.2922 \end{bmatrix}$ | |
| $k = 2$ | $\begin{bmatrix} 0.1959 & -0.0008 \\ 0.0007 & -0.0009 \end{bmatrix}$ | | $\begin{bmatrix} -0.3631 \\ 0.0535 \end{bmatrix}$ | | $\begin{bmatrix} -0.0592 & 0.3209 \end{bmatrix}$ | |
| $k = 3$ | $\begin{bmatrix} 0.1938 & -0.0012 \\ 0.0015 & 0.0004 \end{bmatrix}$ | | $\begin{bmatrix} -0.5746 \\ 0.1815 \end{bmatrix}$ | | $\begin{bmatrix} -0.0369 & 0.2906 \end{bmatrix}$ | |
| $k = 4$ | $\begin{bmatrix} 0.2034 & -0.0029 \\ 0.0028 & 0.0013 \end{bmatrix}$ | | $\begin{bmatrix} -0.6966 \\ 0.2701 \end{bmatrix}$ | | $\begin{bmatrix} -0.0238 & 0.1938 \end{bmatrix}$ | |
| $\vdots$ | $\vdots$ | | $\vdots$ | | $\vdots$ | |

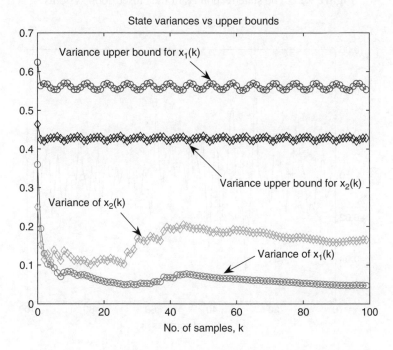

**Figure 9.1** The variance upper bounds and actual variances

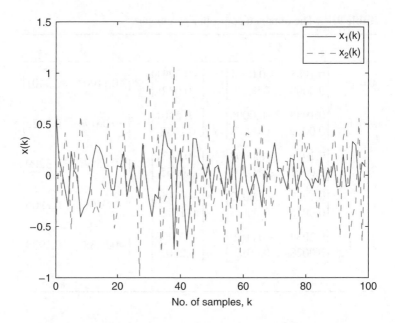

**Figure 9.2**    The state responses of the closed-loop systems

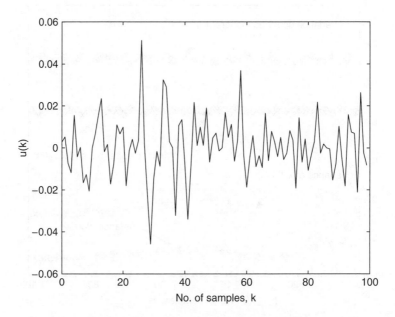

**Figure 9.3**    The control input signal $u(k)$

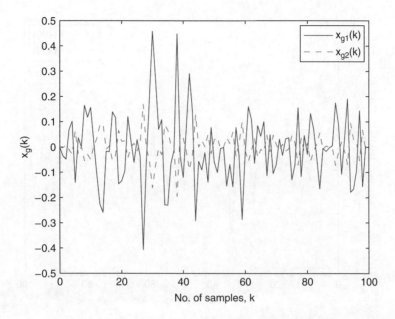

**Figure 9.4** The state responses of the dynamic output feedback controller

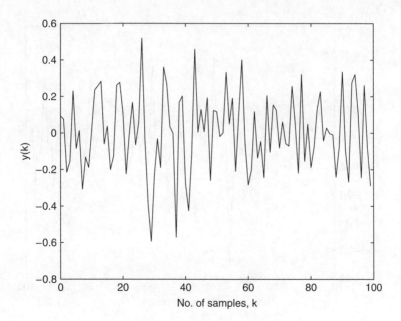

**Figure 9.5** The system output $y(k)$

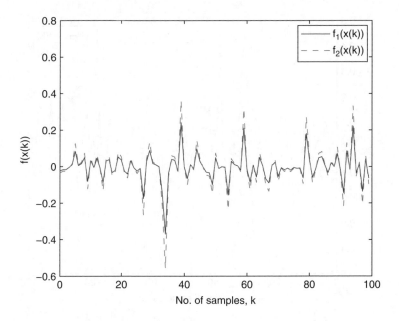

**Figure 9.6**    The stochastic nonlinearity $f(x(k))$

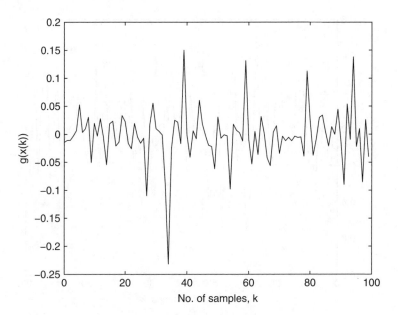

**Figure 9.7**    The stochastic nonlinearity $g(x(k))$

## 9.5 Summary

A multi-objective controller design problem via output feedback for a class of discrete time-varying nonlinear stochastic systems has been discussed in this chapter. The system model considered here with stochastic nonlinearities is widely seen in engineering applications. The control task is to achieve the prescribed $H_\infty$ noise attenuation level and system state covariance constraint simultaneously. The $H_\infty$ and covariance performances have been analyzed separately and then a sufficient condition for the solvability of the addressed controller design problem has been given in terms of the feasibility of a series of RLMIs. Finally, an illustrative example has been provided to show the applicability and effectiveness of the proposed algorithm.

# 10

# Error Variance-Constrained $H_\infty$ Filtering with Degraded Measurements: The Finite-Horizon Case

In this chapter, we aim to solve the robust $H_\infty$ filtering problem for a class of time-varying nonlinear stochastic systems with an error variance constraint. The stochastic nonlinearities considered are quite general, which contain several well-studied stochastic nonlinear systems as special cases. The purpose of the filtering problem is to design a filter that is capable of achieving the pre-specified $H_\infty$ performance and meanwhile guaranteeing a minimized upper bound on the filtering error variance. By means of the adjoint system method, a necessary and sufficient condition for satisfying the $H_\infty$ constraint is first given, expressed as a forward Riccati-like difference equation. Then an upper bound on the variance of the filtering error system is given, guaranteeing that the error variance is not more than a certain value at each sampling instant. The existence condition for the desired filter is established in terms of the feasibility of a set of difference Riccati-like equations, which can be solved forward in time and hence is suitable for online computation. A numerical example is presented finally to show the effectiveness and applicability of the proposed method.

The contributions of this chapter lie in the following parts: (1) For the nonlinear stochastic system, we establish a *forward* Riccati-like difference equation to solve the $H_\infty$ filtering problem, which can be solved forward in time and hence is suitable for online computation. (2) We considered simultaneously the $H_\infty$ and filtering error variance constraints for the *time-varying nonlinear* stochastic system. Compared with the system models in most of the existing literature, which are assumed to be either *time-invariant* or *linear*, the one we study here is much closer to the

*Variance-Constrained Multi-Objective Stochastic Control and Filtering*, First Edition.
Lifeng Ma, Zidong Wang and Yuming Bo.
© 2015 John Wiley & Sons, Ltd. Published 2015 by John Wiley & Sons, Ltd.

actual systems in real-world engineering practice. (3) We converted the solvability of the addressed filtering problem into the feasibility of certain Riccati-like difference equations, whose computing complexity is much lower than that of solving LMI constraints. Hence the proposed method is more suitable for online calculation, which usually requires a high real-time performance.

The rest of this chapter is organized as follows: Section 10.1 formulates the robust filter design problem. Section 10.2 analyzes separately the system $H_\infty$ performance and filtering error variance. Section 10.3 gives the solution to the addressed robust filter design problem. Section 10.4 presents a numerical example. Section 10.5 is the summary.

## 10.1    Problem Formulation

Consider the following nonlinear stochastic system defined in $k \in [0, N-1]$:

$$
\begin{cases}
x(k+1) = (A(k) + \Delta A(k))x(k) + D(k)\omega(k) + f(x(k), k) \\[2mm]
y(k) = \Theta(k)C(k)x(k) + g(x(k), k) \\[2mm]
\qquad = \displaystyle\sum_{j=1}^{m} \theta_j(k)C_j(k)x(k) + g(x(k), k), \\[4mm]
z(k) = L(k)x(k), \quad x_0 = x_0,
\end{cases}
\tag{10.1}
$$

where $x(k) \in \mathbb{R}^n$ is the system state, $y(k) \in \mathbb{R}^m$ is the system measured output, $z(k) \in \mathbb{R}^s$ is a combination of system state to be estimated, $w(k) \in \mathbb{R}^p$ is a zero mean Gaussian white noise with variance $W(k) > 0$, and $A(k)$, $C(k)$, $D(k)$, and $L(k)$ are time-varying matrices with compatible dimensions.

The matrix $\Delta A(k)$ represents the system parameter uncertainty and satisfies

$$
\Delta A(k) = H(k)F(k)E(k), \qquad F^{\mathrm{T}}(k)F(k) \leq I.
\tag{10.2}
$$

The stochastic matrix $\Theta(k)$ describes the phenomenon of multiple measurements degraded in the process of information retrieval from the sensor output. $\Theta(k)$ is defined as

$$
\Theta(k) \triangleq \mathrm{diag}\{\theta_1(k), \theta_2(k), \ldots, \theta_m(k)\},
\tag{10.3}
$$

with $\theta_j(k)(j = 1, 2, \ldots, m)$ being $m$ mutually independent random variables that are also independent from $\omega(k)$. It is assumed that $\theta_j(k)$ has the probabilistic density function $p_j(s)$ on the interval $[0, 1]$ with the mathematical expectation $\bar{\theta}_j(k)$ and variance $(\sigma_j(k))^2$, $\bar{\Theta}(k) \triangleq \mathbb{E}\{\Theta(k)\}$, and

$$
C_j(k) \triangleq \mathrm{diag}\{\underbrace{0, \ldots, 0}_{j-1}, 1, 0, \underbrace{\ldots, 0}_{m-j}\}C(k).
$$

**Remark 10.1** *As discussed in Chapter 7, the stochastic matrix $\Theta(k)$ describes, in the system measurement process with multiple sensors, the working status of these sensors. Notice that $\theta_j(k)$ has the probabilistic density function $p_i(s)$ on the interval*

[0, 1] *and hence the measurement output model in this chapter is more general than those in existing literature, where a Bernoulli distributed stochastic sequence taking values on 0 and 1 is always applied to describe the missing measurement phenomena.*

Denote $h(k) \triangleq h(x(k), k) = \left[ f^{\mathrm{T}}(x(k), k) \ g^{\mathrm{T}}(x(k), k) \right]^{\mathrm{T}}$. The nonlinear functions $f(x(k), k)$ and $g(x(k), k)$ are assumed to satisfy

$$\mathbb{E}\{h(k)|x(k)\} = 0, \quad \mathbb{E}\{h(j)|x(k)\} = 0, \quad \mathbb{E}\{h(k)h^{\mathrm{T}}(j)|x(k)\} = 0, \quad k \neq j, \quad (10.4)$$

$$\mathbb{E}\{h(k)h^{\mathrm{T}}(k)|x(k)\} = \sum_{i=1}^{q} \begin{bmatrix} \Pi_i^{11}(k) & \Pi_i^{12}(k) \\ (\Pi_i^{12}(k))^{\mathrm{T}} & \Pi_i^{22}(k) \end{bmatrix} (x^{\mathrm{T}}(k)\Gamma_i(k)x(k))$$

$$\triangleq \sum_{i=1}^{q} \Pi_i(k)(x^{\mathrm{T}}(k)\Gamma_i(k)x(k)), \quad (10.5)$$

where $\Pi_i^{11}(k)$, $\Pi_i^{22}(k)$, and $\Gamma_i(k)$ $(i = 1, 2, \dots, q)$ are known semi-positive definite matrices and $\Pi_i^{12}(k)$ is a known real-valued matrix with compatible dimensions. Not losing any generality, in this chapter we assume that $\Pi_i(k) = \pi_i(k)\pi_i^{\mathrm{T}}(k)$ and $\Gamma_i(k) = \mu_i(k)\mu_i^{\mathrm{T}}(k)$, with $\pi_i(k)$ and $\mu_i(k)$ being known vectors with appropriate dimensions.

Consider the following type of filter for system (10.1):

$$\begin{cases} \hat{x}(k + 1) = A_f(k)\hat{x}(k) + K_f(k)(y(k) - \bar{\Theta}(k)C(k)\hat{x}(k)), \\ \hat{z}(k) = L(k)\hat{x}(k), \qquad \hat{x}_0 = 0, \end{cases} \quad (10.6)$$

where $A_f(k)$ and $K_f(k)$ are the filter parameters to be designed.

Defining $\eta(k) \triangleq \left[ x^{\mathrm{T}}(k) \ \hat{x}^{\mathrm{T}}(k) \right]^{\mathrm{T}}$, $e(k) \triangleq x(k) - \hat{x}(k)$, and $\tilde{z}(k) \triangleq z(k) - \hat{z}(k) = L(k)e(k)$, we could obtain the filtering error system as follows:

$$\begin{cases} \eta(k + 1) = (\tilde{A}(k) + \Delta\tilde{A}(k))\eta(k) + \tilde{D}(k)\omega(k) + \tilde{G}(k)h(k), \\ \tilde{z}(k) = \tilde{L}(k)\eta(k), \\ \eta_0 = \left[ x_0^{\mathrm{T}} \ \hat{x}_0^{\mathrm{T}} \right]^{\mathrm{T}}, \end{cases} \quad (10.7)$$

where

$$\tilde{A}(k) = \begin{bmatrix} A(k) & 0 \\ K_f(k)\Theta(k)C(k) & A_f(k) - K_f(k)\bar{\Theta}(k)C(k) \end{bmatrix},$$

$$\Delta\tilde{A}(k) = \begin{bmatrix} \Delta A(k) & 0 \\ 0 & 0 \end{bmatrix},$$

$$\tilde{D}(k) = \begin{bmatrix} D(k) \\ 0 \end{bmatrix},$$

$$\tilde{G}(k) = \begin{bmatrix} I & 0 \\ 0 & K_f(k) \end{bmatrix},$$

$$\tilde{L}(k) = \begin{bmatrix} L(k) & -L(k) \end{bmatrix}.$$

We consider the following cost function:

$$J = \mathbb{E}_{\langle h(k),\Theta(k)\rangle} \left\{ \sum_{k=0}^{N-1} \left( \|\tilde{z}(k)\|_2^2 - \gamma^2 \|\omega(k)\|_2^2 \right) \right\} - \gamma^2 \|x_0\|_{S_0}^2, \qquad (10.8)$$

where $\gamma > 0$ is a pre-specified disturbance attenuation level and $S_0 > 0$ is a weighting matrix.

Define the following performance index for system (10.7):

$$\bar{\mathcal{R}}(k) \triangleq \mathbb{E}_{\langle h(k),\Theta(k),\omega(k)\rangle} \{\eta(k)\eta^{\mathrm{T}}(k)\}, \qquad (10.9)$$

and define the performance index for filtering error $e(k)$ as follows:

$$\bar{\mathcal{E}}(k) \triangleq \mathbb{E}_{\langle h(k),\Theta(k),\omega(k)\rangle} \{e(k)e^{\mathrm{T}}(k)\} = Z\bar{\mathcal{R}}(k)Z^{\mathrm{T}}, \qquad (10.10)$$

where $Z = \begin{bmatrix} I & -I \end{bmatrix}$.

This chapter aims to design the filter parameters $A_f(k)$ and $K_f(k)$ such that

(R1)  For a pre-specified $H_\infty$ noise attenuation level $\gamma > 0$, $J < 0$.
(R2)  An upper bound of the performance index $\bar{\mathcal{E}}(k)$ of filtering error defined in (10.10) is first guaranteed, that is

$$\bar{\mathcal{E}}(k) \le \mathcal{E}(k), \qquad (10.11)$$

and then $\mathcal{E}(k)$ is minimized in the sense of matrix norm.

In the following, in order to solve the addressed $H_\infty$ filtering problem for systems with parameter uncertainty we first establish an auxiliary system without parameter uncertainty as follows:

$$\begin{cases} \xi(k+1) = \hat{A}(k)\xi(k) + \bar{D}(k)\bar{\omega}(k) + \bar{G}(k)h(k), \\ \tilde{z}(k) = \bar{L}(k)\xi(k), \quad \xi_0 = \eta_0, \end{cases} \qquad (10.12)$$

where

$$\hat{A}(k) = \tilde{A}(k),$$

$$\bar{G}(k) = \tilde{G}(k),$$

$$\bar{L}(k) = \tilde{L}(k),$$

$$\bar{D}(k) = \begin{bmatrix} \tilde{D}(k) & \epsilon^{-1}(k)\gamma\bar{H}_k \end{bmatrix},$$

$$\bar{E}(k) = \begin{bmatrix} E(k) & 0 \end{bmatrix},$$

$$\bar{H}(k) = \begin{bmatrix} H(k) \\ 0 \end{bmatrix},$$

$$\bar{\omega}(k) = \begin{bmatrix} \omega(k) \\ \epsilon(k)\gamma^{-1}F(k)\bar{E}(k)\xi(k) \end{bmatrix},$$

where $\epsilon(k) \neq 0$ is any known real-valued constant. Such a methodology for eliminating the parameter uncertainty first appeared in Ref. [128]. Since $\xi_0 = \eta_0$, it is clear that $\xi(k) = \eta(k)$ for $k \in [0, \ N-1]$.

Then introduce the following performance formulation:

$$\bar{J} = \mathop{\mathbb{E}}_{\langle h(k), \Theta(k) \rangle} \left\{ \sum_{k=0}^{N-1} \left( \|\bar{z}(k)\|_2^2 + \|\epsilon(k)\bar{E}(k)\xi(k)\|_2^2 - \gamma^2 \|\bar{\omega}(k)\|_2^2 \right) \right\} - \gamma^2 \|\xi_0\|_{S_0}^2.$$

(10.13)

Now we will show that $\bar{J}$ is a tight upper bound of $J$. In fact,

$$\bar{J} - J = \mathop{\mathbb{E}}_{\langle h(k), \Theta(k) \rangle} \left\{ \sum_{k=0}^{N-1} \|\epsilon(k)\bar{E}(k)\xi(k)\|_2^2 - \gamma^2 \sum_{k=0}^{N-1} \left( \|\bar{\omega}(k)\|_2^2 - \|\omega(k)\|_2^2 \right) \right\}$$

$$= \mathop{\mathbb{E}}_{\langle h(k), \Theta(k) \rangle} \left\{ \sum_{k=0}^{N-1} \left( \epsilon(k)^2 \xi^{\mathrm{T}}(k)\bar{E}^{\mathrm{T}}(k)\bar{E}(k)\xi(k) - \gamma^2 \left( \omega^{\mathrm{T}}(k)\omega(k) \right. \right. \right.$$

(10.14)

$$\left. \left. \left. + \gamma^{-2}\epsilon(k)^2 \xi^{\mathrm{T}}(k)\bar{E}^{\mathrm{T}}(k)F^{\mathrm{T}}(k)F(k)\bar{E}(k)\xi(k) - \omega^{\mathrm{T}}(k)\omega(k) \right) \right) \right\}$$

$$= \mathop{\mathbb{E}}_{\langle h(k), \Theta(k) \rangle} \left\{ \sum_{k=0}^{N-1} \|\epsilon(k)\sqrt{I - F^{\mathrm{T}}(k)F(k)}\bar{E}(k)\xi(k)\|_2^2 \right\}$$

$$\geq 0.$$

Hence, $\bar{J} < 0$ implies $J < 0$. Since $F(k)$ can be any arbitrary perturbation matrix satisfying (10.2), $\bar{J}$ is a tight upper bound of $J$.

## 10.2    Performance Analysis

In this section, the $H_\infty$ performance and filtering error variance are first analyzed separately and then several lemmas which will play key roles in the robust filter design procedure in the next section will be proposed.

### 10.2.1    $H_\infty$ Performance Analysis

First denote

$$\bar{A}(k) = \begin{bmatrix} A(k) & 0 \\ K_f(k)\bar{\Theta}(k)C(k) & A_f(k) - K_f(k)\bar{\Theta}(k)C(k) \end{bmatrix},$$

$$\bar{\Gamma}_i(k) = \begin{bmatrix} \Gamma_i(k) & 0 \\ 0 & 0 \end{bmatrix},$$

(10.15)

$$\bar{C}_j(k) = \begin{bmatrix} 0 & 0 \\ K_f(k)C_j(k) & 0 \end{bmatrix}.$$

We consider the following recursion:

$$Q(k) = \bar{A}^{\mathrm{T}}(k)Q(k+1)\bar{A}(k) + \sum_{j=1}^{m} (\sigma_j(k))^2 \bar{C}_j^{\mathrm{T}}(k)Q(k+1)\bar{C}_j(k)$$

$$+ \sum_{i=1}^{q} \bar{\Gamma}_i(k) \operatorname{tr}[\bar{G}^{\mathrm{T}}(k)Q(k+1)\bar{G}(k)\Pi_i(k)] \tag{10.16}$$

$$+ \epsilon(k)^2 \bar{E}^{\mathrm{T}}(k)\bar{E}(k) + \bar{L}^{\mathrm{T}}(k)\bar{L}(k)$$

$$+ \bar{A}^{\mathrm{T}}(k)Q(k+1)\bar{D}(k)\Xi^{-1}(k)\bar{D}^{\mathrm{T}}(k)Q(k+1)\bar{A}(k), \quad Q_N = 0,$$

where we define

$$\Xi(k) = \gamma^2 I - \bar{D}^{\mathrm{T}}(k)Q(k+1)\bar{D}(k). \tag{10.17}$$

**Lemma 10.2.1** *Consider system* (10.12). *Given a positive scalar* $\gamma > 0$, *a sufficient condition for J of* (10.8) *to be negative for all nonzero* $(\omega(k), x_0)$ *is that there exists a solution* $Q(k)$ *to* (10.16) *such that* $\Xi(k) > 0$ *and* $Q_0 < \gamma^2 S_0$.

*Proof.* First define

$$\tilde{J}(k) = \xi(k+1)^{\mathrm{T}}Q(k+1)\xi(k+1) - \xi^{\mathrm{T}}(k)Q(k)\xi(k). \tag{10.18}$$

Then, substituting (10.12) into (10.18) and by the statistical property of $h(k)$ and $\Theta(k)$, we have

$$\mathbb{E}_{\langle h(k), \Theta(k)\rangle} \{\tilde{J}(k)\}$$

$$= \mathbb{E}_{\langle h(k), \Theta(k)\rangle} \left\{ (\hat{A}(k)\xi(k) + \bar{D}(k)\bar{\omega}(k) + \bar{G}(k)h(k))^{\mathrm{T}}Q(k+1) \right.$$

$$\left. \times (\hat{A}(k)\xi(k) + \bar{D}(k)\bar{\omega}(k) + \bar{G}(k)h(k)) \right\} - \xi^{\mathrm{T}}(k)Q(k)\xi(k) \tag{10.19}$$

$$= \xi^{\mathrm{T}}(k)\left( \bar{A}^{\mathrm{T}}(k)Q(k+1)\bar{A}(k) - Q(k) + \sum_{j=1}^{m} (\sigma_j(k))^2 \bar{C}_j^{\mathrm{T}}(k)Q(k+1)\bar{C}_j(k) \right.$$

$$\left. + \sum_{i=1}^{q} \bar{\Gamma}_i(k)\operatorname{tr}[\bar{G}^{\mathrm{T}}(k)Q(k+1)\bar{G}(k)\Pi_i(k)] \right) \xi(k)$$

$$+ 2\xi^{\mathrm{T}}(k)\bar{A}^{\mathrm{T}}(k)Q(k+1)\bar{D}(k)\bar{\omega}(k) + \bar{\omega}^{\mathrm{T}}(k)\bar{D}^{\mathrm{T}}(k)Q(k+1)\bar{D}(k)\bar{\omega}(k).$$

Adding the zero term

$$\|\tilde{z}(k)\|_2^2 + \|\epsilon(k)\bar{E}(k)\xi(k)\|_2^2 - \gamma^2 \|\bar{\omega}(k)\|_2^2$$

$$-(\|\tilde{z}(k)\|_2^2 + \|\epsilon(k)\bar{E}(k)\xi(k)\|_2^2 - \gamma^2 \|\bar{\omega}(k)\|_2^2) \tag{10.20}$$

to $\underset{\langle h(k),\Theta(k)\rangle}{\mathbb{E}} \{\tilde{J}(k)\}$ and then taking the mathematical expectation results in

$$\underset{\langle h(k),\Theta(k)\rangle}{\mathbb{E}} \{\tilde{J}(k)\}$$

$$= \xi^{\mathrm{T}}(k) \left( \bar{A}^{\mathrm{T}}(k)Q(k+1)\bar{A}(k) - Q(k) + \sum_{j=1}^{m} (\sigma_j(k))^2 \bar{C}_j^{\mathrm{T}}(k)Q(k+1)\bar{C}_j(k) \right.$$

$$\left. + \sum_{i=1}^{q} \bar{\Gamma}_i(k)\mathrm{tr}[\bar{G}^{\mathrm{T}}(k)Q(k+1)\bar{G}(k)\Pi_i(k)] + \bar{L}^{\mathrm{T}}(k)\bar{L}(k) + \epsilon(k)^2 \bar{E}^{\mathrm{T}}(k)\bar{E}(k) \right) \xi(k)$$

$$+ 2\xi^{\mathrm{T}}(k)\bar{A}^{\mathrm{T}}(k)Q(k+1)\bar{D}(k)\bar{\omega}(k)$$

$$+ \bar{\omega}^{\mathrm{T}}(k)(-\gamma^2 I + \bar{D}^{\mathrm{T}}(k)Q(k+1)\bar{D}(k))\bar{\omega}(k)$$

$$- \underset{\langle h(k),\Theta(k)\rangle}{\mathbb{E}} \{\|\tilde{z}(k)\|_2^2 + \|\epsilon(k)\bar{E}(k)\xi(k)\|_2^2 - \gamma^2 \|\bar{\omega}(k)\|_2^2\}. \tag{10.21}$$

Completing the squares of $\bar{\omega}(k)$, we obtain the recursion (10.16), (10.17) and

$$\underset{\langle h(k),\Theta(k)\rangle}{\mathbb{E}} \{\tilde{J}(k)\} = - (\bar{\omega}(k) - \bar{\omega}^*(k))^{\mathrm{T}}\Xi(k)(\bar{\omega}(k) - \bar{\omega}^*(k))$$

$$- \underset{\langle h(k),\Theta(k)\rangle}{\mathbb{E}} \{\|\tilde{z}(k)\|_2^2 + \|\epsilon(k)\bar{E}(k)\xi(k)\|_2^2 - \gamma^2 \|\bar{\omega}(k)\|_2^2\}, \tag{10.22}$$

where

$$\bar{\omega}^*(k) = \Xi^{-1}(k)\bar{D}^{\mathrm{T}}(k)Q(k+1)\bar{A}(k)\xi(k). \tag{10.23}$$

Then, taking the sum of both sides of (10.22) with respect to $k$ from 0 to $N-1$ results in

$$\underset{\langle h(k),\Theta(k)\rangle}{\mathbb{E}} \left\{ \sum_{k=0}^{N-1} \tilde{J}(k) \right\} = \underset{\langle h(k)\rangle}{\mathbb{E}} \{\xi_N^{\mathrm{T}}Q_N\xi_N\} - \xi_0^{\mathrm{T}}Q_0\xi_0$$

$$= - \sum_{k=0}^{N-1} \left( (\bar{\omega}(k) - \bar{\omega}^*(k))^{\mathrm{T}}\Xi(k)(\bar{\omega}(k) - \bar{\omega}^*(k)) \right.$$

$$\left. + \underset{\langle h(k),\Theta(k)\rangle}{\mathbb{E}} \{\|\tilde{z}(k)\|_2^2 + \|\epsilon(k)\bar{E}(k)\xi(k)\|_2^2 - \gamma^2 \|\bar{\omega}(k)\|_2^2\} \right). \tag{10.24}$$

Therefore, since $\Xi(k) > 0$, $Q_N = 0$ and $Q_0 - \gamma^2 S_0 < 0$,

$$\bar{J} = \underset{\langle h(k),\Theta(k)\rangle}{\mathbb{E}} \left\{ \sum_{k=0}^{N-1} (\|\tilde{z}(k)\|_2^2 + \|\epsilon(k)\bar{E}(k)\xi(k)\|_2^2 - \gamma^2 \|\bar{\omega}(k)\|_2^2) \right\} - \gamma^2 \|\xi_0\|_{S_0}^2$$

$$\tag{10.25}$$

$$= \xi_0^{\mathrm{T}}Q_0\xi_0 - \sum_{k=0}^{N-1} (\bar{\omega}(k) - \bar{\omega}^*(k))^{\mathrm{T}}\Xi(k)(\bar{\omega}(k) - \bar{\omega}^*(k)) - \gamma^2 \|\xi_0\|_{S_0}^2 < 0,$$

which means $J < 0$. The proof is complete. $\qquad\qquad\square$

**Remark 10.2** *Notice that the Riccati equation in Lemma 10.2.1 is backward in time, which is therefore only suitable for offline computation. However, in engineering practice, it is always required to design a filter capable of online computing. In Ref. [146], such a problem was solved by an adjoint system method; that is, since (10.12) can be used to define a map $\mathcal{L}$ from one Hilbert space $\mathcal{H}_1$ to another $\mathcal{H}_2$, the adjoint system which is used to derive the forward difference equations, can be determined by the adjoint map $\mathcal{L}^*$ from Hilbert space $\mathcal{H}_2$ to $\mathcal{H}_1$. However, in such a procedure, an explicit closed form of solution to system (10.12), namely, $\xi(k)$, is needed. Unfortunately, as discussed in Chapter 2, the nonlinear stochastic system discussed in this book could cover several types of nonlinear systems and we have no information about its explicit form but the statistical property; hence it is impossible to solve the stochastic difference equation to obtain an explicit expression of the solution $\xi(k)$. Therefore, in the following, first we present an auxiliary system that has the same statistical property as system (10.12). Then by means of the adjoint system method, we shall propose a necessary and sufficient condition for $\bar{\bar{J}} < 0$ in a forward form, which can be applied in online computing.*

Construct an auxiliary system for system (10.12) as follows:

$$\begin{cases} \zeta(k+1) = (\hat{A}(k) + M(k)v(k))\zeta(k) + \bar{D}(k)\bar{\bar{\omega}}(k), \\ \bar{\bar{z}}(k) = \bar{L}(k)\zeta(k), \\ \zeta_0 = \xi_0 = \eta_0, \end{cases} \tag{10.26}$$

where $M(k)$ is defined as follows:

$$\begin{aligned} M(k) &\triangleq \begin{bmatrix} M_1(k) & M_2(k) & \cdots & M_{qk} \end{bmatrix}, \\ M_i(k) &= \bar{G}(k)\pi_i(k)\mu_i^{\mathrm{T}}(k)\bar{Z}, \quad i = 1, 2, \dots, q, \\ \bar{Z} &= \begin{bmatrix} I & 0 \end{bmatrix}, \end{aligned} \tag{10.27}$$

and $v(k)$, $\bar{\bar{\omega}}(k)$ are defined as follows:

$$\begin{aligned} v(k) &\triangleq \begin{bmatrix} v_1(k)I & v_2(k)I & \cdots & v_{qk}I \end{bmatrix}^{\mathrm{T}}, \\ \bar{\bar{\omega}}(k) &\triangleq \begin{bmatrix} \omega(k) \\ \epsilon(k)\gamma^{-1}F(k)\bar{E}(k)\zeta(k) \end{bmatrix}, \end{aligned} \tag{10.28}$$

where $v_i(k) \in \mathbb{R}$ ($i = 1, 2, \dots, q$) are $q$ mutually independent zero mean Gaussian white noise with unity variance. Also, $v_i(k)$ are assumed to be independent from $\omega(k)$.

We construct the following cost function:

$$\bar{\bar{J}} = \mathop{\mathbb{E}}_{\langle v(k) \rangle} \left\{ \sum_{k=0}^{N-1} \left( \|\bar{\bar{z}}(k)\|_2^2 + \|\epsilon(k)\bar{E}(k)\zeta(k)\|_2^2 - \gamma^2\|\bar{\bar{\omega}}(k)\|_2^2 \right) - \gamma^2\|\zeta_0\|_{S_0}^2 \right\}. \tag{10.29}$$

In the following, we will show that $\bar{\bar{J}} = \bar{J}$. To this end, since $\xi_0 = \zeta_0$, it is true that

$$\mathop{\mathbb{E}}_{\langle h(k) \rangle} \{\xi(k+1)\} = \mathop{\mathbb{E}}_{\langle v(k) \rangle} \{\zeta(k+1)\}, \quad 0 < k < N-1. \tag{10.30}$$

Define the second moment of system (10.12) and (10.26) respectively as:

$$\mathcal{X}(k+1) \triangleq \mathop{\mathbb{E}}_{\langle h(k)\rangle} \{\xi(k+1)\xi(k+1)^{\mathrm{T}}\},$$

$$\mathcal{Y}(k+1) \triangleq \mathop{\mathbb{E}}_{\langle v(k)\rangle} \{\zeta(k+1)\zeta(k+1)^{\mathrm{T}}\}, \quad 0 < k < N-1,$$

$$\mathcal{X}_0 = \xi_0\xi_0^{\mathrm{T}},$$

$$\mathcal{Y}_0 = \zeta_0\zeta_0^{\mathrm{T}},$$

(10.31)

and define a matrix-valued function by

$$\mathcal{F}(\mathcal{T}) \triangleq (\hat{A}(k) + \Delta\tilde{A}(k))\mathcal{T}(\hat{A}(k) + \Delta\tilde{A}(k))^{\mathrm{T}}$$

$$+ \sum_{i=1}^{q} \bar{G}(k)\Pi_i(k)\bar{G}^{\mathrm{T}}(k)\mathrm{tr}[\bar{\Gamma}_i(k)\mathcal{T}] + \tilde{D}(k)\omega(k)\omega^{\mathrm{T}}(k)\tilde{D}^{\mathrm{T}}(k). \quad (10.32)$$

Now, we could prove that $\mathcal{X}(k) = \mathcal{Y}(k)$, $\forall k \in [0, \ N-1]$. In fact, since $\xi_0 = \zeta_0$, we have $\mathcal{X}_0 = \mathcal{Y}_0$. The following proof is carried out by induction. Suppose that when $0 < k < N-1$, $\mathcal{X}(k) = \mathcal{Y}(k)$. Then at the instant $k+1$, from (10.12) and (10.26), we can obtain

$$\mathcal{X}(k+1) = \mathcal{F}(\mathcal{X}(k)) + \mathop{\mathbb{E}}_{\langle h(k)\rangle} \{(\hat{A}(k) + \Delta\tilde{A}(k))\xi(k)\omega^{\mathrm{T}}(k)\tilde{D}^{\mathrm{T}}(k)$$

$$+ \tilde{D}(k)\omega(k)\xi^{\mathrm{T}}(k)(\hat{A}(k) + \Delta\tilde{A}(k))^{\mathrm{T}}\},$$

$$\mathcal{Y}(k+1) = \mathcal{F}(\mathcal{Y}(k)) + \mathop{\mathbb{E}}_{\langle v(k)\rangle} \{(\hat{A}(k) + \Delta\tilde{A}(k))\zeta(k)\omega^{\mathrm{T}}(k)\tilde{D}^{\mathrm{T}}(k)$$

$$+ \tilde{D}(k)\omega(k)\zeta^{\mathrm{T}}(k)(\hat{A}(k) + \Delta\tilde{A}(k))^{\mathrm{T}}\},$$

(10.33)

and taking note of (10.30), we arrive at $\mathcal{X}(k) = \mathcal{Y}(k)$, $\forall k \in [0, \ N-1]$.

From the definition of norm and the property of matrix trace, we can get

$$\mathbb{E}\{\|\bar{z}(k)\|_2^2\} = \mathrm{tr}[\bar{L}^{\mathrm{T}}(k)\bar{L}(k)\mathcal{X}(k)],$$

$$\mathbb{E}\{\|\epsilon(k)\bar{E}(k)\xi(k)\|_2^2\} = \epsilon(k)^2\mathrm{tr}[\bar{E}^{\mathrm{T}}(k)\bar{E}(k)\mathcal{X}(k)],$$

$$\mathbb{E}\{\|\bar{\omega}(k)\|_2^2\} = \mathrm{tr}[\epsilon(k)^2\gamma^{-2}F(k)\bar{E}(k)\mathcal{X}(k)\bar{E}^{\mathrm{T}}(k)F^{\mathrm{T}}(k)]$$

$$+ \omega^{\mathrm{T}}(k)\omega(k)$$

(10.34)

and

$$\mathbb{E}\{\|\bar{\bar{z}}(k)\|_2^2\} = \mathrm{tr}[\bar{L}^{\mathrm{T}}(k)\bar{L}(k)\mathcal{Y}(k)],$$

$$\mathbb{E}\{\|\epsilon(k)\bar{E}(k)\zeta(k)\|_2^2\} = \epsilon(k)^2\mathrm{tr}[\bar{E}^{\mathrm{T}}(k)\bar{E}(k)\mathcal{Y}(k)],$$

$$\mathbb{E}\{\|\bar{\bar{\omega}}(k)\|_2^2\} = \mathrm{tr}[\epsilon(k)^2\gamma^{-2}F(k)\bar{E}(k)\mathcal{Y}(k)\bar{E}^{\mathrm{T}}(k)F^{\mathrm{T}}(k)]$$

$$+ \omega^{\mathrm{T}}(k)\omega(k).$$

(10.35)

Since $\mathcal{X}(k) = \mathcal{Y}(k)$, we can conclude that $\bar{\bar{J}} = \bar{J}$.

So far, by constructing an auxiliary system and a new cost function, we have successfully transformed the cost function $\bar{J}$ applied on the nonlinear stochastic system (10.12) into an equivalent cost function $\bar{\bar{J}}$ applied on the stochastic system with state-dependent noises (10.26). Therefore, we could apply the stochastic bounded real lemma (Theorem 3.1 in Ref. [147]) to obtain a sufficient and necessary condition expressed by a Riccati equation that can be solved forward in time, satisfying the $H_\infty$ constraint.

**Lemma 10.2.2** *Consider system* (10.26) *and the cost function* $\bar{\bar{J}}$ *defined in* (10.29). *For a given* $\gamma > 0$, *a sufficient and necessary condition for* $\bar{\bar{J}} < 0$ *is that there exists a solution* $\mathcal{Q}(k)$ *to the following matrix-valued equation:*

$$\mathcal{Q}(k+1) = \bar{A}(k)(\mathcal{Q}_k + \bar{\mathcal{Q}}_k)\bar{A}^T(k) + \sum_{j=1}^{m} (\sigma_j(k))^2 \bar{C}_j(k)(\mathcal{Q}_k + \bar{\mathcal{Q}}_k)\bar{C}_j^T(k)$$

$$+ \bar{D}(k)\bar{D}^T(k) + \sum_{i=1}^{q} M_i(k)\mathcal{Q}_k M_i^T(k), \quad \mathcal{Q}_0 = S_0^{-1}, \tag{10.36}$$

*such that*

$$\Phi(k) = \gamma^2 I - \bar{L}(k)\mathcal{Q}_k \bar{L}^T(k) > 0, \quad \forall k \in [0 \quad N-1], \tag{10.37}$$

*where* $\bar{\mathcal{Q}}(k) \triangleq \mathcal{Q}_k \bar{L}^T(k)\Phi^{-1}(k)\bar{L}(k)\mathcal{Q}_k$.

The proof of Lemma 10.2.2 can be carried out in a similar way to the proof of Theorem 3.1 in Ref. [145] by replacing the corresponding parameters and hence is omitted here.

### 10.2.2    System Covariance Analysis

Now we are in a situation to analyze the covariance of system (10.7), and a lemma is present that is important in the robust filter design in the next section. Firstly, from (10.7), $\bar{\mathcal{R}}(k)$ defined in (10.9) satisfies the following Lyapunov equation:

$$\bar{\mathcal{R}}(k+1) = (\bar{A}(k) + \Delta\tilde{A}(k))\bar{\mathcal{R}}_k(\bar{A}(k) + \Delta\tilde{A}(k))^T + \sum_{j=1}^{m} (\sigma_j(k))^2 \bar{C}_j(k)\bar{\mathcal{R}}_k \bar{C}_j^T(k)$$

$$+ \sum_{i=1}^{q} \bar{G}(k)\Pi_i(k)\bar{G}^T(k)\text{tr}[\bar{\Gamma}_i(k)\bar{\mathcal{R}}(k)] + \tilde{D}(k)\mathcal{W}(k)\tilde{D}^T(k). \tag{10.38}$$

The following lemma gives an upper bound of the state covariance of the filtering error system (10.7).

**Lemma 10.2.3** *Let* $\alpha(k) > 0$ *be a sequence of positive scalars. Consider the filtering error system* (10.7). *If the following matrix-valued equation,*

$$\mathcal{R}(k+1) = \bar{A}(k)(\mathcal{R}_k^{-1} - \alpha(k)\bar{E}^{\mathrm{T}}(k)\bar{E}(k))^{-1}\bar{A}^{\mathrm{T}}(k) + \alpha^{-1}(k)\bar{H}(k)\bar{H}^{\mathrm{T}}(k)$$

$$+ \sum_{j=1}^{m} (\sigma_j(k))^2 \bar{C}_j(k)\mathcal{R}_k\bar{C}_j^{\mathrm{T}}(k) + \sum_{i=1}^{q} \bar{G}(k)\Pi_i(k)\bar{G}^{\mathrm{T}}(k)\mathrm{tr}[\bar{\Gamma}_i(k)\mathcal{R}(k)]$$

$$+ \tilde{D}(k)\mathcal{W}(k)\tilde{D}^{\mathrm{T}}(k), \tag{10.39}$$

*has a positive definite solution* $\mathcal{R}(k)$ *such that*

$$\alpha^{-1}(k)I - \bar{E}(k)\mathcal{R}(k)\bar{E}^{\mathrm{T}}(k) > 0, \quad k \in [0, \ N-1], \tag{10.40}$$

*then* $\mathcal{R}(k)$ *is an upper bound of covariance of system* (10.7), *namely* $\bar{\mathcal{R}}(k)$.

The proof of Lemma 10.2.3 can be done according to the proof of Lemma 10.2.1 of this chapter and Theorem 1 in Ref. [148]. The latter, however, did not take the phenomenon of degraded measurements into account. By the technique we have developed in the proof of Lemma 10.2.1 above for dealing with the stochastic matrix $\Theta(k)$, we could easily obtain (10.39) and (10.40) and hence the detailed proof is omitted here.

## 10.3   Robust Filter Design

Now we are ready to give our main result, the designing algorithm of the robust filter (10.6) for the nonlinear stochastic system (10.1).

**Theorem 10.3.1**  *Given* $\gamma > 0$. *Let* $\alpha(k) > 0$ *be a sequence of positive scalars. If the following set of matrix-valued difference equations,*

$$\left\{ \begin{array}{l} \mathcal{Q}(k+1) = \bar{A}(k)(\mathcal{Q}_k + \bar{\mathcal{Q}}_k)\bar{A}^{\mathrm{T}}(k) + \displaystyle\sum_{j=1}^{m} (\sigma_j(k))^2 \bar{C}_j(k)(\mathcal{Q}_k + \bar{\mathcal{Q}}_k)\bar{C}_j^{\mathrm{T}}(k) \\[4mm] \qquad\qquad + \bar{D}(k)\bar{D}^{\mathrm{T}}(k) + \displaystyle\sum_{i=1}^{q} M_i(k)\mathcal{Q}_k M_i^{\mathrm{T}}(k), \quad \mathcal{Q}_0 = S_0^{-1}, \\[4mm] P(k+1) = A(k)(P^{-1}(k) - \alpha(k)E^{\mathrm{T}}(k)E(k))^{-1}A^{\mathrm{T}}(k) + \alpha^{-1}(k)H(k)H^{\mathrm{T}}(k) \\[4mm] \qquad\qquad + \displaystyle\sum_{i=1}^{q} \Pi_i^{11}(k)\mathrm{tr}[\Gamma_i(k)P(k)] + D(k)\mathcal{W}(k)D^{\mathrm{T}}(k), \quad P_0 = \mathcal{P}_0, \\[4mm] \mathcal{E}(k+1) = A(k)(\mathcal{E}^{-1}(k) - \alpha(k)E^{\mathrm{T}}(k)E(k))^{-1}A^{\mathrm{T}}(k) + \alpha^{-1}(k)H(k)H^{\mathrm{T}}(k) \\[4mm] \qquad\qquad + \displaystyle\sum_{i=1}^{q} \Pi_i^{11}(k)\mathrm{tr}[\Gamma_i(k)P(k)] + D(k)\mathcal{W}(k)D^{\mathrm{T}}(k) \\[4mm] \qquad\qquad - \mathcal{M}(k)\Omega^{-1}(k)\mathcal{M}^{\mathrm{T}}(k), \quad \mathcal{E}_0 = \mathcal{E}_0, \end{array} \right. \tag{10.41}$$

*where*

$$\mathcal{M}(k) \triangleq \sum_{i=1}^{q} \Pi_i^{12}(k)\mathrm{tr}[\Gamma_i(k)\mathcal{P}(k)] + A(k)(\mathcal{E}^{-1}(k) - \alpha(k)E^{\mathrm{T}}(k)E(k))^{-1}C^{\mathrm{T}}(k)\bar{\Theta}(k)$$

*has the solution* $(\mathcal{Q}(k), \mathcal{P}(k) > 0, \mathcal{E}(k) > 0)$ *such that*

$$\begin{cases} \gamma^2 I - \bar{L}(k)\mathcal{Q}_k\bar{L}^{\mathrm{T}}(k) > 0, \\ \alpha^{-1}(k)I - E(k)\mathcal{P}(k)E^{\mathrm{T}}(k) > 0, \\ \mathcal{P}(k) - \mathcal{E}(k) > 0, \end{cases} \tag{10.42}$$

*then there exists a filter of form (10.6) that is able to achieve the pre-specified* $H_\infty$ *performance, and meanwhile the upper bound of the filtering error variance can be minimized in the sense of matrix norm. Moreover, the filter parameters can be parameterized by*

$$\begin{cases} K_f(k) = \left( A(k)(\mathcal{E}^{-1}(k) - \alpha(k)E^{\mathrm{T}}(k)E(k))^{-1}C^{\mathrm{T}}(k)\bar{\Theta}(k) \right. \\ \qquad\qquad \left. + \sum_{i=1}^{q} \Pi_i^{12}(k)\mathrm{tr}[\Gamma_i(k)\mathcal{P}(k)] \right) \Omega^{-1}(k), \\ A_f(k) = A(k) + (A(k) - K_f(k)\bar{\Theta}(k)C(k))\mathcal{E}(k)E^{\mathrm{T}}(k) \\ \qquad\qquad \times (\alpha^{-1}(k)I - E(k)\mathcal{E}(k)E^{\mathrm{T}}(k))^{-1}E(k), \end{cases} \tag{10.43}$$

*where*

$$\Omega(k) = \bar{\Theta}(k)C(k)(\mathcal{E}^{-1}(k) - \alpha(k)E^{\mathrm{T}}(k)E(k))^{-1}C^{\mathrm{T}}(k)\bar{\Theta}(k)$$

$$+ \sum_{i=1}^{q} \Pi_i^{22}(k)\mathrm{tr}[\Gamma_i(k)\mathcal{P}(k)] + \sum_{j=1}^{m} \sigma_j^2(k)C_j(k)\mathcal{P}(k)C_j^{\mathrm{T}}(k). \tag{10.44}$$

*Proof.* First, according to Lemma 10.2.2, the $H_\infty$ performance is achieved. Then, by the method developed in Ref. [148], we suppose that $\mathcal{R}(k)$ has a form of

$$\mathcal{R}(k) = \begin{bmatrix} \mathcal{P}(k) & \mathcal{P}(k) - \mathcal{E}(k) \\ \mathcal{P}(k) - \mathcal{E}(k) & \mathcal{P}(k) - \mathcal{E}(k) \end{bmatrix}. \tag{10.45}$$

Substituting (10.43) to (10.44) into $\mathcal{P}(k+1)$ and $\mathcal{E}(k+1)$ of (10.41) and by some tedious but straightforward manipulations, we can find that $\mathcal{R}(k)$ is a solution to the matrix-valued function (10.39). Therefore, according to Lemma 10.2.3, $\mathcal{R}(k)$ is an upper bound of the covariance of filtering error system (10.7). Now, the rest of the proof is to show that the obtained $A_f(k)$ and $K_f(k)$ in (10.43) minimize the filtering error variance $\mathcal{E}(k)$ in the matrix norm sense. To this end,

$$\mathcal{E}(k+1) = \begin{bmatrix} I & -I \end{bmatrix} \mathcal{R}(k+1) \begin{bmatrix} I & -I \end{bmatrix}^{\mathrm{T}}$$

$$= \mathcal{A}(k)(\mathcal{R}^{-1}(k) - \alpha(k)\bar{E}^{\mathrm{T}}(k)\bar{E}(k))^{-1}\mathcal{A}^{\mathrm{T}}(k)$$

$$+ \alpha^{-1}(k)H(k)H^{\mathrm{T}}(k) + \sum_{j=1}^{m} \sigma_j^2(k)K_f(k)C_j(k)P(k)C_j^{\mathrm{T}}(k)K_f^{\mathrm{T}}(k)$$

$$+ \sum_{i=1}^{q} \Pi_i^{11}(k)\mathrm{tr}[\Gamma_i(k)P(k)] - K_f(k)\sum_{i=1}^{q} \Pi_i^{21}(k)\mathrm{tr}[\Gamma_i(k)P(k)] \qquad (10.46)$$

$$- \sum_{i=1}^{q} \Pi_i^{12}(k)\mathrm{tr}[\Gamma_i(k)P(k)]K_f^{\mathrm{T}}(k) + K_f(k)\sum_{i=1}^{q} \Pi_i^{22}(k)\mathrm{tr}[\Gamma_i(k)P(k)]K_f^{\mathrm{T}}(k)$$

$$+ D(k)\mathcal{W}(k)D^{\mathrm{T}}(k),$$

where

$$\mathcal{A}(k) \triangleq \begin{bmatrix} A(k) - K_f(k)\bar{\Theta}(k)C(k) & -A_f(k) + K_f(k)\bar{\Theta}(k)C(k) \end{bmatrix}.$$

In order to find the filter parameters minimizing the filtering error variance, we take the first variation of (10.46) with respect to $A_{kf}$ and $K_{kf}$:

$$\frac{\partial \mathcal{E}(k+1)}{\partial A_f(k)} = 2\mathcal{A}(k)(\mathcal{R}^{-1}(k) - \alpha(k)\bar{E}^{\mathrm{T}}(k)\bar{E}(k))^{-1}\begin{bmatrix} 0 & -I \end{bmatrix}^{\mathrm{T}} = 0,$$

$$\frac{\partial \mathcal{E}(k+1)}{\partial K_f(k)} = 2\mathcal{A}(k)(\mathcal{R}^{-1}(k) - \alpha(k)\bar{E}^{\mathrm{T}}(k)\bar{E}(k))^{-1}\begin{bmatrix} -\bar{\Theta}(k)C(k) & \bar{\Theta}(k)C(k) \end{bmatrix}^{\mathrm{T}}$$

$$- 2\sum_{i=1}^{q} \Pi_i^{12}(k)\mathrm{tr}[\Gamma_i(k)P(k)] + 2K_f(k)\sum_{i=1}^{q} \Pi_i^{22}(k)\mathrm{tr}[\Gamma_i(k)P(k)]$$

$$+ 2\sum_{j=1}^{m} \sigma_j(k)^2 K_f(k)C_j(k)P(k)C_j^{\mathrm{T}}(k) = 0. \qquad (10.47)$$

Therefore, $A_f(k)$ can be obtained by

$$A_f(k) = A(k) + (A(k) - K_f(k)\bar{\Theta}(k)C(k))\mathcal{E}(k)E^{\mathrm{T}}(k)$$

$$\times (\alpha^{-1}(k)I - E(k)P(k)E^{\mathrm{T}}(k))^{-1}E(k)$$

$$\times (I - (\mathcal{E}(k) - P(k))E^{\mathrm{T}}(k)(\alpha^{-1}(k)I - E(k)P(k)E^{\mathrm{T}}(k))^{-1}E(k))^{-1}$$

$$= A(k) + (A(k) - K_f(k)\bar{\Theta}(k)C(k))\mathcal{E}(k)$$

$$\times E^{\mathrm{T}}(k)E(k)(\alpha^{-1}(k)I - P(k)E^{\mathrm{T}}(k)E(k))^{-1}$$

$$\times (I - (\mathcal{E}(k) - P(k))E^{\mathrm{T}}(k)E(k)(\alpha^{-1}(k)I - E(k)P(k)E^{\mathrm{T}}(k))^{-1})^{-1}$$

$$= A(k) + (A(k) - K_f(k)\bar{\Theta}(k)C(k))\mathcal{E}(k)$$

$$\times E^{\mathrm{T}}(k)E(k)(\alpha^{-1}(k)I - \mathcal{E}(k)E^{\mathrm{T}}(k)E(k))^{-1}$$

$$
\begin{aligned}
&= A(k) + (A(k) - K_f(k)\bar{\Theta}(k)C(k))\mathcal{E}(k) \\
&\quad \times E^{\mathrm{T}}(k)(\alpha^{-1}(k)I - E(k)\mathcal{E}(k)E^{\mathrm{T}}(k))^{-1}E(k),
\end{aligned}
\tag{10.48}
$$

which is exactly the form in (10.43).

Next, by some tedious but straightforward manipulations, and taking (10.47) into consideration, we can find the parametric expression of parameter $K_f(k)$ as follows:

$$
K_f(k) = \Bigg( A(k)(\mathcal{E}^{-1}(k) - \alpha(k)E^{\mathrm{T}}(k)E(k))^{-1}C^{\mathrm{T}}(k)\bar{\Theta}(k)
$$
$$
+ \sum_{i=1}^{q} \Pi_i^{12}(k)\mathrm{tr}[\Gamma_i(k)\mathcal{P}(k)] \Bigg) \Omega^{-1}(k),
\tag{10.49}
$$

where

$$
\Omega(k) = \bar{\Theta}(k)C(k)(\mathcal{E}^{-1}(k) - \alpha(k)E^{\mathrm{T}}(k)E(k))^{-1}C^{\mathrm{T}}(k)\bar{\Theta}(k)
$$
$$
+ \sum_{i=1}^{q} \Pi_i(k)^{22}\mathrm{tr}[\Gamma_i(k)\mathcal{P}(k)] + \sum_{j=1}^{m} \sigma_j^2(k)C_j(k)\mathcal{P}(k)C_j^{\mathrm{T}}(k),
\tag{10.50}
$$

which is exactly the same form in (10.43) and (10.44). The proof is complete.    □

**Remark 10.3** *Theorem 10.3.1 proposed an algorithm for seeking the suboptimal filter parameters via solving a set of Riccati-like difference equations recursively. Moreover, notice that all three difference equations are in a form that can be solved forward in time; hence the proposed algorithm is suitable for online computing. The possible conservatism caused by introducing the scalars $\alpha(k)$ can be significantly reduced if we choose the proper values of $\alpha(k)$.*

## 10.4   Numerical Example

Consider a nonlinear stochastic system with the following parameters:

$$
A(k) = \begin{bmatrix} -0.26 & 0.15 \\ 0.4 - 0.25\sin(2k) & 0.14 \end{bmatrix}, \quad D(k) = \begin{bmatrix} 0.5 \\ 1 \end{bmatrix},
$$

$$
C(k) = \begin{bmatrix} 1.2 - 0.4\sin(0.1k) & 0 \\ 0 & 0.15 \end{bmatrix}, \quad L(k) = \begin{bmatrix} 0.12\sin(0.5k) & 0.25 \end{bmatrix},
$$

$$
H(k) = \begin{bmatrix} 0.65 \\ -0.75 \end{bmatrix}, \quad F(k) = \sin(0.5k), \quad E(k) = \begin{bmatrix} 0 & 0.13 \end{bmatrix}.
$$

The stochastic nonlinear functions are taken to be

$$f(x(k), k) = \begin{bmatrix} 0.4 \\ 0.5 \end{bmatrix} (0.3 \text{ sign}[x_1(k)]x_1(k)\varsigma_1(k) + 0.4 \text{ sign}[x_2(k)]x_2(k)\varsigma_2(k)),$$

$$g(x(k), k) = \begin{bmatrix} 0.6 \\ 0.8 \end{bmatrix} (0.3 \text{ sign}[x_1(k)]x_1(k)\varsigma_1(k) + 0.4 \text{ sign}[x_2(k)]x_2(k)\varsigma_2(k))$$

where $x_i(k)$ is the $i$th component of $x(k)$ and $\varsigma_i(k)$ is a zero mean independent Gaussian white noise process with unity covariances, which is also assumed to be independent from $\omega(k)$. It is easy to check that $f(x(k), k)$ and $g(x(k), k)$ satisfy

$$\mathbb{E}\left\{ \begin{bmatrix} f(x(k), k) \\ g(x(k), k) \end{bmatrix} \middle| x(k) \right\} = 0,$$

$$\mathbb{E}\left\{ \begin{bmatrix} f(x(k), k) \\ g(x(k), k) \end{bmatrix} [f^{\mathrm{T}}(x_j, j) \ g^{\mathrm{T}}(x_j, j)] \middle| x(k) \right\} = 0, \qquad k \neq j,$$

$$\mathbb{E}\left\{ \begin{bmatrix} f(x(k), k) \\ g(x(k), k) \end{bmatrix} [f^{\mathrm{T}}(x(k), k) \ g^{\mathrm{T}}(x(k), k)] \middle| x(k) \right\}$$

$$= \begin{bmatrix} 0.4 \\ 0.5 \\ 0.6 \\ 0.8 \end{bmatrix} \begin{bmatrix} 0.4 \\ 0.5 \\ 0.6 \\ 0.8 \end{bmatrix}^{\mathrm{T}} x^{\mathrm{T}}(k) \left( \begin{bmatrix} 0.3 \\ 0 \end{bmatrix} \begin{bmatrix} 0.3 \\ 0 \end{bmatrix}^{\mathrm{T}} + \begin{bmatrix} 0 \\ 0.4 \end{bmatrix} \begin{bmatrix} 0 \\ 0.4 \end{bmatrix}^{\mathrm{T}} \right) x(k).$$

Hence,

$$\Pi^{11}(k) = \begin{bmatrix} 0.16 & 0.20 \\ 0.20 & 0.25 \end{bmatrix}, \quad \Pi^{12}(k) = \begin{bmatrix} 0.24 & 0.32 \\ 0.30 & 0.40 \end{bmatrix},$$

$$\Pi^{22}(k) = \begin{bmatrix} 0.36 & 0.48 \\ 0.48 & 0.64 \end{bmatrix}, \quad \mu_1 = \begin{bmatrix} 0.3 \\ 0 \end{bmatrix}, \quad \mu_2 = \begin{bmatrix} 0 \\ 0.4 \end{bmatrix}.$$

Set the initial condition by $x_0 = [0.5 \quad 0]^{\mathrm{T}}$. In addition, we assume $\Theta_j(k)$ $(j = 1, 2)$ is time-invariant. In order to illustrate the effectiveness of our results for measurement degraded cases, we consider the following two cases:

**Case 1**: The probabilistic density functions of $\theta_1$ and $\theta_2$ in $[0, 1]$ described by

$$p_1(s_1) = \begin{cases} 0.8 & s_1 = 0 \\ 0.1 & s_1 = 0.5 \\ 0.1 & s_1 = 1 \end{cases} \quad \text{and} \quad p_2(s_2) = \begin{cases} 0.7 & s_2 = 0 \\ 0.2 & s_2 = 0.5 \\ 0.1 & s_2 = 1 \end{cases}$$

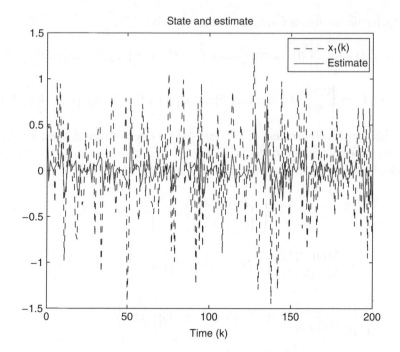

**Figure 10.1**    The state $x_1(k)$ and its estimate for Case 1

from which the expectations and variances can be easily calculated as $\bar{\theta}_1 = 0.15, \bar{\theta}_2 = 0.2, \sigma_1^2 = 0.1025$, and $\sigma_2^2 = 0.11$.

***Case 2***: The probabilistic density functions of $\theta_1$ and $\theta_2$ in $[0, 1]$ described by

$$p_1(s_1) = \begin{cases} 0, & s_1 = 0 \\ 0.1, & s_1 = 0.5 \\ 0.9, & s_1 = 1 \end{cases} \quad \text{and} \quad p_2(s_2) = \begin{cases} 0.1, & s_2 = 0 \\ 0.1, & s_2 = 0.5 \\ 0.8, & s_2 = 1 \end{cases}$$

with the expectations and variances obtained as $\bar{\theta}_1 = 0.95, \bar{\theta}_2 = 0.85, \sigma_1^2 = 0.0225$, and $\sigma_2^2 = 0.1025$.

We choose $\gamma = 1$ as the prescribed noise attenuation level. Selecting $\alpha(k) = 0.5$ and using Theorem 10.3.1 with the initial conditions

$$\hat{x}_0 = \begin{bmatrix} 0.5 \\ 0 \end{bmatrix}, \quad S_0 = \begin{bmatrix} 5 & 0 \\ 0 & 5 \end{bmatrix}, \quad \mathcal{P}_0 = \begin{bmatrix} 3 & 0 \\ 0 & 3 \end{bmatrix}, \quad \mathcal{E}_0 = \begin{bmatrix} 2 & 0 \\ 0 & 2 \end{bmatrix},$$

the filter parameters can be obtained by (10.43) via solving (10.41). The simulation results are shown in Figures 10.1 to 10.9, where we can see clearly that (1) the actual variances for the states stay below their upper bounds and (2) as the measurement degraded phenomenon becomes severe, both the variances and their upper bounds will increase. This means that the less true signal we can obtain from the measured output, the lower the estimate precision will be, which is reasonable.

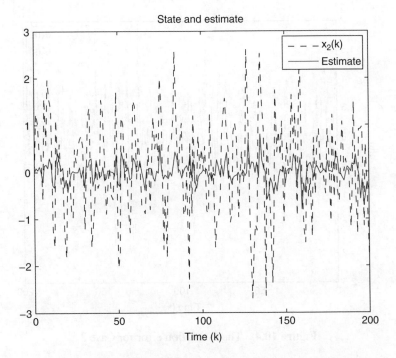

**Figure 10.2**    The state $x_2(k)$ and its estimate for Case 1

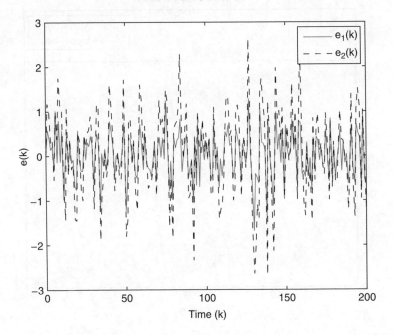

**Figure 10.3**    The estimation error for Case 1

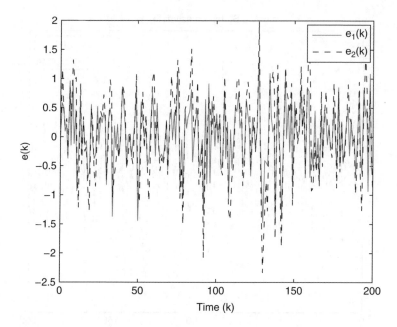

**Figure 10.4**    The estimation error for Case 2

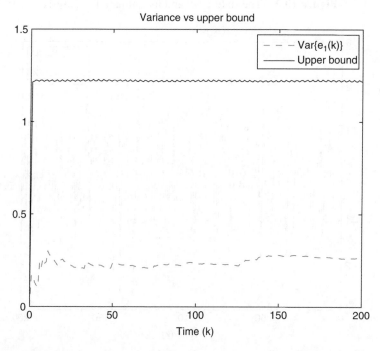

**Figure 10.5**    The comparison of the actual variance of $e_1(k)$ and its upper bound for Case 1

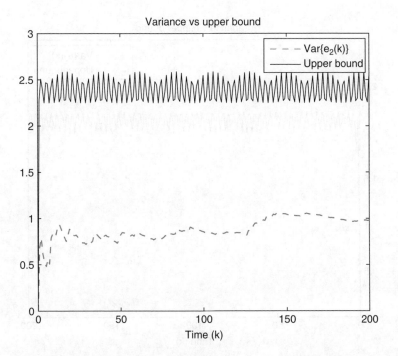

**Figure 10.6**  The comparison of the actual variance of $e_2(k)$ and its upper bound for Case 1

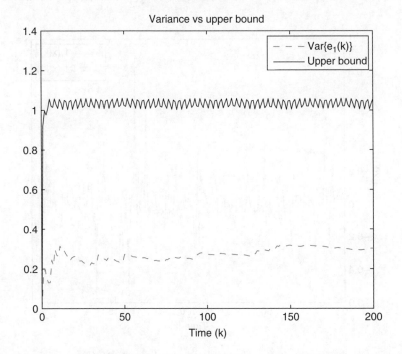

**Figure 10.7**  The comparison of the actual variance of $e_1(k)$ and its upper bound for Case 2

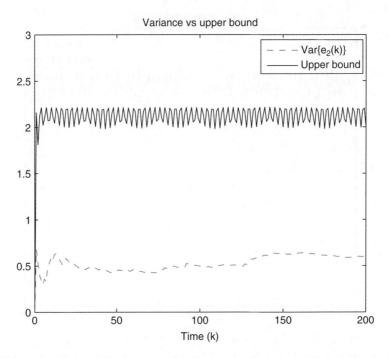

**Figure 10.8**    The comparison of the actual variance of $e_2(k)$ and its upper bound for Case 2

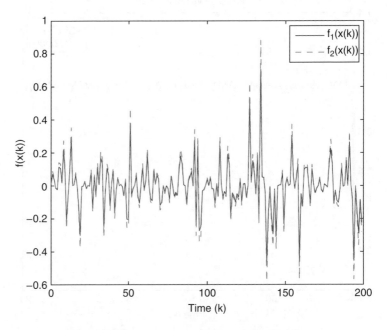

**Figure 10.9**    The stochastic nonlinearity $f(x(k))$

## 10.5 Summary

In this chapter, we have studied the robust filtering problem for a class of nonlinear stochastic systems with both $H_\infty$ noise attenuation level and filtering error variance constraints. The measurement degraded phenomenon has also been taken into consideration. A sufficient condition for the existence of a desired filter has been given in terms of the solvability of a set of Riccati-like difference equations. The simulation results show the effectiveness and applicability of the proposed algorithm. It is worth pointing out that one of the future research topics would be to study more general nonlinear systems. The corresponding results will appear in the near future.

# 11

# Mixed $H_2/H_\infty$ Control with Randomly Occurring Nonlinearities: The Finite-Horizon Case

As the stochastic systems have been playing more and more important roles in control theory and applications, a special case of the multi-objective design problem, the mixed $H_2/H_\infty$ control and filtering for stochastic systems have recently stirred much research interest and some initial results have been available in the literature. Several analysis and design methods including the linear matrix inequality approach and game theory strategy have been developed for such a problem and have made a significant impact to both theoretical advance and practical application of the mixed $H_2/H_\infty$ control and filtering for linear stochastic systems. Nevertheless, when it comes to the nonlinear stochastic system, the reported results are relatively fewer due mainly to the complicated dynamics of the nonlinearities. Moreover, most of the existing literature has focused on the *time-invariant* nonlinear stochastic systems despite the fact that real-world systems are usually time-varying, that is, all the system parameters are explicitly dependent on time. In fact, for a truly time-varying nonlinear stochastic system, we would be more interested in the transient behaviors over a *finite time-interval*, e.g., the mixed $H_2/H_\infty$ performance over a finite horizon. Unfortunately, for *time-varying* stochastic *nonlinear* systems, the $H_2/H_\infty$ control as well as filtering problems have not been adequately investigated yet.

On the other hand, in many real-world engineering practice, due to random abrupt changes of the environmental circumstances, the system may be corrupted by nonlinear disturbances in a *probabilistic* way. As opposed to the traditional deterministic

*Variance-Constrained Multi-Objective Stochastic Control and Filtering*, First Edition.
Lifeng Ma, Zidong Wang and Yuming Bo.
© 2015 John Wiley & Sons, Ltd. Published 2015 by John Wiley & Sons, Ltd.

disturbances, such nonlinear disturbances could be either existent or nonexistent at a specific time and the existence is governed by certain Bernoulli distributed white sequences with known probabilities. For example, in the maneuvering target tracking problem, it is quite common for the targets to maneuver in a nonlinear way, occasionally in order to get rid of the potential track. In this case, the randomly occurring nonlinearities (RONs) can be ideally applied to describe such a maneuvering behavior. Another case that can be perfectly interpreted by RONs is the networked control systems (NCSs). In NCSs, due to the limited bandwidth, the network inevitably suffers from the poor stability and variational reliability that give rise to probabilistic nonlinear disturbances. Such phenomena of RONS have already been verified by many experiments. Unfortunately, up to now, the phenomena of RONs have not gained adequate research attention, especially in the context of mixed $H_2/H_\infty$ control problems.

Summarizing the discussion made in this chapter, we aim to deal with the mixed $H_2/H_\infty$ control problem for a class of stochastic time-varying systems with nonlinearities. The nonlinearities are described by statistical means and could cover several kinds of well-studied nonlinearities as special cases. The occurrence of the addressed nonlinearities is governed by two sequences of Bernoulli distributed white sequences with known probabilities. Such nonlinearities are named as randomly occurring nonlinearities as they appear in a probabilistic way. The purpose of the problem under investigation is to design a controller such that the closed-loop system achieves the pre-specified $H_\infty$ disturbance attenuation level and, at the same time, the energy of system output is minimized when the worst-case disturbance happens. A sufficient condition is given for the existence of the desired controller by means of the solvability of certain coupled matrix equations. By resorting to the game theory approach, an algorithm is developed to obtain the controller gain at each sampling instant. A numerical example is presented to show the effectiveness and applicability of the proposed method. The main contributions of this chapter lie in the following aspects: (1) a unified framework is established to solve the mixed $H_2/H_\infty$ control problem when the system is both *time-varying* and *nonlinear*; (2) the controller design technique developed in this chapter is presented in terms of the solvability of certain Riccati-like difference equations, which have a much lower computation complexity than the commonly used LMI technique, and is therefore more convenient and effective for practical design; and (3) the phenomena of randomly occurring nonlinearity are considered in the mixed $H_2/H_\infty$ control problem. As a conclusion, the control problem considered in this chapter is significant in both the theoretical and practical senses.

The rest of the chapter is organized as follows: In Section 11.1, the mixed $H_2/H_\infty$ control problem for a class of discrete time-varying nonlinear stochastic systems is formulated. A stochastic Bounded Real Lemma (BRL) is given in Section 11.2. Section 11.3 proposes the state feedback design algorithm. In Section 11.4, a numerical example is presented to show the effectiveness of the proposed algorithm. Section 11.5 gives the summary.

## 11.1 Problem Formulation

Consider the following discrete-time nonlinear stochastic system defined on $k \in [0, N-1]$:

$$\begin{cases} x(k+1) = A(k)x(k) + B(k)u(k) + D(k)\omega(k) + \alpha(k)f(x(k), u(k), k), \\ \quad\quad y(k) = C(k)x(k) + \beta(k)g(x(k), u(k), k), \quad\quad x(0) = x_0, \end{cases} \quad (11.1)$$

where $x(k) \in \mathbb{R}^n$ is the system state, $u(k) \in \mathbb{R}^m$ is the control input, $y(k) \in \mathbb{R}^r$ is the system output, $\omega(k) \in \mathbb{R}^p$ is the external disturbance belonging to $l_2([0, N-1], \mathbb{R}^p)$, and $A(k), B(k), C(k)$, and $D(k)$ are time-varying matrices with appropriate dimensions.

For notation simplicity, let $h(k) \triangleq \left[ f^T(x(k), u(k), k) \quad g^T(x(k), u(k), k) \right]^T$. For all $x(k)$ and $u(k)$, $h(k)$ is assumed to satisfy:

$$\mathbb{E}\{h(k)|x(k)\} = 0, \quad (11.2)$$

$$\mathbb{E}\{h(k)h^T(j)|x(k)\} = 0, \quad k \neq j, \quad (11.3)$$

$$\mathbb{E}\{h(k)h^T(k)|x(k)\} = \sum_{i=1}^{q} \begin{bmatrix} \Pi_i^{11}(k) & \Pi_i^{12}(k) \\ * & \Pi_i^{22}(k) \end{bmatrix}$$

$$\times \mathbb{E}\{(x^T(k)\Gamma_i(k)x(k) + u^T(k)\Xi_i(k)u(k))\}, \quad (11.4)$$

where $\Pi_i^{11}(k) \geq 0$, $\Pi_i^{22}(k) \geq 0$, $\Gamma_i(k) \geq 0$, $\Xi_i(k) \geq 0$, and $\Pi_i^{12}(k)$ $(i = 1, 2, \ldots, q)$ are known matrices with compatible dimensions.

**Remark 11.1** *As discussed in Ref. [31], the nonlinear descriptions* (11.2) *to* (11.4) *can characterize several well-studied nonlinear stochastic systems, such as systems with state-dependent multiplicative noises, nonlinear systems with random sequences whose powers depend on a sector-bound nonlinear function of the state, nonlinear systems with a random sequence whose power depends on the sign of a nonlinear function of the state, to name just a few. It is worth pointing out, as in Ref. [149], that a mixed multi-objective controller has been designed for a type of discrete-time system with state-dependent noise, which is a special case of the stochastic systems we study in this chapter. Moreover, in this chapter, we also consider the phenomena of RONs, which will be illustrated in detail in the following section. Such a model is applied to describe the way in which the time-varying working conditions affect the system structure. Compared with Ref. [149], the model we study in this chapter is much closer to engineering practice, not only in the type of nonlinearities but also in the way these nonlinearities affect the system.*

**Remark 11.2** *It is well known that nonlinearities exist universally in engineering practice. A phenomenon that deserves particular attention is that in many real-time applications, such as target tracking and networked control systems (NCSs), the nonlinearities are not persistent all the time, but are intermittent stochastically. In other words, the nonlinearities occur at random time points in a probabilistic* *way. To be specific, let us take the NCS as an example. In practice, many kinds of nonlinearities occurring in networked control systems are caused by the unpredictable changes of network environments/conditions including network-induced congestion, network-induced delays, data packet dropouts, etc. Furthermore, the probability of the occurrence of such network-induced events can be estimated through statistical tests. Therefore, when the network operates well, there will be no negative effects from the environment on the NCSs and the random nonlinearities will not occur; otherwise one or some of the network-induced problems (e.g., delay, disorder, congestion, dropout) will give rise to the nonlinear disturbances/influences. Overall, the randomly changing network conditions would result in the random occurrence of nonlinearities. In the following, the Bernoulli distributed white sequences are exploited to illustrate such a randomly occurring phenomenon, i.e., the randomly occurring nonlinearities (RONs).*

The random variables $\alpha(k) \in \mathbb{R}$ and $\beta(k) \in \mathbb{R}$, which account for the phenomena of RONs, are assumed to be independent from $h(k)$ and take values of 0 and 1, with

$$\text{Prob}\{\alpha(k) = 1\} = \mu, \qquad \text{Prob}\{\alpha(k) = 0\} = 1 - \mu, \qquad (11.5)$$

$$\text{Prob}\{\beta(k) = 1\} = \nu, \qquad \text{Prob}\{\beta(k) = 0\} = 1 - \nu. \qquad (11.6)$$

Hence it can be easily obtained that

$$\mathbb{E}\{\alpha(k)\} = \mathbb{E}\{\alpha^2(k)\} = \mu \qquad (11.7)$$

and

$$\mathbb{E}\{\beta(k)\} = \mathbb{E}\{\beta^2(k)\} = \nu. \qquad (11.8)$$

**Remark 11.3** *The Bernoulli distributed white sequences $\alpha(k)$ and $\beta(k)$ are employed to describe the randomly occurring stochastic nonlinearities. Such a phenomenon often occurs with changes of the environmental circumstances, for example, random failures and repairs of the components, changes in the interconnections of subsystems, sudden environment changes, and modification of their types and/or intensity. Obviously, if $\alpha(k) = 0$ ($\beta(k) = 0$), there is no nonlinearity in the state (output) equation; if $\alpha(k) = 1$ ($\beta(k) = 1$), there is a nonlinear disturbance with first and second moments specified by (11.2) to (11.4).*

For system (11.1), given $\gamma > 0$, we consider the following performance index:

$$J_1 \triangleq \mathbb{E} \left\{ \sum_{k=0}^{N-1} \left( \|y(k)\|_{R(k)}^2 - \gamma^2 \|\omega(k)\|_{W(k)}^2 \right) \right\} - \gamma^2 \|x_0\|_{S(k)}^2, \qquad (11.9)$$

where $R(k) > 0$, $W(k) > 0$, and $S(k) > 0$ are weighting matrices.

In this chapter, our aim is to determine a state feedback control law $u(k) = K(k)x(k)$ such that:

(R1) For a pre-specified $H_\infty$ disturbance attenuation level $\gamma > 0$, $J_1 < 0$ holds.
(R2) When there exists a worst-case disturbance $\omega^*(k)$, the output energy defined by $J_2 \triangleq \mathbb{E}\left\{ \sum_{k=0}^{N-1} \|y(k)\|^2 \right\}$ is minimized.

## 11.2 $H_\infty$ Performance

In this section, we first analyze the $H_\infty$ performance for the time-varying stochastic system with randomly occurring nonlinearities. Then a sufficient condition is given for satisfying the pre-specified $H_\infty$ requirement. To this end, setting $u(k) = 0$, we obtain the corresponding unforced system for system (11.1) as follows:

$$\begin{cases} x(k + 1) = A(k)x(k) + D(k)\omega(k) + \alpha(k)f(x(k), k), \\ \quad\quad y(k) = C(k)x(k) + \beta(k)g(x(k), k), \quad\quad x(0) = x_0. \end{cases} \tag{11.10}$$

We now consider the following recursion:

$$Q(k) = A^T(k)Q(k + 1)A(k) + \sum_{i=1}^{q} \Gamma_i(k)(\text{tr}[\mu Q(k + 1)\Pi_i^{11}(k)] + \text{tr}[\nu R(k)\Pi_i^{22}(k)])$$

$$+ (D(k)^T Q(k + 1)A(k))^T \Theta^{-1}(k)(D(k)^T Q(k + 1)A(k))$$

$$+ C^T(k)R(k)C(k), \tag{11.11}$$

with $Q(N) = 0$ and

$$\Theta(k) \triangleq \gamma^2 W(k) - D^T(k)Q(k + 1)D(k). \tag{11.12}$$

**Lemma 11.2.1** *Consider the system (11.10) with given positive scalar $\gamma > 0$ and weighting matrices $W(k)$, $R(k)$, and $S(k)$. A necessary and sufficient condition for $J_1$ in (11.9) to be negative for all nonzero $(\{\omega(k)\}, x_0)$ is that there exists a solution $Q(k)$ to (11.11) such that $\Theta(k) > 0$ and $Q(0) < \gamma^2 S(k)$.*

*Proof.*
**Sufficiency**
By defining

$$\tilde{J}(k) \triangleq x^T(k + 1)Q(k + 1)x(k + 1) - x^T(k)Q(k)x(k) \tag{11.13}$$

we have

$$\mathbb{E}\left\{\tilde{J}(k)\right\} = \mathbb{E}\left\{ (A(k)x(k) + D(k)\omega(k) + \alpha(k)f(x(k), k))^T Q(k + 1) \right.$$

$$\times (A(k)x(k) + D(k)\omega(k) + \alpha(k)f(x(k), k))\} - x^T(k)Q(k)x(k)$$

$$= x^T(k)\left( A^T(k)Q(k + 1)A(k) - Q(k)\right.$$

$$+ \sum_{i=1}^{q} \Gamma_i(k) \mathrm{tr}[\mu Q(k+1)\Pi_i^{11}(k)] \Bigg) x(k)$$

$$+ 2x^{\mathrm{T}}(k)A^{\mathrm{T}}(k)Q(k+1)D(k)\omega(k)$$

$$+ \omega(k)^{\mathrm{T}}D^{\mathrm{T}}(k)Q(k+1)D(k)\omega(k). \tag{11.14}$$

Adding the following zero term

$$\|y(k)\|_{R(k)}^2 - \gamma^2 \|\omega(k)\|_{W(k)}^2 - (\|y(k)\|_{R(k)}^2 - \gamma^2 \|\omega(k)\|_{W(k)}^2) \tag{11.15}$$

to the right side of (11.14) and then taking the mathematical expectation results in

$$\mathbb{E}\left\{\tilde{J}(k)\right\}$$

$$= x^{\mathrm{T}}(k)\left(A(k)^{\mathrm{T}}Q(k+1)A(k) - Q(k) + C(k)^{\mathrm{T}}R(k)C(k)\right.$$

$$\left. + \sum_{i=1}^{q} \Gamma_i(k)\left(\mathrm{tr}[\mu Q(k+1)\Pi_i^{11}(k)] + \mathrm{tr}[\nu R(k)\Pi_i^{22}(k)]\right)\right)x(k) \tag{11.16}$$

$$+ 2x^{\mathrm{T}}(k)A^{\mathrm{T}}(k)Q(k+1)D(k)\omega(k)$$

$$+ \omega(k)^{\mathrm{T}}(-\gamma^2 W(k) + D^{\mathrm{T}}(k)Q(k+1)D(k))\omega(k)$$

$$- \mathbb{E}\left\{\|y(k)\|_{R(k)}^2 - \gamma^2\|\omega(k)\|_{W(k)}^2\right\}.$$

By applying the completing squares method, we have

$$\mathbb{E}\left\{\tilde{J}(k)\right\} = -(\omega(k) - \omega^*(k))^{\mathrm{T}}\Theta(k)(\omega(k) - \omega^*(k))$$

$$- \mathbb{E}\left\{\|y(k)\|_{R(k)}^2 - \gamma^2\|\omega(k)\|_{W(k)}^2\right\}, \tag{11.17}$$

where

$$\omega^*(k) = \Theta^{-1}(k)D^{\mathrm{T}}(k)Q(k+1)A(k)x(k). \tag{11.18}$$

Then, it follows that

$$\mathbb{E}\left\{\sum_{k=0}^{N-1}\tilde{J}(k)\right\} = \mathbb{E}\left\{x^{\mathrm{T}}(N)Q(N)x(N)\right\} - x_0^{\mathrm{T}}Q(0)x_0$$

$$= -\sum_{k=0}^{N-1}\left((\omega(k) - \omega^*(k))^{\mathrm{T}}\Theta(k)(\omega(k) - \omega^*(k))\right. \tag{11.19}$$

$$\left. + \mathbb{E}\left\{\|y(k)\|_{R(k)}^2 - \gamma^2\|\omega(k)\|_{W(k)}^2\right\}\right).$$

Since $\Theta(k) > 0$, $Q(N) = 0$, and $Q(0) - \gamma^2 S(k) < 0$, we obtain

$$
J_1 = \mathbb{E}\left\{ \sum_{k=0}^{N-1} \left( \|y(k)\|_{R(k)}^2 - \gamma^2 \|\omega(k)\|_{W(k)}^2 \right) \right\} - \gamma^2 \|x_0\|_{S(k)}^2
$$

$$
= x_0^{\mathrm{T}} Q(0) x_0 - \sum_{k=0}^{N-1} (\omega(k) - \omega^*(k))^{\mathrm{T}} \Theta(k)(\omega(k) - \omega^*(k)) - \gamma^2 \|x_0\|_{S(k)}^2 < 0. \quad (11.20)
$$

**Necessity**

First, considering (11.20), we select $\omega(k) = \omega^*(k)$ for $\forall k \in [0, N)$ and $x_0 \neq 0$. It is clear that $Q(0) < \gamma^2 S(k)$ if $J_1 < 0$.

We now proceed to prove that if $J_1 < 0$, then there exist $Q(k)$ solving the matrix equations (11.11) such that $\Theta(k) > 0$ is satisfied for all nonzero $(\{\omega(k)\}, x_0)$. In fact, if $\Theta(k)$ are nonsingular, we can always solve $Q(k)$ for (11.11) with the known final condition $Q(N) = 0$ recursively backward in time $k$. Therefore, the proof of necessity is equivalent to proving the following proposition:

$$
J_1 < 0 \Rightarrow \lambda_\kappa(\Theta(k)) > 0 \quad \forall k \in [0, N), \qquad \kappa = 1, 2, \dots, n. \quad (11.21)
$$

The rest of the proof is carried out by contradiction. The inverse proposition of (11.21) is that we can find, from $J_1 < 0$, at least one eigenvalue of $\Theta(k)$ that is either equal to 0 or negative at some time point $k^* \in [0, N)$. For simplicity, we denote such an eigenvalue of $\Theta(k)$ at time $k^*$ as $\lambda°(k^*) \triangleq \lambda°(\Theta(k^*))$. In the following, we shall use such a non-positive $\lambda°(k^*)$ to reveal that there exist certain $(\{\omega(k)\}, x_0) \neq 0$ such that $J_1 \geq 0$. In fact, if $\lambda°(k^*) < 0$, we can choose

$$
\omega(k) = \begin{cases} \omega^*(k^*) + \varphi°(k^*), & k = k^*, \\ \omega^*(k), & \text{otherwise}, \end{cases} \quad (11.22)
$$

where $\varphi°(k^*)$ is the eigenvector of $\Theta(k^*)$ with respect to $\lambda°(k^*)$. Hence, we have

$$
J_1 = -\sum_{k=0}^{N-1} (\omega(k) - \omega^*(k))^{\mathrm{T}} \Theta(k)(\omega(k) - \omega^*(k)) - x_0^{\mathrm{T}}(\gamma^2 S(k) - Q(0))x_0
$$

$$
= -\sum_{k=0, k \neq k^*}^{N-1} (\omega(k) - \omega^*(k))^{\mathrm{T}} \Theta(k)(\omega(k) - \omega^*(k)) \quad (11.23)
$$

$$
- x_0^{\mathrm{T}}(\gamma^2 S(k) - Q(0))x_0 - (\omega(k^*) - \omega^*(k^*))^{\mathrm{T}} \Theta(k^*)(\omega(k^*) - \omega^*(k^*))
$$

$$
= -(\varphi°(k^*))^{\mathrm{T}} \Theta(k^*)\varphi°(k^*) - x_0^{\mathrm{T}}(\gamma^2 S(k) - Q(0))x_0
$$

$$
= -\lambda°(k^*)\|\varphi°(k^*)\|^2 - x_0^{\mathrm{T}}(\gamma^2 S(k) - Q(0))x_0.
$$

We can see from the above that if $x_0$ is chosen such that $\|x_0\|^2$ is sufficiently small, then $J_1 > 0$ is true for all $(\{\omega(k)\}, x_0) \neq 0$, which contradicts the condition $J_1 < 0$. Therefore, all the eigenvalues of $\Theta(k)$ should be non-negative.

Next, if $\lambda^\circ(k^*) = 0$, we choose $x_0 = 0$ and

$$
\omega(k) = \begin{cases} \varphi^\circ(k^*), & k = k^*, \\ \omega^*(k), & k^* < k \le N - 1, \\ 0, & 0 \le k < k^*. \end{cases}
\tag{11.24}
$$

Then

$$
\begin{aligned}
J_1 = \mathbb{E} &\left\{ \sum_{k=0}^{k^*-1} \left( \|y(k)\|^2 - \gamma^2 \|\omega(k)\|^2 \right) \right\} - x_0^{\mathrm{T}}(\gamma^2 S(k) - Q(0))x_0 \\
&- \sum_{k=k^*+1}^{N-1} (\omega(k) - \omega^*(k))^{\mathrm{T}} \Theta(k)(\omega(k) - \omega^*(k)) \\
&- (\omega(k^*) - \omega^*(k^*))^{\mathrm{T}} \Theta(k^*)(\omega(k^*) - \omega^*(k^*)).
\end{aligned}
\tag{11.25}
$$

Since $x_0 = 0$ and $\omega(k) = 0$ when $0 \le k < k^*$, we can see from (11.10) that $\mathbb{E}\left\{ \|y_k\|^2 \right\} = 0$. Therefore,

$$
J_1 = -(\varphi^\circ(k^*))^{\mathrm{T}} \Theta(k^*) \varphi^\circ(k^*) = -\lambda^\circ(k^*) \| \varphi^\circ(k^*) \|^2 = 0,
\tag{11.26}
$$

which also contradicts the condition $J_1 < 0$. Therefore, we can conclude that $\Theta(k) > 0$. The proof is now complete.              □

So far, we have analyzed the system's $H_\infty$ performance in terms of the solvability of a backward Riccati-like equation. The presented lemma will enable us to develop an algorithm to obtain the multi-objective controller parameters, which will be illustrated in detail in the following section.

## 11.3 Mixed $H_2/H_\infty$ Controller Design

In this section, we shall design a state feedback controller for system (11.1) such that the design objectives (R1) and (R2) are satisfied simultaneously. It turns out that the solvability of the addressed mixed $H_2/H_\infty$ control problem can be determined by the solvability of certain coupled backwards Riccati-type matrix equations. A computational algorithm is proposed in the sequel to solve such a set of matrix equations.

### 11.3.1 State-Feedback Controller Design

Implementing $u(k) = K(k)x(k)$ to system (11.1), we consider the following closed-loop stochastic nonlinear system over the finite horizon $[0, \ N - 1]$:

$$
\begin{cases} x(k+1) = (A(k) + B(k)K(k))x(k) + D(k)\omega(k) + \alpha(k)f(x(k), K(k)x(k), k), \\ y(k) = C(k)x(k) + \beta(k)g(x(k), K(k)x(k), k), \qquad x(0) = x_0. \end{cases}
\tag{11.27}
$$

The following theorem presents a sufficient condition for the closed-loop system (11.27) to guarantee both the $H_\infty$ and $H_2$ performances at the same time.

**Theorem 11.3.1** *For a pre-specified $H_\infty$ performance index $\gamma > 0$ and properly chosen weighting matrices $W(k)$, $R(k)$, and $S(k)$, there exists a state feedback controller $u(k) = K(k)x(k)$ for system (11.27) such that the requirements (R1) and (R2) are achieved simultaneously if the following coupled matrix equations have solutions $Q(k)$ and $P(k)$ ($0 < k \leq N - 1$):*

$$
\begin{cases}
Q(k) = (A(k) + B(k)K(k))^{\mathrm{T}}Q(k+1)(A(k) + B(k)K(k)) + C^{\mathrm{T}}(k)R(k)C(k) \\
\qquad + \sum_{i=1}^{q} \left(\Gamma_i(k) + K(k)^{\mathrm{T}}\Xi_i(k)K(k)\right) \left(\mathrm{tr}[\mu Q(k+1)\Pi_i^{11}(k)] + \mathrm{tr}[\nu R(k)\Pi_i^{22}(k)]\right) \\
\qquad + (D^{\mathrm{T}}(k)Q(k+1)(A(k) + B(k)K(k)))^{\mathrm{T}} \\
\qquad \times \Theta^{-1}(k)\left(D^{\mathrm{T}}(k)Q(k+1)(A(k) + B(k)K(k))\right), \\
\Theta(k) = \gamma^2 W(k) - D(k)^{\mathrm{T}}Q(k+1)D(k) > 0, \\
Q(N) = 0,\, Q(0) < \gamma^2 S(k),
\end{cases}
\tag{11.28}
$$

$$
\begin{cases}
P(k) = \bar{A}^{\mathrm{T}}(k)P(k+1)\bar{A}(k) + C^{\mathrm{T}}(k)C(k) \\
\qquad + \sum_{i=1}^{q} \Gamma_i(k)\left(\mathrm{tr}[\mu P(k+1)\Pi_i^{11}(k)] + \mathrm{tr}[\nu\Pi_i^{22}(k)]\right) \\
\qquad - \bar{A}^{\mathrm{T}}(k)P(k+1)B(k)\Omega^{-1}(k)B^{\mathrm{T}}(k)P(k+1)\bar{A}(k), \\
\Omega(k) = \sum_{i=1}^{q} \Xi_i(k)(\mathrm{tr}[\mu P(k+1)\Pi_i(k)^{11} + \mathrm{tr}[\nu\Pi_i^{22}(k)]) + B^{\mathrm{T}}(k)P(k+1)B(k), \\
P(N) = 0,
\end{cases}
\tag{11.29}
$$

*where $\bar{A}(k) \triangleq A(k) + D(k)T(k)$ and the matrix $T(k)$ and the feedback gain $K(k)$ can be recursively computed by*

$$
\begin{cases}
T(k) = \Theta^{-1}(k)\left(D^{\mathrm{T}}(k)Q(k+1)(A(k) + B(k)K(k))\right), \\
K(k) = -\Omega^{-1}(k)B^{\mathrm{T}}(k)P(k+1)(A(k) + D(k)T(k)).
\end{cases}
\tag{11.30}
$$

*Proof.* According to Lemma 11.2.1, the closed-loop system (11.27) achieves $J_1 < 0$ if there exists a solution $Q(k)$ to equation (11.28) such that $\Theta(k) > 0$. If $Q(k)$ exists, the worst-case disturbance can be expressed by $\omega^*(k) = T(k)x(k)$. Moreover, $J_{1\max} = x_0^{\mathrm{T}}Q(0)x_0$ when $\omega(k) = \omega^*(k)$.

In the following stage, under the situation of a worst-case disturbance, we shall try to find out the parametric expression of $K(k)$ that is able to minimize the output energy

$J_2$. To this end, when the worst-case disturbance happens, the original system (11.1) can be formulated by

$$\begin{cases} x(k+1) = (A(k) + D(k)T(k))x(k) + B(k)u(k) + \alpha(k)f(x(k), u(k), k), \\ y(k) = C(k)x(k) + \beta(k)g(x(k), u(k), k), \qquad x(0) = x_0. \end{cases} \tag{11.31}$$

Similar to the derivation of Lemma 11.2.1, we first define

$$\tilde{J}_{2k} \triangleq x^T(k+1)P(k+1)x(k+1) - x^T(k)P(k)x(k). \tag{11.32}$$

Then we have

$$\begin{aligned} \mathbb{E}\left\{\tilde{J}_{2k}\right\} =& \mathbb{E}\left\{(\bar{A}(k)x(k) + B(k)u(k) + \alpha(k)f(x(k), u(k), k))^T P(k+1)\right. \\ & \times (\bar{A}(k)x(k) + B(k)u(k) + \alpha(k)f(x(k), u(k), k))\left.\right\} - x^T(k)P(k)x(k) \\ =& x^T(k)\left(\bar{A}^T(k)P(k+1)\bar{A}(k) - P(k) + \sum_{i=1}^{q}\Gamma_i(k)\mathrm{tr}[\mu P(k+1)\Pi_i^{11}(k)]\right)x(k) \\ & + u^T(k)\left(B^T(k)P(k+1)B(k) + \sum_{i=1}^{q}\Xi_i(k)\mathrm{tr}[\mu P(k+1)\Pi_i^{11}(k)]\right)u(k) \\ & + 2x^T(k)\bar{A}^T(k)P(k+1)B(k)u(k). \end{aligned} \tag{11.33}$$

Then it follows that

$$\begin{aligned} & \mathbb{E}\left\{\tilde{J}_{2k} + \|y(k)\|^2 - \|y(k)\|^2\right\} \\ =& x^T(k)\left(\bar{A}(k)^T P(k+1)\bar{A}(k) - P(k) + C(k)^T C(k)\right. \\ & \left. + \sum_{i=1}^{q}\Gamma_i(k)\mathrm{tr}[\mu P(k+1)\Pi_i^{11}(k)] + \sum_{i=1}^{q}\Gamma_i(k)\mathrm{tr}[\nu\Pi_i^{22}(k)]\right)x(k) \\ & + u^T(k)\left(B(k)^T P(k+1)B(k)\right. \\ & \left. + \sum_{i=1}^{q}\Xi_i(k)\mathrm{tr}[\mu P(k+1)\Pi_i^{11}(k)] + \sum_{i=1}^{q}\Xi_i(k)\mathrm{tr}[\nu\Pi_i^{22}(k)]\right)u(k) \\ & + 2x^T(k)\bar{A}^T(k)P(k+1)B(k)u(k) - \mathbb{E}\left\{\|y(k)\|^2\right\}. \end{aligned} \tag{11.34}$$

Completing the squares of $u(k)$, it is obtained that

$$\begin{aligned} & \mathbb{E}\left\{\tilde{J}_{2k} + \|y(k)\|^2 - \|y(k)\|^2\right\} \\ =& \mathbb{E}\left\{(u(k) - u^*(k))^T\Omega(k)(u(k) - u^*(k)) - \|y(k)\|^2\right\}, \end{aligned} \tag{11.35}$$

where $\Omega(k)$ is defined in (11.29) and $u^*(k) = K(k)x(k)$ with $K(k)$ defined in (11.30). Furthermore, we have

$$
\begin{aligned}
J_2 &= \mathbb{E}\left\{ \sum_{k=0}^{N-1} \|y(k)\|^2 \right\} \\
&= \mathbb{E}\left\{ \sum_{k=0}^{N-1} (u(k) - u^*(k))^{\mathrm{T}}\Omega(k)(u(k) - u^*(k)) \right\} \\
&\quad + x_0^{\mathrm{T}}P(0)x_0 - x^{\mathrm{T}}(N)P(N)x(N).
\end{aligned}
\tag{11.36}
$$

Since $P(N) = 0$, it is easy to see that, when $u(k) = u^*(k)$, the following relationship

$$
J_{2\min} = x_0^{\mathrm{T}}P(0)x_0 \tag{11.37}
$$

is true, which means that

$$
J_1(u^*(k), \omega^*(k)) \geq J_1(u^*(k), \omega(k)), \quad J_2(u^*(k), \omega^*(k)) \leq J_2(u(k), \omega^*(k)). \tag{11.38}
$$

We can now conclude that $(u^*(k), \omega^*(k))$ is a Nash equilibrium we intend to find. The proof is complete. $\qquad\square$

Notice that the solvability of the addressed controller design problem is presented as the feasibility of four coupled matrix-valued equations. In the following subsection, we propose a numerical recursive algorithm to obtain the value of the required controller gain $K(k)$.

### 11.3.2 Computational Algorithm

The following algorithm shows how to solve the addressed problem with known parameters and terminal values $Q(N)$ and $P(N)$.

**Algorithm of Mixed $H_2/H_\infty$ Control Problem**

- *Step 1.* Select the proper initial values. Set $k = N - 1$; then $Q(k + 1) = Q(N)$ and $P(k + 1) = P(N)$ are known.
- *Step 2.* Calculate $\Theta(k)$ and $\Omega(k)$ with known $Q(k + 1)$ and $P(k + 1)$ via the second equation of (11.28) and (11.29) respectively.
- *Step 3.* With the obtained $\Theta(k)$ and $\Omega(k)$, solve the following coupled matrix equations:

$$
\begin{cases}
T(k) = \Theta^{-1}(k)\left(D^{\mathrm{T}}(k)Q(k + 1)(A(k) + B(k)K(k))\right), \\
K(k) = -\Omega(k)^{-1}B^{\mathrm{T}}(k)P(k + 1)(A(k) + D(k)T(k))
\end{cases}
\tag{11.39}
$$

to obtain the values of $T(k)$ and $K(k)$.

- *Step 4.* Solve the first equation of (11.28) and (11.29) to get $Q(k)$ and $P(k)$ respectively.
- *Step 5.* If $k = 0$, EXIT. Else, set $k = k - 1$ and go to Step 2.

**Remark 11.4** *By the recursive algorithm developed above, we can obtain the controller gain $K(k)$ step by step at every sampling instant. However, in Step 2, one needs to check whether the obtained $\Theta(k)$ is positive definite or not. If not, then the addressed problem is not solvable. Moreover, in Step 5, when the recursion is finished, we should also check if the initial condition $Q(0) < \gamma^2 S(k)$ is satisfied with the obtained $Q(0)$. If not, the addressed problem is not solvable as well, and then we need to change the pre-specified parameter values and repeat the iterative algorithm again.*

**Remark 11.5** *Unfortunately, so far, there have been very few satisfactory results on the convergence of the proposed algorithm. Actually the algorithm relies largely on the system parameters and the selection of the initial values. If, at a certain time instant, the values of the controller parameters do not exist, we can either change the initial values or relax the constraints on the pre-specified performance indices in order to obtain a feasible solution. It is worth pointing out that one of our possible research topics in future is to develop a more applicable numerical algorithm than the one given in this chapter.*

## 11.4   Numerical Example

Let us use the networked control systems as the background to illustrate the numerical example. There are different ways to define the quality-of-service (QoS) for the NCS. In this chapter, we take into account two of the most popular QoS measures: (1) the point-to-point network allowable data dropout rate that is used to indicate the probability of data packet dropout in data transmission leading to the randomly occurring nonlinearities and (2) the point-to-point network throughput that is used to indicate how fast the signal can be sampled and sent as a packet through the network.

For the system considered in this chapter, the sampling period $h$ and the data dropout rate $\rho$ determine the control performance. We assume that the data are single-packet transmitted and different data packet have the same length $L$, and the network throughput distributed by the packet scheduler is $Q_{i(k)}$ in $\mathrm{T} \in [i_k h, i(k+1)h)$. The network allowable data dropout rate is related to the packet scheduler, backlog controller, and algorithm complex of loss dropper policy. In this chapter, we use the stochastic variables $\alpha(k)$ and $\beta(k)$ to quantify the randomly occurring nonlinearities $f(x(k), u(k), k)$ and $g(x(k), u(k), k)$. On the other hand, the sampling period $h$ is decided by the network throughput $Q_{i(k)}$ and the number of sensors. A small sampling period can have good control performance, but can induce network congestion and improve the data dropout rate.

Since there has been a rich body of literature studying the appropriate sampling method [150], in this chapter we assume that: (1) the sensors are clock-driven, the controller and actuators are event-driven, and the network exists only between sensor and controller; (2) data, either from measurement or from control, are transmitted with a single packet and full-state variables are available for measurements; (3) for the case of nonorder sequences, the time stamping technique is applied to choose the latest message; and (4) a sampled-data model can be obtained through online measurement such as sending probing data packet to measure network characteristics and QoS scheduling.

Now consider the system (11.1) that describes a special type of nonlinear stochastic model where the network-induced nonlinearities occur in a probabilistic way. Let the stochastic nonlinearities $f(x(k), u(k), k)$ and $g(x(k), u(k), k)$ have the following bilinear form:

$$f(x(k), u(k), k) = \sum_{i=1}^{2} \begin{bmatrix} a_i(k) \\ b_i(k) \end{bmatrix} (l_i(k)x_1(k)\zeta_1(k) + m_i(k)x_2(k)\zeta_2(k) + n_i(k)u(k)\zeta_3(k)),$$

$$g(x(k), u(k), k) = \sum_{i=1}^{2} \varrho_i(k)(l_i(k)x_1(k)\zeta_1(k) + m_i(k)x_2(k)\zeta_2(k) + n_i(k)u(k)\zeta_3(k)),$$

where $x_1(k)$ and $x_2(k)$ stand for the first and second components of the system state respectively and $\zeta_j(k)$ ($j = 1, 2, 3$) represents three mutually independent Gaussian white noise sequence; $a_i(k)$, $b_i(k)$, $\varrho_i(k)$, $l_i(k)$, $m_i(k)$, and $n_i(k)$ are time-varying coefficients.

Let the exogenous disturbance input be $\omega(k) = \exp(-k/35) \times n(k)$, where $n(k)$ is uniformly distributed over $[-0.5, 0.5]$. The probability is assumed to be $\delta = 0.9$.

Now consider the following discrete time-varying nonlinear stochastic system with time-varying parameters as follows:

For $k = 0$,

$$A(0) = \begin{bmatrix} 0.5 & -0.1 \\ -0.3 & 0.07 \end{bmatrix}, \qquad B(0) = \begin{bmatrix} 0.2 \\ -0.6 \end{bmatrix}, \qquad D(0) = \begin{bmatrix} 0.2 \\ -0.8 \end{bmatrix},$$

$$H(0) = 0.1, \qquad S(0) = I, \qquad C(0) = \begin{bmatrix} 0.5 & -0.6 \end{bmatrix},$$

$$R(0) = I, \qquad W(0) = I,$$

$$\mu = 0.16, \qquad v = 0.25,$$

$$a_1(0) = 0.6324, \qquad a_2(0) = 0, \qquad b_1(0) = 0, \qquad b_2(0) = 0.7746,$$

$$\varrho_1(0) = 0.6, \qquad \varrho_2(0) = 0.8, \qquad l_1(0) = l_2(0) = 0.4472,$$

$$m_1(0) = m_2(0) = 0.7071, \qquad n_1(0) = 0.7, \qquad n_2(0) = 0.5477.$$

Hence, we can easily obtain that

$$\Gamma_1(0) = \Gamma_2(0) = \begin{bmatrix} 0.2 & 0 \\ 0 & 0.5 \end{bmatrix}, \qquad \Pi_1^{11}(0) = \begin{bmatrix} 0.4 & 0 \\ 0 & 0 \end{bmatrix},$$

$$\Pi_2^{11}(0) = \begin{bmatrix} 0 & 0 \\ 0 & 0.6 \end{bmatrix}, \qquad \Pi_1^{22}(0) = 0.36, \qquad \Pi_2^{22}(0) = 0.64,$$

$$\Xi_1(0) = 0.49, \qquad \Xi_2(0) = 0.3.$$

For $k = 1$,

$$A(1) = \begin{bmatrix} 0.4 & -0.08 \\ -0.1 & 0.05 \end{bmatrix}, \qquad B(1) = \begin{bmatrix} 0.3 \\ -0.6 \end{bmatrix}, \qquad D(1) = \begin{bmatrix} 0.1 \\ -0.5 \end{bmatrix},$$

$$H(1) = 0.1, \qquad S(1) = I, \qquad C(1) = \begin{bmatrix} 0.3 & -0.1 \end{bmatrix},$$

$$R(1) = I, \qquad W(1) = I,$$

$$\mu = 0.16, \qquad v = 0.25,$$

$$a_1(1) = 0.8367, \qquad a_2(1) = 0, \qquad b_1(1) = 0, \qquad b_2(1) = 0.4472,$$

$$\varrho_1(1) = 0.3162, \qquad \varrho_2(1) = 0.5477, \qquad l_1(1) = l_2(1) = 0.3162,$$

$$m_1(1) = m_2(1) = 0.5477, \qquad n_1(1) = n_2(1) = 0.3162.$$

Hence, we can easily obtain that

$$\Gamma_1(1) = \Gamma_2(1) = \begin{bmatrix} 0.1 & 0 \\ 0 & 0.3 \end{bmatrix}, \qquad \Pi_1^{11}(1) = \begin{bmatrix} 0.7 & 0 \\ 0 & 0 \end{bmatrix},$$

$$\Pi_2^{11}(1) = \begin{bmatrix} 0 & 0 \\ 0 & 0.2 \end{bmatrix}, \qquad \Pi_1^{22}(1) = 0.1, \qquad \Pi_2^{22}(1) = 0.3,$$

$$\Xi_1(1) = 0.1, \qquad \Xi_2(1) = 0.1.$$

For $k = 2$,

$$A(2) = \begin{bmatrix} 0.2 & -0.05 \\ -0.1 & 0.08 \end{bmatrix}, \qquad B(2) = \begin{bmatrix} 0.03 \\ -0.5 \end{bmatrix}, \qquad D(2) = \begin{bmatrix} 0.05 \\ -0.8 \end{bmatrix},$$

$$H(2) = 0.1, \qquad S(2) = I, \qquad C(2) = \begin{bmatrix} 0.1 & -0.2 \end{bmatrix},$$

$$R(2) = I, \qquad W(2) = I,$$

$$\mu = 0.16, \qquad v = 0.25,$$

$$a_1(2) = 0.8944, \qquad a_2(2) = 0, \qquad b_1(2) = 0, \qquad b_2(2) = 0.7071,$$

$$\varrho_1(2) = 0.8944, \qquad \varrho_2(2) = 0.7071, \qquad l_1(2) = l_2(2) = 0.4472,$$

$$m_1(2) = m_2(2) = 0.3162, \qquad n_1(2) = n_2(2) = 0.6324.$$

Hence, we can easily obtain that

$$\Gamma_1(2) = \Gamma_2(2) = \begin{bmatrix} 0.2 & 0 \\ 0 & 0.1 \end{bmatrix}, \qquad \Pi_1^{11}(2) = \begin{bmatrix} 0.8 & 0 \\ 0 & 0 \end{bmatrix},$$

$$\Pi_2^{11}(2) = \begin{bmatrix} 0 & 0 \\ 0 & 0.5 \end{bmatrix}, \qquad \Pi_1^{22}(2) = 0.8, \qquad \Pi_2^{22}(2) = 0.5,$$

$$\Xi_1(2) = 0.4, \qquad \Xi_2(2) = 0.4.$$

For $k = 3$,

$$A(3) = \begin{bmatrix} 0.3 & -0.1 \\ -0.1 & 0.2 \end{bmatrix}, \qquad B(3) = \begin{bmatrix} 0.1 \\ -0.2 \end{bmatrix}, \qquad D(3) = \begin{bmatrix} 0.03 \\ -0.5 \end{bmatrix},$$

$$H(3) = 0.1, \qquad S(3) = I, \qquad C(3) = \begin{bmatrix} 0.2 & -0.5 \end{bmatrix},$$

$$R(3) = I, \qquad W(3) = I,$$

$$\mu = 0.16, \qquad v = 0.25,$$

$$a_1(3) = 0.7071, \qquad a_2(3) = 0, \qquad b_1(3) = 0, \qquad b_2(3) = 0.7071,$$

$$\varrho_1(3) = 0.6324, \qquad \varrho_2(3) = 0.7746, \qquad l_1(3) = l_2(3) = 0.3162,$$

$$m_1(3) = m_2(3) = 0.3162, \qquad n_1(3) = 0.4472, \qquad n_2(3) = 0.5477.$$

Hence, we can easily obtain that

$$\Gamma_1(3) = \Gamma_2(3) = \begin{bmatrix} 0.1 & 0 \\ 0 & 0.1 \end{bmatrix}, \qquad \Pi_1^{11}(3) = \begin{bmatrix} 0.5 & 0 \\ 0 & 0 \end{bmatrix},$$

$$\Pi_2^{11}(3) = \begin{bmatrix} 0 & 0 \\ 0 & 0.5 \end{bmatrix}, \qquad \Pi_1^{22}(3) = 0.4, \qquad \Pi_2^{22}(3) = 0.6,$$

$$\Xi_1(3) = 0.2, \qquad \Xi_2(3) = 0.3.$$

Choosing $\gamma = 1$, we apply the *Algorithm of Mixed $H_2/H_\infty$ Control Problem* from the instant $k = 3$ to $k = 0$ with $Q(4) = 0$ and $P(4) = 0$:

- Set $k = 3$.
- Then we obtain

$$\Theta(3) = 1, \qquad \Omega(3) = 0.05.$$

- With obtained $\Theta(3)$ and $\Omega(3)$, we have

$$T(3) = \begin{bmatrix} 0 & 0 \end{bmatrix}, \qquad K(3) = \begin{bmatrix} 0 & 0 \end{bmatrix}.$$

- With known $\Theta(3)$, $\Omega(3)$, $T(3)$, and $K(3)$, we have

$$Q(3) = \begin{bmatrix} 0.0650 & -0.1000 \\ -0.1000 & 0.2750 \end{bmatrix}, \qquad P(3) = \begin{bmatrix} 0.0650 & -0.1000 \\ -0.1000 & 0.2750 \end{bmatrix}.$$

By similar recursions, it can be obtained that

$$\Theta(2) = 0.8158,$$

$$\Omega(2) = 0.2139,$$

$$T(2) = \begin{bmatrix} -0.0161 & 0.0091 \end{bmatrix}, \quad K(2) = \begin{bmatrix} -0.4518 & 0.2559 \end{bmatrix},$$

$$Q(2) = \begin{bmatrix} 0.11222 & -0.0373 \\ -0.0373 & 0.0852 \end{bmatrix}, \quad P(2) = \begin{bmatrix} 0.0840 & -0.0203 \\ -0.0213 & 0.0756 \end{bmatrix},$$

$$\Theta(1) = 0.9738, \quad \Omega(1) = 0.0758,$$

$$T(1) = \begin{bmatrix} 0.0068 & -0.0075 \end{bmatrix}, \quad K(1) = \begin{bmatrix} -0.2709 & 0.0769 \end{bmatrix},$$

$$Q(1) = \begin{bmatrix} 0.1144 & -0.0327 \\ -0.0327 & 0.0452 \end{bmatrix}, \quad P(1) = \begin{bmatrix} 0.0683 & -0.0186 \\ -0.0198 & 0.0402 \end{bmatrix},$$

$$\Theta(0) = 0.9560, \quad \Omega(0) = 0.1509,$$

$$T(0) = \begin{bmatrix} 0.0336 & -0.0073 \end{bmatrix}, \quad K(0) = \begin{bmatrix} -0.1461 & 0.0291 \end{bmatrix},$$

$$Q(0) = \begin{bmatrix} 0.3401 & -0.3078 \\ -0.3078 & 0.4925 \end{bmatrix}, \quad P(0) = \begin{bmatrix} 0.2870 & -0.2964 \\ -0.2967 & 0.4883 \end{bmatrix}.$$

## 11.5 Summary

In this chapter, the mixed $H_2/H_\infty$ controller design problem has been dealt with for a class of nonlinear stochastic systems with randomly occurring nonlinearities that are characterized by two Bernoulli distributed white sequences with known probabilities. The stochastic nonlinearities addressed cover several well-studied nonlinearities in the literature. For the multi-objective controller design problem, a stochastic-type bounded real lemma has been derived. With the bounded real lemma, the sufficient condition of the solvability of the mixed $H_2/H_\infty$ control problem has been established by means of the solvability of four coupled matrix-valued equations. Then, a recursive algorithm has been developed to obtain the value of the feedback controller step by step at every sampling instant. A numerical example has been given to show the effectiveness of the proposed method.

# 12

# Mixed $H_2/H_\infty$ Control with Markovian Jump Parameters and Probabilistic Sensor Failures: The Finite-Horizon Case

Recently, quite a lot of research interest has been devoted to Markovian jump systems (MJSs) whose dynamics may experience abrupt changes in their structure and parameters caused by phenomena such as component failures or repairs, changing subsystem interconnections, and environmental disturbance. It is worth mentioning that most of the literature relevant to MJSs have been about linear systems, and the respective results for nonlinear Markovian jump systems have been scattered. Several methods including the linear matrix inequality approach and the T-S fuzzy algorithm have been employed to solve the analysis and design problem for Markovian jump systems. However, it should be pointed out that most of the algorithms proposed in the existing literature are applicable to *time-invariant* systems only. In fact, the majority of real-world engineering systems exhibit the *time-varying* nature and the desired control performance index is set over a finite horizon to reflect the transient system behavior. Nevertheless, the finite-horizon control problem for general time-varying nonlinear Markovian jump systems in the presence of positivistic sensor failures has yet not been investigated, due probably to the mathematical complexity.

Motivated by the above discussion, in this chapter it is our objective to deal with the mixed $H_2/H_\infty$ control problem over a finite horizon for a class of nonlinear Markovian jump systems with both stochastic nonlinearities and probabilistic sensor failures. The stochastic nonlinearities described by statistical means could cover several types of well-studied nonlinearities, and the failure probability for each sensor is governed by an individual random variable satisfying a certain probability distribution over a given

*Variance-Constrained Multi-Objective Stochastic Control and Filtering*, First Edition.
Lifeng Ma, Zidong Wang and Yuming Bo.
© 2015 John Wiley & Sons, Ltd. Published 2015 by John Wiley & Sons, Ltd.

interval for each mode. The purpose of the addressed problem is to design a controller such that the closed-loop system achieves the expected $H_2$ performance requirements with a guaranteed $H_\infty$ disturbance attenuation level. The solvability of the addressed control problem is expressed as the feasibility of certain coupled matrix equations. The controller gain at each time instant $k$ can be obtained by solving the corresponding set of matrix equations. A numerical example is given to illustrate the effectiveness and applicability of the proposed algorithm. The main contributions of this chapter lie in the following two aspects: (1) by virtue of the recursive Riccati equation method, the multi-objective control problem is solved for a class of nonlinear time-varying Markovian jump systems, which are much more general than those investigated in the existing literature, and (2) both the stochastic nonlinearities and the sensor failures have been taken into consideration in order better to reflect the engineering practice, especially within networked environments.

The rest of the chapter is arranged as follows: Section 12.1 formulates the mixed $H_2/H_\infty$ control problem for the time-varying Markovian jump systems, where the stochastic nonlinearities and sensor failures embedded in the model are properly justified. In Section 12.2, the $H_\infty$ performance is analyzed in terms of the Riccati equation method. Section 12.3 gives the methodology to solve the addressed multi-objective control problem and outlines the computational algorithm to recursively obtain the required parameters. A numerical example is presented in Section 12.4 to show the effectiveness and applicability of the proposed algorithm. Section 12.5 draws the summary.

## 12.1 Problem Formulation

Consider the following nonlinear discrete time-varying Markovian jump system defined on $k \in [0, N - 1]$:

$$x(k + 1) = A_{r(k)}(k)x(k) + B_{r(k)}(k)u(k) + f_{r(k)}(x(k), u(k), k) + D_{r(k)}(k)\omega(k), \quad (12.1)$$

with the measurement equation

$$y(k) = \Theta(k)C_{r(k)}(k)x(k) + g_{r(k)}(x(k), u(k), k) + E_{r(k)}(k)\omega(k)$$

$$= \sum_{\tau=1}^{m} \theta^{(\tau)}(k)C_{r(k)}^{(\tau)}(k) + g_{r(k)}(x(k), u(k), k) + E_{r(k)}(k)\omega(k), \quad (12.2)$$

where $x(k) \in \mathbb{R}^n$ is the state, $u(k) \in \mathbb{R}^{n1}$ is the control input, $y(k) \in \mathbb{R}^m$ is the output, $\omega(k) \in \mathbb{R}^{n2}$ is the external disturbance belonging to $l_2$, and $A_{r(k)}(k)$, $B_{r(k)}(k)$, $C_{r(k)}(k)$, $D_{r(k)}(k)$, and $E_{r(k)}(k)$ are known real-valued time-varying matrices with appropriate dimensions; $r(k)$ is a time-varying Markov chain taking values in a finite set $S = \{1, 2, \ldots, S\}$ ($S \in \mathbb{I}^+$), with the transition probability matrix $\mathcal{P}(k) \triangleq \{p_{ij}(k)\}$ being defined as follows:

$$p_{ij}(k) = \text{Prob}(r(k + 1) = j | r(k) = i), \quad (12.3)$$

where $0 \leq p_{ij}(k) \leq 1$, $\forall i, j \in S$, and $\sum_{j=1}^{S} p_{ij}(k) = 1$.

The stochastic matrix $\Theta(k)$ describes the phenomenon of multiple measurements degraded in the process of information retrieval from the sensor output, that is,

$$\Theta(k) \triangleq \text{diag}\left\{\theta^{(1)}(k), \theta^{(2)}(k), \dots, \theta^{(m)}(k)\right\}, \tag{12.4}$$

with $\theta^{(\tau)}(k)$ ($\tau = 1, 2, \dots, m$) being $m$ independent random variables. It is assumed that $\theta^{(\tau)}(k)$ is independent of $r(k)$ and has the probability density function on the interval $[0, 1]$ with mathematical expectation $\bar{\theta}^{(\tau)}(k)$ and variance $(\sigma^{(\tau)}(k))^2$. $C_{r(k)}^{(\tau)}(k) \triangleq$ $\text{diag}\{\underbrace{0, \dots, 0}_{\tau-1}, 1, \underbrace{0, \dots, 0}_{m-\tau}\} C_{r(k)}(k)$ and $\bar{\Theta}(k) \triangleq \mathbb{E}\{\Theta(k)\}$.

**Remark 12.1** *The description of measurement degradation phenomenon given in (12.4) is quite general and much closer to the engineering reality than those employed in the existing literature. In such a model, when $\theta^{(\tau)}(k) = 1$, it means that the $\tau$th sensor is in good condition at time $k$; otherwise there might be partial or complete sensor failure. To be more specific, when $\theta^{(\tau)}(k) = 0$, the measurements are completely missing; when $0 < \theta^{(\tau)}(k) < 1$, the output signals are measured with reduced gains that lead to degraded measurements. Therefore, the model (12.2) to (12.4) offers a comprehensive means to reflect system complexities such as nonlinearities, stochasticity, and data degraded from multiple sensors. It should be pointed out that, in most available publications, the data missing phenomenon has simply been modeled by a single Bernoulli sequence; see, for example, Ref. [54].*

Denote $f_{r(k)}(k) \triangleq f_{r(k)}(x(k), u(k), k)$ and $g_{r(k)}(k) \triangleq g_{r(k)}(x(k), u(k), k)$. The nonlinear stochastic functions $f_{r(k)}(k)$ and $g_{r(k)}(k)$ are assumed to have the following first moments for all $x(k)$, $u(k)$, and $k$:

$$\mathbb{E}\left\{ \begin{bmatrix} f_{r(k)}(k) \\ g_{r(k)}(k) \end{bmatrix} \middle| (x(k), u(k)) \right\} = 0, \tag{12.5}$$

with the covariance given by

$$\mathbb{E}\left\{ \begin{bmatrix} f_{r(k)}(k) \\ g_{r(k)}(k) \end{bmatrix} \begin{bmatrix} f_{r(j)}^{\mathrm{T}}(j) & g_{r(j)}^{\mathrm{T}}(j) \end{bmatrix} \middle| (x(k), u(k)) \right\} = 0, \qquad k \neq j, \tag{12.6}$$

and

$$\mathbb{E}\left\{ \begin{bmatrix} f_{r(k)}(k) \\ g_{r(k)}(k) \end{bmatrix} \begin{bmatrix} f_{r(k)}^{\mathrm{T}}(k) & g_{r(k)}^{\mathrm{T}}(k) \end{bmatrix} \middle| (x(k), u(k)) \right\}$$

$$= \sum_{l=1}^{q} \begin{bmatrix} \Omega_{r(k)}^{(l)}(k) & \Lambda_{r(k)}^{(l)}(k) \\ \left(\Lambda_{r(k)}^{(l)}(k)\right)^{\mathrm{T}} & \Xi_{r(k)}^{(l)}(k) \end{bmatrix} \left( x^{\mathrm{T}}(k)\Gamma_{r(k)}^{(l)}(k)x(k) + u^{\mathrm{T}}(k)\Upsilon_{r(k)}^{(l)}(k)u(k) \right), \tag{12.7}$$

where $\Omega_{r(k)}^{(l)}(k) > 0$, $\Lambda_{r(k)}^{(l)}(k)$, $\Xi_{r(k)}^{(l)}(k) > 0$, $\Gamma_{r(k)}^{(l)}(k) > 0$ and $\Upsilon_{r(k)}^{(l)}(k) > 0$ ($l = 1, 2, \dots, q$) are known matrices for each $r(k) \in S$.

**Remark 12.2** *As discussed in Ref. [31], the stochastic nonlinearities $f_{r(k)}(k)$ and $g_{r(k)}(k)$ account for several classes of well-studied nonlinear systems, such as the system with state-dependent multiplicative noises and the system whose state has power dependent on the sector bound (or sign) of the nonlinear state function of the state.*

For each $r(k) = i \in \mathcal{S}$, we denote $A_{r(k)}(k) = A_i(k)$, $B_{r(k)}(k) = B_i(k)$, $C_{r(k)}(k) = C_i(k)$, $D_{r(k)}(k) = D_i(k)$, $E_{r(k)}(k) = E_i(k)$, $\Omega_{r(k)}^{(l)}(k) = \Omega_i^{(l)}(k)$, $\Lambda_{r(k)}^{(l)}(k) = \Lambda_i^{(l)}(k)$, $\Xi_{r(k)}^{(l)}(k) = \Xi_i^{(l)}(k)$, $\Gamma_{r(k)}^{(l)}(k) = \Gamma_i^{(l)}(k)$, and $\Upsilon_{r(k)}^{(l)}(k) = \Upsilon_i^{(l)}(k)$.

Before stating the problem to be addressed, let us first introduce the $H_\infty$ specification. For a pre-specified $\gamma > 0$, consider the following $H_\infty$ performance index for system (12.1):

$$J_\infty \triangleq \mathbb{E}\left\{ \sum_{k=0}^{N-1} \left( \|y(k)\|^2 - \gamma^2 \|\omega(k)\|^2 \right) \right\} - \gamma^2 x_0^\mathsf{T} W_i x_0, \qquad (12.8)$$

where $W_i > 0$ $(i \in \mathcal{S})$ are weighting matrices.

In this chapter, our aim is to design state feedback controller $u(k) = K_i(k)x(k)$ such that:

(R1) For the pre-specified $H_\infty$ disturbance attenuation level $\gamma$, the inequality $J_\infty < 0$ holds for all nonzero $\left( \{\omega(k), \}, x_0 \right)$.

(R2) When there exists a worst-case disturbance $\omega^*(k)$, the output energy defined by

$$J_2 \triangleq \mathbb{E}\left\{ \sum_{k=0}^{N-1} \|y(k)\|^2 \right\}$$

is minimized.

## 12.2   $H_\infty$ Performance

In this section, the $H_\infty$ performance of the system will be analyzed first. A lemma that will play a vital role in the multi-objective controller design procedure is presented to guarantee the pre-specified $H_\infty$ disturbance attenuation level. To this end, setting $u(k) = 0$, we obtain the corresponding unforced system as follows:

$$\begin{cases} x(k+1) = A_i(k)x(k) + f_i(x(k), k) + D_i(k)\omega(k), \\ y(k) = \Theta(k)C_i(k)x(k) + g_i(x(k), k) + E_i(k)\omega(k). \end{cases} \qquad (12.9)$$

The following lemma gives a necessary and sufficient condition such that system (12.9) satisfies the $H_\infty$ requirement.

**Lemma 12.2.1** *Consider system* (12.9). *Let the disturbance attenuation level $\gamma > 0$ and positive definite matrices $W_i$ $(i \in \mathcal{S})$ be given. If, with the final condition*

$R_i(N) = 0$, there exist solutions $\{R_i(k)\}$ $(i \in \mathcal{S}, 0 \leq k < N)$ to the following matrix equation:

$$R_i(k) = A_i^{\mathrm{T}}(k)\bar{R}_i(k+1)A_i(k) + C_i^{\mathrm{T}}(k)\bar{\Theta}^{\mathrm{T}}(k)\bar{\Theta}(k)C_i(k)$$

$$+ \sum_{\tau=1}^{m} \left(\sigma^{(\tau)}(k)\right)^2 \left(C_i^{(\tau)}(k)\right)^{\mathrm{T}} C_i^{(\tau)}(k) + \bar{D}_i^{\mathrm{T}}(k)\Phi_i^{-1}(k)\bar{D}_i(k) \qquad (12.10)$$

$$+ \sum_{l=1}^{q} \Gamma_i^{(l)}(k)(\mathrm{tr}[\bar{R}_i(k+1)\Omega_i^{(l)}(k)] + \mathrm{tr}[\Xi_i^{(l)}(k)]),$$

such that

$$\begin{cases} \Phi_i(k) = \gamma^2 I - D_i^{\mathrm{T}}(k)\bar{R}_i(k+1)D_i(k) - E_i^{\mathrm{T}}(k)E_i(k) > 0, \\ R_i(0) < \gamma^2 W_i, \end{cases} \qquad (12.11)$$

where

$$\bar{R}_i(k+1) = \sum_{j=1}^{s} p_{ij}R_j(k+1),$$

$$\bar{D}_i(k) = D_i^{\mathrm{T}}(k)\bar{R}_i(k+1)A_i(k) + E_i^{\mathrm{T}}(k)\bar{\Theta}(k)C_i(k),$$

then the prescribed $H_\infty$ disturbance attenuation level in the design task (R1) can be achieved for all nonzero $(\{\omega(k)\}, x_0)$.

*Proof.*
**Sufficiency**
By defining

$$\tilde{J}_i(k) = x^{\mathrm{T}}(k+1)R_{r(k+1)}(k+1)x(k+1) - x^{\mathrm{T}}(k)R_i(k)x(k) \qquad (12.12)$$

and taking the statistical property of $f_i(x(k), k)$ and $g_i(x(k), k)$ into account, we have

$$\mathbb{E}\{\tilde{J}_i(k)\}$$

$$= \mathbb{E}\left\{ (A_i(k)x(k) + f_i(x(k), k) + D_i(k)\omega(k))^{\mathrm{T}}\bar{R}_i(k+1) \right.$$

$$\left. \times (A_i(k)x(k) + f_i(x(k), k) + D_i(k)\omega(k)) \right\} - x^{\mathrm{T}}(k)R_i(k)x(k)$$

$$= x^{\mathrm{T}}(k)\left( A_i^{\mathrm{T}}(k)\bar{R}_i(k+1)A_i(k) - R_i(k) + \sum_{l=1}^{q} \Gamma_i^{(l)}(k)\mathrm{tr}[\bar{R}_i(k+1)\Omega_i^{(l)}(k)] \right)x(k)$$

$$+ 2x^{\mathrm{T}}(k)A_i^{\mathrm{T}}(k)\bar{R}_i(k+1)D_i(k)\omega(k) + \omega^{\mathrm{T}}(k)D_i^{\mathrm{T}}(k)\bar{R}_i(k+1)D_i(k)\omega(k). \qquad (12.13)$$

Adding the following zero term

$$\|y(k)\|^2 - \gamma^2\|\omega(k)\|^2 - (\|y(k)\|^2 - \gamma^2\|\omega(k)\|^2) \qquad (12.14)$$

to both sides of (12.13) and then taking the mathematical expectation, we have

$$
\begin{aligned}
\mathbb{E}\{\tilde{J}_i(k)\} \\
= x^{\mathrm{T}}(k) &\left( A_i^{\mathrm{T}}(k)\bar{R}_i(k+1)A_i(k) - R_i(k) + C_i^{\mathrm{T}}(k)\bar{\Theta}^{\mathrm{T}}(k)\bar{\Theta}(k)C_i(k) \right. \\
&+ \sum_{\tau=1}^{m} \left(\sigma^{(\tau)}(k)\right)^2 \left(C_i^{(\tau)}(k)\right)^{\mathrm{T}} C_i^{(\tau)}(k) \\
&\left.+ \sum_{l=1}^{q} \Gamma_i^{(l)}(k)\left(\mathrm{tr}[\bar{R}_i(k+1)\Omega_i^{(l)}(k)] + \mathrm{tr}[\Xi_i^{(l)}(k)]\right)\right) x(k) \\
&+ 2x^{\mathrm{T}}(k)\left(A_i^{\mathrm{T}}(k)\bar{R}_i(k+1)D_i(k) + C_i^{\mathrm{T}}(k)\bar{\Theta}^{\mathrm{T}}(k)E_i(k)\right)\omega(k) \\
&+ \omega^{\mathrm{T}}(k)\left(-\gamma^2 I + D_i^{\mathrm{T}}(k)\bar{R}_i(k+1)D_i(k) + E_i^{\mathrm{T}}(k)E_i(k)\right)\omega(k) \\
&- \mathbb{E}\{\|y(k)\|^2 - \gamma^2\|\omega(k)\|^2\}.
\end{aligned}
\tag{12.15}
$$

By the "completing squares" method, we obtain

$$
\mathbb{E}\{\tilde{J}_i(k)\} = -(\omega(k) - \omega^*(k))^{\mathrm{T}}\Phi_i(k)(\omega(k) - \omega^*(k)) - \mathbb{E}\{\|y(k)\|^2 - \gamma^2\|\omega(k)\|^2\},
\tag{12.16}
$$

where

$$
\omega^*(k) = \Phi_i^{-1}(k)\left(D_i^{\mathrm{T}}(k)\bar{R}_i(k+1)A_i(k) + E_i^{\mathrm{T}}(k)\bar{\Theta}(k)C_i(k)\right)x(k).
\tag{12.17}
$$

Then, it can be easily verified that

$$
\begin{aligned}
\mathbb{E}\left\{\sum_{k=0}^{N-1}\tilde{J}_i(k)\right\} &= \mathbb{E}\left\{x_N^{\mathrm{T}}R_i(N)x_N\right\} - x_0^{\mathrm{T}}R_i(0)x_0 \\
&= -\sum_{k=0}^{N-1}\left((\omega(k) - \omega^*(k))^{\mathrm{T}}\Phi_i(k)(\omega(k) - \omega^*(k))\right. \\
&\quad \left.- \mathbb{E}\{\|y(k)\|^2 - \gamma^2\|\omega(k)\|^2\}\right).
\end{aligned}
\tag{12.18}
$$

Since $\Phi_i(k) > 0$, $R_i(0) < \gamma^2 W_i$, and by the final condition $R_i(N) = 0$, it follows immediately that

$$
\begin{aligned}
J_\infty &= \mathbb{E}\left\{\sum_{k=0}^{N-1}(\|y(k)\|^2 - \gamma^2\|\omega(k)\|^2)\right\} - \gamma^2 x_0^{\mathrm{T}} W_i x_0 \\
&= -\sum_{k=0}^{N-1}(\omega(k) - \omega^*(k))^{\mathrm{T}}\Phi_i(k)(\omega(k) - \omega^*(k)) + x_0^{\mathrm{T}}(R_i(0) - \gamma^2 W_i)x_0 < 0,
\end{aligned}
\tag{12.19}
$$

which means that the pre-specified $H_\infty$ performance is satisfied. The proof of sufficiency is complete. $\qquad\square$

**Necessity**

Firstly, considering (12.19), we select $\omega(k) = \omega^*(k)$ for $\forall k \in [0, N)$ and $x_0 \neq 0$. It is clear that $R_i(0) < \gamma^2 W_i$ if $J_\infty < 0$.

We now proceed to prove that if $J_\infty < 0$, then there exist $R_i(k)$ ($i \in \mathcal{S}, 0 \leq k < N$) solving the matrix equations (12.10) such that $\Phi_i(k) > 0$ is satisfied for all nonzero ($\{\omega(k)\}, x_0$). In fact, if $\Phi_i(k)$ ($i \in \mathcal{S}, 0 \leq k < N$) are nonsingular, we can always solve $R_i(k)$ for (12.10) with the known final condition $R_i(N) = 0$ recursively backward in time $k$. Therefore, the proof of necessity is equivalent to proving the following proposition:

$$J_\infty < 0 \Rightarrow \lambda_\kappa(\Phi_i(k)) > 0 \quad \forall i \in \mathcal{S} \ \& \ k \in [0, N), \qquad \kappa = 1, 2, \ldots, n. \qquad (12.20)$$

The rest of the proof is carried out by contradiction. The inverse proposition of (12.20) is that we can find, from $J_\infty < 0$, at least one eigenvalue of $\Phi_i(k)$ that is either equal to 0 or negative for some $i' \in \mathcal{S}$ at some time point $k^* \in [0, N)$. For simplicity, we denote such an eigenvalue of $\Phi_i(k)$ for mode $i'$ at time $k^*$ as $\lambda^\circ(i', k^*) = \lambda^\circ(\Phi_{i'}(k^*))$. In the following, we shall use such a non-positive $\lambda^\circ(i', k^*)$ to reveal that there exist certain ($\{\omega(k)\}, x_0) \neq 0$ such that $J_\infty \geq 0$. In fact, if $\lambda^\circ(i', k^*) < 0$, we can choose

$$\omega(k) = \begin{cases} \omega^*(k^*) + \varphi^\circ(i', k^*), & k = k^*, \\ \omega^*(k), & \text{otherwise,} \end{cases} \qquad (12.21)$$

where $\varphi^\circ(i', k^*)$ is the eigenvector of $\Phi_{i'}(k^*)$ with respect to $\lambda^\circ(i', k^*)$. Hence, we have

$$J_\infty = -\sum_{k=0}^{N-1} (\omega(k) - \omega^*(k))^T \Phi_i(k) (\omega(k) - \omega^*(k)) - x_0^T (\gamma^2 W_i - R_i(0)) x_0$$

$$= -\sum_{k=0, k \neq k^*}^{N-1} (\omega(k) - \omega^*(k))^T \Phi_i(k) (\omega(k) - \omega^*(k)) - x_0^T (\gamma^2 W_i - R_i(0)) x_0$$

$$\quad - (\omega(k^*) - \omega^*(k^*))^T \Phi_{i'}(k^*) (\omega(k^*) - \omega^*(k^*))$$

$$= -\left(\varphi^\circ(i', k^*)\right)^T \Phi_{i'}(k^*) \varphi^\circ(i', k^*) - x_0^T (\gamma^2 W_i - R_i(0)) x_0$$

$$= -\lambda^\circ(i', k^*) \|\varphi^\circ(i', k^*)\|^2 - x_0^T (\gamma^2 W_i - R_i(0)) x_0. \qquad (12.22)$$

We can see from above that if $x_0$ is chosen such that $\|x_0\|^2$ is sufficiently small, then $J_\infty > 0$ is true for all ($\{\omega(k)\}, x_0) \neq 0$, which contradicts the condition $J_\infty < 0$. Therefore, all the eigenvalues of $\Phi_i(k)$ should be non-negative.

Next, if $\lambda^\circ(i', k^*) = 0$, we choose $x_0 = 0$ and

$$\omega(k) = \begin{cases} \varphi^\circ(i', k^*), & k = k^*, \\ \omega^*(k), & k^* < k \leq N - 1, \\ 0, & 0 \leq k < k^*, \end{cases} \qquad (12.23)$$

Then

$$
J_\infty = \mathbb{E} \left\{ \sum_{k=0}^{k^*-1} \left( \|y(k)\|^2 - \gamma^2 \|\omega(k)\|^2 \right) \right\} - x_0^{\mathrm{T}}(\gamma^2 W_i - R_i(0))x_0
$$

$$
- \sum_{k=k^*+1}^{N-1} (\omega(k) - \omega^*(k))^{\mathrm{T}} \Phi_i(k)(\omega(k) - \omega^*(k)) \tag{12.24}
$$

$$
- (\omega(k^*) - \omega^*(k^*))^{\mathrm{T}} \Phi_{i'}(k^*)(\omega(k^*) - \omega^*(k^*)).
$$

Since $x_0 = 0$ and $\omega(k) = 0$ when $0 \le k < k^*$, we can see from (12.9) that $\mathbb{E}\{\|y_k\|^2\} = 0$. Therefore,

$$
J_\infty = - (\varphi^\circ(i', k^*))^{\mathrm{T}} \Phi_{i'}(k^*)\varphi^\circ(i', k^*) = -\lambda^\circ(i', k^*)\|\varphi^\circ(i', k^*)\|^2 = 0, \tag{12.25}
$$

which also contradicts the condition $J_\infty < 0$. Therefore, we can conclude that $\Phi_i(k) > 0$. The proof is now complete.  □

So far, we have analyzed the system's $H_\infty$ performance in terms of the solvability of a backward Riccati-like equation. The presented lemma will enable us to develop an algorithm to obtain the multi-objective controller parameters, which will be illustrated in detail in the following section.

## 12.3  Mixed $H_2/H_\infty$ Controller Design

In this section, a state feedback control law will be designed for system (12.1) such that the closed-loop system is able to achieve the pre-specified $H_\infty$ performance with the output energy being minimized in case the worst disturbance happens. It will be shown that the desired controller parameters can be obtained by solving certain coupled backward Riccati-like matrix equations recursively.

### 12.3.1  Controller Design

In this subsection, we aim to find the sufficient condition of the existence of the desired multi-objective controller. To this end, implementing $u(k) = K_i(k)x(k)$ to system (12.1), we get the following closed-loop nonlinear stochastic Markovian jump systems over the finite horizon $[0, N-1]$:

$$
\begin{cases} x(k+1) = (A_i(k) + B_i(k)K_i(k))\, x(k) + f_i(x(k), K_i(k)x(k), k) + D_i(k)\omega(k), \\ y(k) = \Theta(k)C_i(k)x(k) + g_i(x(k), K_i(k)x(k), k) + E_i(k)\omega(k). \end{cases} \tag{12.26}
$$

The following theorem presents a sufficient condition for the existence of the desired controllers, which ensures both $H_\infty$ and $H_2$ performances of the closed-loop system (12.26).

**Theorem 12.3.1** *Let $\gamma > 0$ and $W_i$ ($i \in \mathcal{S}$) be given. There exist state feedback controller $u(k) = K_i(k)x(k)$ for system (12.1) such that the design goals (R1) and (R2) are achieved if the following coupled matrix equations have solutions $(R_i(k), K_i(k))$ and $(Q_i(k), T_i(k))$ ($i \in \mathcal{S}, 0 \le k \le N-1$):*

$$
\begin{cases}
\begin{aligned}
R_i(k) =\ & (A_i(k) + B_i(k)K_i(k))^{\mathrm{T}} \bar{R}_i(k+1)(A_i(k) + B_i(k)K_i(k)) \\
& + C_i^{\mathrm{T}}(k)\bar{\Theta}^{\mathrm{T}}(k)\bar{\Theta}(k)C_i(k) + \sum_{l=1}^{q} (\Gamma_i^{(l)}(k) + K_i^{\mathrm{T}}(k)\Upsilon_i^{(l)}(k)K_i(k)) \\
& \times (tr[\bar{R}_i(k+1)\Omega_i^{(l)}(k)] + tr[\Xi_i^{(l)}(k)]) \\
& + \sum_{\tau=1}^{m} (\sigma^{(\tau)}(k))^2 (C_i^{(\tau)}(k))^{\mathrm{T}} C_i^{(\tau)}(k) + \bar{K}_i^{\mathrm{T}}(k)\Phi_i^{-1}(k)\bar{K}_i(k),
\end{aligned} \\
R_i(N) = 0,
\end{cases}
\tag{12.27}
$$

$$
\begin{cases}
\begin{aligned}
Q_i(k) =\ & (A_i(k) + D_i(k)T_i(k))^{\mathrm{T}} \bar{Q}_i(k+1)(A_i(k) + D_i(k)T_i(k)) \\
& + \sum_{l=1}^{q} \Gamma_i^{(l)}(k)tr[\bar{Q}_i(k+1)\Omega_i^{(l)}(k)] \\
& + (\bar{\Theta}(k)C_i(k) + E_i(k)T_i(k))^{\mathrm{T}} (\bar{\Theta}(k)C_i(k) + E_i(k)T_i(k)) \\
& + \bar{T}_i^{\mathrm{T}}(k)\Psi_i^{-1}(k)\bar{T}_i(k) + \sum_{\tau=1}^{m} (\sigma^{(\tau)}(k))^2 (C_i^{(\tau)}(k))^{\mathrm{T}} C_i^{(\tau)}(k) \\
& + \sum_{l=1}^{q} \Gamma_i^{(l)}(k)tr[\Xi_i^{(l)}(k)],
\end{aligned} \\
Q_i(N) = 0,
\end{cases}
\tag{12.28}
$$

*with*

$$
\begin{cases}
\Phi_i(k) = \gamma^2 I - D_i^{\mathrm{T}}(k)\bar{R}_i(k+1)D_i(k) - E_i^{\mathrm{T}}(k)E_i(k) > 0, \\
R_i(0) < \gamma^2 W_i,
\end{cases}
\tag{12.29}
$$

*where*

$$
\bar{Q}_i(k+1) = \sum_{j=1}^{S} p_{ij} Q_j(k+1),
$$

$$
\Psi_i(k) = B_i^{\mathrm{T}}(k)\bar{Q}_i(k+1)B_i(k) + \sum_{l=1}^{q} \Upsilon_i^{(l)}(k)(tr[\bar{Q}_i(k+1)\Omega_i^{(l)}(k)] + tr[\Xi_i^{(l)}(k)]),
$$

$$
\bar{K}_i(k) = D_i^{\mathrm{T}}(k)\bar{R}_i(k+1)(A_i(k) + B_i(k)K_i(k)) + E_i^{\mathrm{T}}(k)\bar{\Theta}(k)C_i(k),
$$

$$
\bar{T}_i(k) = B_i^{\mathrm{T}}(k)\bar{Q}_i(k+1)(A_i(k) + D_i(k)T_i(k)).
$$

*In this case, the matrices* $T_i(k)$ *and the feedback gains* $K_i(k)$ *can be computed by*

$$\begin{cases} K_i(k) = -\Psi_i^{-1}(k)B_i^T(k)\bar{Q}_i(k+1)\left(A_i(k)+D_i(k)T_i(k)\right), \\ T_i(k) = \Phi_i^{-1}(k)\left(D_i^T(k)\bar{R}_i(k+1)\left(A_i(k)+B_i(k)K_i(k)\right)+E_i^T(k)\bar{\Theta}(k)C_i(k)\right). \end{cases}$$

$$(12.30)$$

*Proof.* First, from Lemma 12.2.1, we can easily see that if there exist solutions $R_i(k)$ to (12.27) such that $\Phi_i(k) > 0$ and $R_i(0) < \gamma^2 W_i$, then system (12.26) achieves the pre-specified $H_\infty$ performance. Moreover, the worst-case disturbance can be expressed by $\omega^*(k) = T_i(k)x(k)$.

In the next stage, we shall proceed to deal with the $H_2$ constraint when the worst-case disturbance happens and then determine the desired feedback gains $K_i(k)$ minimizing the output energy. To this end, under the situation of the worst-case disturbance, the original system (12.1) can be rewritten as follows:

$$\begin{cases} x(k+1) = (A_i(k)+D_i(k)T_i(k))\,x(k)+B_i(k)u(k)+f_i(x(k),u(k),k), \\ y(k) = (\Theta(k)C_i(k)+E_i(k)T_i(k))x(k)+g_i(x(k),u(k),k). \end{cases}$$

$$(12.31)$$

In order to obtain the parametric expression of $K_i(k)$, we first define

$$\tilde{J}_{2i}(k) \triangleq x^T(k+1)Q_{r(k+1)}(k+1)x(k+1) - x^T(k)Q_i(k)x(k). \tag{12.32}$$

For convenience, we denote

$$\bar{A}_i(k) = A_i(k)+D_i(k)T_i(k), \qquad \bar{C}_i(k) = \bar{\Theta}(k)C_i(k)+E_i(k)T_i(k).$$

Then, using the statistical properties of $f_i(x(k),u(k),k)$ and $g_i(x(k),u(k),k)$, it can be seen that

$$\mathbb{E}\{\tilde{J}_{2i}(k)\}$$

$$= \mathbb{E}\{(\bar{A}_i(k)x(k)+B_i(k)u(k)+f_i(x(k),u(k),k))^T$$

$$\times \bar{Q}_i(k+1)(\bar{A}_i(k)x(k)+B_i(k)u(k)+f_i(x(k),u(k),k))\} - x^T(k)Q_i(k)x(k)$$

$$= x^T(k)\left(\bar{A}_i^T(k)\bar{Q}_i(k+1)\bar{A}_i(k) - Q_i(k) + \sum_{l=1}^{q}\Gamma_i^{(l)}(k)\text{tr}[\bar{Q}_i(k+1)\Omega_i^{(l)}(k)]\right)x(k)$$

$$+ u^T(k)\left(B_i^T(k)\bar{Q}_i(k+1)B_i(k) + \sum_{l=1}^{q}\Upsilon_i^{(l)}(k)\text{tr}[\bar{Q}_i(k+1)\Omega_i^{(l)}(k)]\right)u(k)$$

$$+ 2x^T(k)\bar{A}_i(k)\bar{Q}_i(k+1)B_i(k)u(k). \tag{12.33}$$

Then, it follows that

$$\mathbb{E}\{\tilde{J}_{2i}(k) + \|y(k)\|^2 - \|y(k)\|^2\}$$

$$= x^{\mathrm{T}}(k)\left(\bar{A}_i^{\mathrm{T}}(k)\bar{Q}_i(k+1)\bar{A}_i(k) - Q_i(k) + \sum_{l=1}^{q}\Gamma_i^{(l)}(k)\mathrm{tr}[\bar{Q}_i(k+1)\Omega_i^{(l)}(k)]\right.$$

$$\left.+\bar{C}_i^{\mathrm{T}}(k)\bar{C}_i(k) + \sum_{\tau=1}^{m}(\sigma^{(\tau)}(k))^2(C_i^{(\tau)}(k))^{\mathrm{T}}C_i^{(\tau)}(k) + \sum_{l=1}^{q}\Gamma_i^{(l)}(k)\mathrm{tr}[\Xi_i^{(l)}(k)]\right)x(k)$$

$$+ u^{\mathrm{T}}(k)\left(B_i^{\mathrm{T}}(k)\bar{Q}_i(k+1)B_i(k)\right.$$

$$\left.+ \sum_{l=1}^{q}\Upsilon_i^{(l)}(k)\left(\mathrm{tr}[\bar{Q}_i(k+1)\Omega_i^{(l)}(k)] + \mathrm{tr}[\Xi_i^{(l)}(k)]\right)\right)u(k)$$

$$+ 2x^{\mathrm{T}}(k)\bar{A}_i^{\mathrm{T}}(k)\bar{Q}_i(k+1)B_i(k)u(k) - \mathbb{E}\{\|y(k)\|^2\}. \tag{12.34}$$

Completing the squares of $u(k)$, we have

$$\mathbb{E}\{\tilde{J}_{2i}(k) + \|y(k)\|^2 - \|y(k)\|^2\}$$
$$= \mathbb{E}\{(u(k) - u^*(k))^{\mathrm{T}}\Psi_i(k)(u(k) - u^*(k)) - \|y(k)\|^2\}, \tag{12.35}$$

where $\Psi_i(k)$ is defined in (12.28) and $u^*(k) = K_i(k)x(k)$ with $K_i(k)$ being defined in (12.30).

Now we consider the closed-loop $H_2$ performance $J_2$. From the derivation above, it is true that

$$J_2 = \mathbb{E}\left\{\sum_{k=0}^{N-1}\|y(k)\|^2\right\} = \mathbb{E}\left\{\sum_{k=0}^{N-1}(u(k) - u^*(k))^{\mathrm{T}}\Psi_i(k)(u(k) - u^*(k))\right\}$$

$$+ x_0^{\mathrm{T}}Q_i(0)x_0 - x^{\mathrm{T}}(N)Q_i(N)x(N). \tag{12.36}$$

Noticing the final condition $Q_i(N) = 0$ and the fact that $\Psi_i(k) > 0$ is always satisfied, it is easy to see that $J_2$ is minimized when $u(k) = u^*(k)$ and

$$J_{\min} = x_0^{\mathrm{T}}Q_i(0)x_0. \tag{12.37}$$

Therefore, the control design goals (R1) and (R2) are achieved simultaneously. The proof is complete. □

**Remark 12.3** *So far, a unified framework has been established to solve the mixed $H_2/H_\infty$ control problem for the Markovian jump system with both stochastic* *nonlinearities and missing measurements. It is worth noting that the proposed design technique is presented in terms of the solvability of certain Riccati-like difference equations that have a much lower computation complexity than the commonly used linear matrix inequality (LMI) constraints, and is therefore more convenient and effective for the practical design.*

## 12.3.2 Computational Algorithm

Noticing that the feedback controller gain is involved in the proposed coupled Riccati-like equations, in this subsection, we shall present a recursive algorithm to obtain the numerical value of $K_i(k)$ by solving the corresponding equations iteratively.

### Multi-Objective Controller Design Algorithm

*Step 1.* Set $k = N - 1, r(N - 1) = i \in \mathcal{S}$. Then $R_i(N) = 0$ and $Q_i(N) = 0$ are available.
*Step 2.* Compute $\Phi_i(k)$ and $\Psi_i(k)$ via the second equations of (12.27) and (12.28) respectively.
*Step 3.* With the obtained $\Phi_i(k)$ and $\Psi_i(k)$, solve equation (12.30) to obtain the values of $T_i(k)$ and $K_i(k)$.
*Step 4.* Solve the first equations of (12.27) and (12.28) to get $R_i(k)$ and $Q_i(k)$ respectively.
*Step 5.* If $k = 0$, stop. Else, set $k = k - 1$ and go to Step 2.

**Remark 12.4** *In the proposed* Multi-Objective Controller Design Algorithm, *a recursive controller design scheme is presented. At each time point, a desired feedback gain is obtained such that both the $H_2$ and $H_\infty$ constraints are achieved. Noticing that the existence of the desired controller is expressed in terms of the solvability of coupled backward Riccati-like equations, the proposed numerical algorithm is also backward in time and can only be applied in off-line design.*

## 12.4 Numerical Example

In this section, a numerical example is presented to demonstrate the effectiveness of the method proposed in this chapter.

Set $N = 5, \gamma = 1.5, W_i = 0.2I, \mathcal{S} = \{1, 2\}$, and the final condition $Q_{r(5)}(5) = 0$ and $R_{r(5)}(5) = 0$. The transition probability matrix $\mathcal{P}(k)$ of $r(k)$ is

$$\mathcal{P}(0) = \begin{bmatrix} 0.10 & 0.90 \\ 0.60 & 0.40 \end{bmatrix}, \quad \mathcal{P}(1) = \begin{bmatrix} 0.30 & 0.70 \\ 0.50 & 0.50 \end{bmatrix},$$

$$\mathcal{P}(2) = \begin{bmatrix} 0.20 & 0.80 \\ 0.65 & 0.35 \end{bmatrix}, \quad \mathcal{P}(3) = \begin{bmatrix} 0.45 & 0.55 \\ 0.70 & 0.30 \end{bmatrix},$$

$$\mathcal{P}(4) = \begin{bmatrix} 0.15 & 0.85 \\ 0.30 & 0.70 \end{bmatrix}.$$

Let the probability of data-packet dropout be defined as follows:

$$\mathcal{P}(\theta(k)) = \begin{cases} 0.8 & \theta(k) = 0, \\ 0.1 & \theta(k) = 0.5, \\ 0.1 & \theta(k) = 1, \end{cases}$$

from which the expectation and variance can be easily calculated as $\bar{\theta}(k) = 0.15$ and $\sigma^2(k) = 0.1025$.

For mode $i = 1$, the parameters of system (12.1) are as follows:

$$A_1(0) = \begin{bmatrix} 0.50 & -0.10 \\ -0.30 & 0.70 \end{bmatrix}, \quad A_1(1) = \begin{bmatrix} 0.20 & 0.80 \\ 0.35 & -0.50 \end{bmatrix},$$

$$A_1(2) = \begin{bmatrix} -0.75 & 0.10 \\ 0.30 & 0.08 \end{bmatrix}, \quad A_1(3) = \begin{bmatrix} 0.10 & 0.45 \\ 0.70 & -0.60 \end{bmatrix},$$

$$A_1(4) = \begin{bmatrix} 0.55 & -0.30 \\ 0.25 & -0.10 \end{bmatrix},$$

$$B_1(0) = \begin{bmatrix} 0.25 \\ -0.30 \end{bmatrix}, \quad B_1(1) = \begin{bmatrix} 0.50 \\ 0.15 \end{bmatrix},$$

$$B_1(2) = \begin{bmatrix} -0.60 \\ 0.4 \end{bmatrix}, \quad B_1(3) = \begin{bmatrix} 0.80 \\ -0.15 \end{bmatrix},$$

$$B_1(4) = \begin{bmatrix} -0.35 \\ 0.25 \end{bmatrix},$$

$$C_1(0) = \begin{bmatrix} 0.25 & 0.45 \end{bmatrix}, \quad C_1(1) = \begin{bmatrix} -0.80 & 0.15 \end{bmatrix},$$

$$C_1(2) = \begin{bmatrix} 0.30 & -0.95 \end{bmatrix}, \quad C_1(3) = \begin{bmatrix} 0.80 & -0.25 \end{bmatrix},$$

$$C_1(4) = \begin{bmatrix} 0.15 & -0.65 \end{bmatrix},$$

$$D_1(0) = \begin{bmatrix} 0.15 \\ 0.20 \end{bmatrix}, \quad D_1(1) = \begin{bmatrix} -0.10 \\ 0.30 \end{bmatrix},$$

$$D_1(2) = \begin{bmatrix} 0.55 \\ -0.25 \end{bmatrix}, \quad D_1(3) = \begin{bmatrix} 0.05 \\ 0.35 \end{bmatrix},$$

$$D_1(4) = \begin{bmatrix} 0.25 \\ 0.10 \end{bmatrix},$$

$$E_1(0) = 0.20, \quad E_1(1) = -0.50, \quad E_1(2) = -0.30,$$

$$E_1(3) = 0.80, \quad E_1(4) = -0.15, \quad \bar{\Theta}_1 = 0.15,$$

$$\Omega_1(0) = \begin{bmatrix} 0.40 & 0 \\ 0 & 0 \end{bmatrix}, \quad \Omega_1(1) = \begin{bmatrix} 0.70 & 0 \\ 0 & 0 \end{bmatrix},$$

$$\Omega_1(2) = \begin{bmatrix} 0.80 & 0 \\ 0 & 0 \end{bmatrix}, \quad \Omega_1(3) = \begin{bmatrix} 0.80 & 0 \\ 0 & 0 \end{bmatrix},$$

$$\Omega_1(4) = \begin{bmatrix} 0.50 & 0 \\ 0 & 0 \end{bmatrix},$$

$$\Xi_1(0) = 0.36, \quad \Xi_1(1) = 0.10, \quad \Xi_1(2) = 0.80,$$

$$\Xi_1(3) = 0.80, \quad \Xi_1(4) = 0.40,$$

$$\Gamma_1(0) = \begin{bmatrix} 0.20 & 0 \\ 0 & 0.50 \end{bmatrix}, \quad \Gamma_1(1) = \begin{bmatrix} 0.10 & 0 \\ 0 & 0.30 \end{bmatrix},$$

$$\Gamma_1(2) = \begin{bmatrix} 0.20 & 0 \\ 0 & 0.10 \end{bmatrix}, \quad \Gamma_1(3) = \begin{bmatrix} 0.20 & 0 \\ 0 & 0.10 \end{bmatrix},$$

$$\Gamma_1(4) = \begin{bmatrix} 0.10 & 0 \\ 0 & 0.10 \end{bmatrix},$$

$$\Upsilon_1(0) = 0.49, \quad \Upsilon_1(1) = 0.10, \quad \Upsilon_1(2) = 0.40,$$

$$\Upsilon_1(3) = 0.40, \quad \Upsilon_1(4) = 0.20.$$

For mode $i = 2$, the parameters of system (12.1) are as follows:

$$A_2(0) = \begin{bmatrix} -0.20 & 0.30 \\ -0.10 & 0.80 \end{bmatrix}, \quad A_2(1) = \begin{bmatrix} 0.55 & 0.80 \\ -0.35 & 0.45 \end{bmatrix},$$

$$A_2(2) = \begin{bmatrix} 0.25 & -0.10 \\ 0.90 & -0.70 \end{bmatrix}, \quad A_2(3) = \begin{bmatrix} 0.20 & 0.65 \\ 0.40 & -0.30 \end{bmatrix},$$

$$A_2(4) = \begin{bmatrix} 0.80 & -0.35 \\ 0.15 & -0.90 \end{bmatrix},$$

$$B_2(0) = \begin{bmatrix} 0.40 \\ -0.35 \end{bmatrix}, \quad B_2(1) = \begin{bmatrix} 0.60 \\ 0.20 \end{bmatrix},$$

$$B_2(2) = \begin{bmatrix} -0.80 \\ 0.50 \end{bmatrix}, \quad B_2(3) = \begin{bmatrix} 0.30 \\ -0.35 \end{bmatrix},$$

$$B_2(4) = \begin{bmatrix} -0.15 \\ 0.85 \end{bmatrix},$$

$$C_2(0) = \begin{bmatrix} 0.15 & 0.60 \end{bmatrix}, \qquad C_2(1) = \begin{bmatrix} -0.20 & 0.70 \end{bmatrix},$$

$$C_2(2) = \begin{bmatrix} 0.50 & -0.20 \end{bmatrix}, \qquad C_2(3) = \begin{bmatrix} 0.75 & -0.15 \end{bmatrix},$$

$$C_2(4) = \begin{bmatrix} 0.80 & -0.25 \end{bmatrix},$$

$$D_2(0) = \begin{bmatrix} 0.50 \\ 0.10 \end{bmatrix}, \qquad D_2(1) = \begin{bmatrix} -0.20 \\ 0.15 \end{bmatrix},$$

$$D_2(2) = \begin{bmatrix} 0.50 \\ -0.85 \end{bmatrix}, \qquad D_2(3) = \begin{bmatrix} 0.70 \\ 0.25 \end{bmatrix},$$

$$D_2(4) = \begin{bmatrix} 0.95 \\ 0.10 \end{bmatrix},$$

$$E_2(0) = 0.50, \quad E_2(1) = -0.70, \quad E_2(2) = -0.40,$$

$$E_2(3) = 0.20, \quad E_2(4) = -0.10, \quad \bar{\Theta}_2 = 0.15,$$

$$\Omega_2(0) = \begin{bmatrix} 0.80 & 0 \\ 0 & 0 \end{bmatrix}, \quad \Omega_2(1) = \begin{bmatrix} 0.40 & 0 \\ 0 & 0 \end{bmatrix},$$

$$\Omega_2(2) = \begin{bmatrix} 0.50 & 0 \\ 0 & 0 \end{bmatrix}, \quad \Omega_2(3) = \begin{bmatrix} 0.80 & 0 \\ 0 & 0 \end{bmatrix},$$

$$\Omega_2(4) = \begin{bmatrix} 0.70 & 0 \\ 0 & 0 \end{bmatrix},$$

$$\Xi_2(0) = 0.80, \quad \Xi_2(1) = 0.36, \quad \Xi_2(2) = 0.40,$$

$$\Xi_2(3) = 0.80, \quad \Xi_2(4) = 0.10,$$

$$\Gamma_2(0) = \begin{bmatrix} 0.20 & 0 \\ 0 & 0.10 \end{bmatrix}, \quad \Gamma_2(1) = \begin{bmatrix} 0.20 & 0 \\ 0 & 0.50 \end{bmatrix},$$

$$\Gamma_2(2) = \begin{bmatrix} 0.10 & 0 \\ 0 & 0.10 \end{bmatrix}, \quad \Gamma_2(3) = \begin{bmatrix} 0.20 & 0 \\ 0 & 0.10 \end{bmatrix},$$

$$\Gamma_2(4) = \begin{bmatrix} 0.10 & 0 \\ 0 & 0.30 \end{bmatrix},$$

$$\Upsilon_2(0) = 0.40, \quad \Upsilon_2(1) = 0.49, \quad \Upsilon_2(2) = 0.20,$$

$$\Upsilon_2(3) = 0.40, \quad \Upsilon_2(4) = 0.10.$$

**Table 12.1**    The random mode

| Time | $k = 0$ | $k = 1$ | $k = 2$ | $k = 3$ | $k = 4$ |
|------|---------|---------|---------|---------|---------|
| Mode | $i = 2$ | $i = 1$ | $i = 2$ | $i = 2$ | $i = 1$ |

Assume that the Markovian jump mode $r(k)$ is switched as shown in Table 12.1. Using the developed computational algorithm, we can check the feasibility of the coupled Riccati-like equations and then calculate the desired controller parameters, which are shown as follows:

$$R_2(0) = \begin{bmatrix} 0.50 & -0.04 \\ -0.04 & 1.30 \end{bmatrix}, \quad R_1(1) = \begin{bmatrix} 0.77 & -0.21 \\ -0.21 & 0.76 \end{bmatrix},$$

$$R_2(2) = \begin{bmatrix} 0.77 & -0.38 \\ -0.38 & 0.38 \end{bmatrix}, \quad R_2(3) = \begin{bmatrix} 0.99 & -0.10 \\ -0.10 & 0.52 \end{bmatrix},$$

$$R_1(4) = \begin{bmatrix} 0.10 & -0.10 \\ -0.10 & 0.50 \end{bmatrix},$$

$$Q_2(0) = \begin{bmatrix} 0.23 & -0.10 \\ -0.10 & 0.73 \end{bmatrix}, \quad Q_1(1) = \begin{bmatrix} 0.14 & -0.02 \\ -0.02 & 0.66 \end{bmatrix},$$

$$Q_2(2) = \begin{bmatrix} 0.26 & -0.16 \\ -0.16 & 0.19 \end{bmatrix}, \quad Q_2(3) = \begin{bmatrix} 0.25 & -0.01 \\ -0.01 & 0.14 \end{bmatrix},$$

$$Q_1(4) = \begin{bmatrix} 0.04 & -0.01 \\ -0.01 & 0.10 \end{bmatrix},$$

$$K_2(0) = \begin{bmatrix} -0.01 & 0.27 \end{bmatrix}, K_1(1) = \begin{bmatrix} -0.12 & -1.41 \end{bmatrix},$$

$$K_2(2) = \begin{bmatrix} -0.18 & 0.21 \end{bmatrix}, K_2(3) = \begin{bmatrix} 0.01 & -0.08 \end{bmatrix},$$

$$K_1(4) = \begin{bmatrix} 0 & 0 \end{bmatrix},$$

$$T_2(0) = \begin{bmatrix} -0.04 & 0.11 \end{bmatrix}, T_1(1) = \begin{bmatrix} 0.05 & -0.09 \end{bmatrix},$$

$$T_2(2) = \begin{bmatrix} -0.14 & 0.15 \end{bmatrix}, T_2(3) = \begin{bmatrix} 0.03 & 0.11 \end{bmatrix},$$

$$T_1(4) = \begin{bmatrix} -0.01 & 0.02 \end{bmatrix},$$

## 12.5   Summary

In this chapter, the mixed $H_2/H_\infty$ control problem has been solved for a class of nonlinear stochastic Markovian jump systems with sensor failures. The stochastic nonlinearities considered could cover several well-studied nonlinearities and the model of sensor failures is much more general than those in the existing literature. The existence of the desired multi-objective controller is expressed in terms of the solvability of certain coupled Riccati-like equations. The numerical values of the feedback controller could be obtained by solving the set of equations via the proposed Multi-Objective Controller Design Algorithm. An illustrative example has been presented to show the effectiveness and applicability of the proposed design strategy.

# 13

# Robust Variance-Constrained $H_\infty$ Control with Randomly Occurring Sensor Failures: The Finite-Horizon Case

In this chapter, the robust variance-constrained $H_\infty$ control problem is investigated for a class of nonlinear stochastic systems with possible sensor failures that are occurring in a random way. Such a phenomenon is called randomly occurring sensor failures (ROSFs). The nonlinearities described by statistical means could cover several well-studied nonlinearities, and the occurrence of the sensor failures is governed by a Bernoulli distributed random variable. The purpose of the addressed problem is to design an output-feedback controller such that, for certain systems with ROSFs, (1) the closed-loop system meets the desired $H_\infty$ performance over a finite horizon and (2) the state covariance is not more than a pre-specified upper bound at each time point. A sufficient condition for the existence of the desired controller is given and a computing algorithm is developed to achieve the aforementioned requirements simultaneously by means of a recursive linear matrix inequalities (RLMIs) approach. An illustrative simulation example is provided to show the applicability of the proposed algorithm.

The rest of the chapter is arranged as follows: Section 13.1 formulates the control problem for the nonlinear stochastic systems with ROSFs. In Section 13.2, the $H_\infty$ and variance performance indices are firstly analyzed separately, and then a theorem is proposed to combine the two performance indices in a unified framework via the RLMIs approach. Section 13.3 gives the methodology to solve the addressed control problem and outlines the computational algorithm to gain the required controller parameters recursively. A numerical example is presented in Section 13.4 to show the effectiveness of the proposed algorithm. Section 13.5 gives a summary.

*Variance-Constrained Multi-Objective Stochastic Control and Filtering*, First Edition.
Lifeng Ma, Zidong Wang and Yuming Bo.
© 2015 John Wiley & Sons, Ltd. Published 2015 by John Wiley & Sons, Ltd.

## 13.1 Problem Formulation

Consider the following time-varying nonlinear stochastic system defined on $k \in [0, N]$:

$$\begin{cases} x(k+1) = A(k)x(k) + B(k)u(k) + f(x(k), k) + D(k)w(k), \\ \quad y(k) = \beta(C(k)x(k)) + g(x(k), k) + E(k)w(k), \end{cases} \tag{13.1}$$

where $x(k) \in \mathbb{R}^n$, $y(k) \in \mathbb{R}^m$, and $u(k) \in \mathbb{R}^p$ represent the system state, output and control input respectively; $w(k) \in \mathbb{R}^r$ represents the Gaussian white sequence with covariance $W(k)$. $A(k)$, $B(k)$, $C(k)$, $D(k)$, and $E(k)$ are known real-valued time-varying matrices with appropriate dimensions.

The nonlinear stochastic functions $f(x(k), k)$ and $g(x(k), k)$ are assumed to have the following first moment for all $x(k)$ and $k$:

$$\mathbb{E}\left\{ \begin{bmatrix} f(x(k), k) \\ g(x(k), k) \end{bmatrix} \middle| x(k) \right\} = 0, \tag{13.2}$$

with the covariance given by

$$\mathbb{E}\left\{ \begin{bmatrix} f(x(k), k) \\ g(x(k), k) \end{bmatrix} [f^{\mathrm{T}}(x(j), j) \ g^{\mathrm{T}}(x(j), j)] \middle| x(k) \right\} = 0, \quad k \neq j, \tag{13.3}$$

and

$$\mathbb{E}\left\{ \begin{bmatrix} f(x(k), k) \\ g(x(k), k) \end{bmatrix} [f^{\mathrm{T}}(x(k), k) \ g^{\mathrm{T}}(x(k), k)] \middle| x(k) \right\}$$

$$= \sum_{i=1}^{q} \Pi^i(k)(x^{\mathrm{T}}(k)\Gamma_i(k)x(k))$$

$$= \sum_{i=1}^{q} \begin{bmatrix} \Pi^i_{11}(k) & \Pi^i_{12}(k) \\ (\Pi^i_{12})^{\mathrm{T}}(k) & \Pi^i_{22}(k) \end{bmatrix} (x^{\mathrm{T}}(k)\Gamma_i(k)x(k)), \tag{13.4}$$

where $\Pi^i_{mn}$ and $\Gamma_i$ $(m, n = 1, 2; i = 1, 2, \ldots, q)$ are known positive definite matrices with compatible dimensions.

**Remark 13.1** *The stochastic nonlinearities $f(x(k), k)$ and $g(x(k), k)$ described by equations (13.2) to (13.4) are frequently seen in both theoretical study and engineering practice, and could account for several classes of well-studied nonlinear systems as discussed in Ref. [31].*

The nonlinear function $\beta(\cdot)$, which illustrates the randomly occurring sensor failures, is defined as follows:

$$\beta(C(k)x(k)) = \theta(k)\sigma(C(k)x(k)) + (1 - \theta(k))C(k)x(k), \tag{13.5}$$

where the sensor failures function $\sigma(\cdot)$ has the following form:

$$\sigma(C(k)x(k)) = \Psi(x(k))C(k)x(k), \tag{13.6}$$

with $\Psi(x(k))$ being a nonlinear time-varying matrix of $x(k)$. In the rest of this chapter, without any confusion, we denote $\Psi(k) \triangleq \Psi(x(k))$ for brevity of notation.

In this chapter, the sensor failures matrix $\Psi(k)$ is assumed to have the following form:

$$\Psi(k) = \text{diag}\{\psi_1(k), \psi_2(k), \ldots, \psi_m(k)\},$$
$$0 \le \psi_{il}(k) \le \psi_i(k) \le \psi_{iu}(k) < \infty, \quad \psi_{il}(k) < 1, \psi_{iu}(k) \ge 1, \tag{13.7}$$

with $\psi_{il}(k)$ and $\psi_{iu}(k)$ being known constants characterizing the lower and upper bounds on $\psi_i(k)$ respectively.

**Remark 13.2** *The function* (13.6) *with* (13.7) *is introduced to describe the constraints imposed on the output sampling when the ROSFs are occurring, where $\psi_i(k)(i = 1, 2, \ldots, m)$ represents the possible failures occurring on the ith sensing channel. Specifically, when $\psi_i = 1$, there will be no sensor failures, while $0 < \psi_i(k) < 1$ means there might be possible data missing, measurements degradation, or inaccurate signal amplification during the process of output sampling.*

The random variable $\theta(k)$ in (13.5), which accounts for the ROSFs, is assumed to be independent of $w(k)$ as well as the stochastic nonlinearities $f(x(k), k)$ and $g(x(k), k)$, and is a Bernoulli distributed sequence taking values on 0 and 1 with

$$\text{Prob}\{\theta(k) = 1\} = \mathbb{E}\{\theta(k)\} = \bar{\theta}, \tag{13.8}$$

where $\bar{\theta}$ is a known positive constant. Hence, it can be easily found that

$$\kappa_\theta^2 = \mathbb{E}\{(\theta(k) - \bar{\theta})^2\} = (1 - \bar{\theta})\bar{\theta}. \tag{13.9}$$

**Remark 13.3** *The Bernoulli distributed random variable $\theta(k)$ is employed to illustrate the phenomenon of ROSFs. To be specific, when $\theta(k) = 0$, the sensor will not be confronted with failures; otherwise certain constraints might be imposed on the output signal due to the sensor failures. Compared to the traditional models, which treat the sensor failures in a deterministic way, such a model proposed in this chapter could better illustrate the systems in real-world engineering practice due to the wide existence of time-varying working conditions.*

Let $\Psi(k) = H(k) + H(k)N(k)$, where

$$H(k) = \text{diag}\{h_1(k), h_2(k), \ldots, h_m(k)\}, \quad h_i(k) = \frac{\psi_{il}(k) + \psi_{iu}(k)}{2},$$
$$N(k) = \text{diag}\{n_1(k), n_2(k), \ldots, n_m(k)\}, \quad n_i(k) = \frac{\psi_i(k) - h_i(k)}{h_i(k)}. \tag{13.10}$$

Denoting    $L(k) = \text{diag}\{l_1(k), l_2(k), \ldots, l_m(k)\}$    with    $l_i(k) = (\psi_{iu}(k) - \psi_{il}(k))/$ $(\psi_{iu}(k) + \psi_{il}(k))$, we obtain

$$N^{\mathrm{T}}(k)N(k) \le L^{\mathrm{T}}(k)L(k) \le I. \tag{13.11}$$

So far, we have converted the constraints on the system output imposed by the sensor failures into certain norm-bounded parameter uncertainties. Then the original system (13.1) can be reformulated as follows:

$$\begin{cases} x(k+1) = A(k)x(k) + B(k)u(k) + f(x(k), k) + D(k)w(k), \\ \quad y(k) = \hat{C}(k)x(k) + \theta(k)H(k)N(k)C(k)x(k) + g(x(k), k) + E(k)w(k), \end{cases} \tag{13.12}$$

where $\hat{C}(k) = (\theta(k)H(k) + (1 - \theta(k))I)C(k)$. In the following, we shall give the design objective of this chapter. Firstly, we define an $H_\infty$ performance index for system (13.12) as follows:

$$J_\infty = \mathbb{E}\left\{ \sum_{k=0}^{N-1} \left( \|y(k)\|^2 - \gamma^2\|\omega(k)\|^2 \right) \right\} - \gamma^2\|x(0)\|_\Omega^2, \tag{13.13}$$

where $\gamma > 0$ is a given positive scalar and $\Omega > 0$ is a known weighting matrix.

In this chapter, it is our objective to determine an output feedback controller of the following form:

$$\begin{cases} x_g(k+1) = A_g(k)x_g(k) + B_g(k)y(k), \\ \quad u(k) = C_g(k)x_g(k), \end{cases} \tag{13.14}$$

such that, for the closed-loop system, given a pre-specified $H_\infty$ disturbance attenuation level $\gamma > 0$, $J_\infty < 0$ holds for all nonzero $w(k)$ and $x(0)$.

Before giving our main results, we need firstly to reconstruct the system (13.12) so as to obtain a system without parameter uncertainty $N(k)$ as follows:

$$\begin{cases} x(k+1) = A(k)x(k) + B(k)u(k) + f(x(k), k) + \hat{D}(k)\hat{D}(k), \\ \quad y(k) = \hat{C}(k)x(k) + g(x(k), k) + \hat{E}(k)\hat{w}(k), \end{cases} \tag{13.15}$$

where

$$\hat{w}(k) = \begin{bmatrix} w(k) \\ \gamma^{-1}\varepsilon(k)N(k)C(k)x(k) \end{bmatrix}, \quad \hat{D}(k) = \begin{bmatrix} D(k) & 0 \end{bmatrix},$$

$$\hat{E}(k) = \begin{bmatrix} E(k) & \gamma\theta(k)\varepsilon^{-1}(k)H(k) \end{bmatrix},$$

with $\varepsilon(k) \ne 0$ being any known real-valued constant.

We now consider the following performance index:

$$\bar{J}_\infty = \mathbb{E}\left\{ \sum_{k=0}^{N-1} \left( \|y(k)\|^2 + \|\varepsilon(k)L(k)C(k)x(k)\|^2 - \gamma^2\|\hat{\omega}(k)\|^2 \right) \right\} - \gamma^2\|x(0)\|_\Omega^2. \tag{13.16}$$

Next, we shall show that $\bar{J}_\infty$ is an upper bound of $J_\infty$. To this end, subtracting (13.13) from (13.16) leads to

$$
\begin{aligned}
\bar{J}_\infty - J_\infty &= \mathbb{E}\left\{ \sum_{k=0}^{N-1} \|\varepsilon(k)L(k)C(k)x(k)\|^2 - \gamma^2 \left(\|\hat{\omega}(k)\|^2 - \|\omega(k)\|^2\right)\right\} \\
&= \mathbb{E}\left\{ \sum_{k=0}^{N-1} \|\varepsilon(k)L(k)C(k)x(k)\|^2 - \|\varepsilon(k)N(k)C(k)x(k)\|^2 \right\} \\
&= \mathbb{E}\left\{ \sum_{k=0}^{N-1} \left( \varepsilon^2(k)x^{\mathrm{T}}(k)C^{\mathrm{T}}(k)\left(L^{\mathrm{T}}(k)L(k) - N^{\mathrm{T}}(k)N(k)\right)C(k)x(k)\right)\right\} \\
&\geq 0.
\end{aligned}
\tag{13.17}
$$

Thus, it is easy to see that $\bar{J}_\infty$ is an upper bound of $J_\infty$. It is worth noting that since $N(k)$ can be any arbitrary matrix corresponding to any arbitrary sensor failure $\bar{J}_\infty$ is a tight upper bound of $J_\infty$.

Applying the full order dynamic output controller (13.14) to system (13.15), we have the following closed-loop system:

$$
\begin{cases}
\xi(k+1) = A_f(k)\xi(k) + G_f(k)h(k) + D_f(k)\hat{w}(k), \\
y(k) = C_f(k)\xi(k) + g(x(k), k) + \hat{E}(k)\hat{w}(k),
\end{cases}
\tag{13.18}
$$

where

$$
\xi(k) = \begin{bmatrix} x(k) \\ x_g(k) \end{bmatrix}, \quad h(k) = \begin{bmatrix} f(x(k), k) \\ g(x(k), k) \end{bmatrix},
$$

$$
A_f(k) = \begin{bmatrix} A(k) & B(k)C_g(k) \\ B_g(k)\hat{C}(k) & A_g(k) \end{bmatrix},
$$

$$
G_f(k) = \begin{bmatrix} I & 0 \\ 0 & B_g(k) \end{bmatrix}, \quad D_f(k) = \begin{bmatrix} \hat{D}(k) \\ B_g(k)\hat{E}(k) \end{bmatrix},
$$

$$
C_f(k) = \begin{bmatrix} \hat{C}(k) & 0 \end{bmatrix}.
$$

In this chapter, it is our objective to design a finite-horizon dynamic output feedback controller of form (13.14) such that the following two requirements are satisfied simultaneously:

(R1) For given $\gamma > 0$, $\Omega > 0$, and the initial state $x(0)$, the $H_\infty$ performance index

$$
\bar{J}_\infty < 0
\tag{13.19}
$$

is achieved for all stochastic nonlinearities and all possible randomly occurring sensor failures.

(R2) For a sequence of specified positive definite matrices $\{\Theta(k)\}_{0<k\leq N}$, at each sampling instant $k$, the system state satisfies

$$X(k) \triangleq \mathbb{E}\{x(k)x^{\mathrm{T}}(k)\} \leq \Theta(k), \quad \forall k. \tag{13.20}$$

The control problem addressed above is referred to as the robust $H_\infty$ output feedback control problem for nonlinear discrete time-varying stochastic systems with covariance constraints.

## 13.2 $H_\infty$ and Covariance Performance Analysis

In this section, the $H_\infty$ and covariance performances will be firstly analyzed separately for the time-varying stochastic system (13.18). Then, in terms of recursive matrix inequalities, a theorem is proposed to incorporate the two performance indices within a unified framework.

### 13.2.1 $H_\infty$ Performance

In this subsection, a sufficient condition is given for the closed-loop system (13.18) to guarantee the prescribed $H_\infty$ and covariance performance indices simultaneously. The following assignments are made for notational convenience:

$$\bar{C}_f(k) = \begin{bmatrix} \bar{C}(k) & 0 \end{bmatrix}, \quad \bar{C}(k) = (\bar{\theta}H(k) + (1-\bar{\theta})I)C(k),$$

$$\check{C}_f(k) = \begin{bmatrix} \check{C}(k) & 0 \end{bmatrix}, \quad \check{C}(k) = \kappa_\theta(H(k) - I)C(k),$$

$$\bar{A}_f(k) = \begin{bmatrix} A(k) & B(k)C_g(k) \\ B_g(k)\bar{C}(k) & A_g(k) \end{bmatrix}, \quad \check{A}_f(k) = \begin{bmatrix} 0 & 0 \\ B_g(k)\check{C}(k) & 0 \end{bmatrix},$$

$$\bar{E}(k) = \begin{bmatrix} E(k) & \gamma\bar{\theta}\varepsilon^{-1}(k)H(k) \end{bmatrix}, \quad \check{E}(k) = \begin{bmatrix} 0 & \gamma\kappa_\theta\varepsilon^{-1}(k)H(k) \end{bmatrix},$$

$$\bar{D}_f(k) = \begin{bmatrix} \hat{D}(k) \\ B_g(k)\bar{E}(k) \end{bmatrix}, \quad \check{D}_f(k) = \begin{bmatrix} 0 \\ B_g(k)\check{E}(k) \end{bmatrix},$$

$$\bar{\Omega} = \mathrm{diag}\{\Omega, 0\}, \quad Z = \begin{bmatrix} I & 0 \end{bmatrix}.$$

The following theorem gives a sufficient condition for system (13.18) to achieve the $H_\infty$ requirement.

**Theorem 13.2.1** *Given a positive scalar $\gamma > 0$ and a matrix $\Omega > 0$. Let the feedback controller gains $A_g(k)$, $B_g(k)$, and $C_g(k)$ be given. If there exist a sequence of positive definite matrices $\{P(k)\}_{1\leq k\leq N}$ with $P(0) \leq \gamma^2\bar{\Omega}$ satisfying the following matrix inequality*

$$\Lambda(k) \triangleq \begin{bmatrix} \Lambda_{11}(k) & \Lambda_{12}(k) \\ \Lambda_{12}^{\mathrm{T}}(k) & \Lambda_{22}(k) \end{bmatrix} < 0, \tag{13.21}$$

*where*

$$\Lambda_{11}(k) = \bar{A}_f^{\mathrm{T}}(k)P(k+1)\bar{A}_f(k) + \check{A}_f^{\mathrm{T}}(k)P(k+1)\check{A}_f(k) - P(k) + \bar{C}_f^{\mathrm{T}}(k)\bar{C}_f(k) + \check{C}_f^{\mathrm{T}}(k)\check{C}_f(k)$$

$$+ \sum_{i=1}^{q} Z^{\mathrm{T}}\Gamma_i(k)Z(\mathrm{tr}[G_f(k)^{\mathrm{T}}P(k+1)G_f(k)\Pi^i(k)] + \mathrm{tr}[\Pi_{22}^i(k)])$$

$$+ \varepsilon^2(k)Z^{\mathrm{T}}C^{\mathrm{T}}(k)L^{\mathrm{T}}(k)L(k)C(k)Z,$$

$$\Lambda_{12}(k) = \bar{A}_f^{\mathrm{T}}(k)P(k+1)\bar{D}_f(k) + \check{A}_f^{\mathrm{T}}(k)P(k+1)\check{D}_f(k) + \bar{C}_f^{\mathrm{T}}(k)\bar{E}(k) + \check{C}_f^{\mathrm{T}}(k)\check{E}(k),$$

$$\Lambda_{22}(k) = -\gamma^2 I + \bar{D}_f^{\mathrm{T}}(k)P(k+1)\bar{D}_f(k) + \check{D}_f^{\mathrm{T}}(k)P(k+1)\check{D}_f(k) + \bar{E}^{\mathrm{T}}(k)\bar{E}(k) + \check{E}^{\mathrm{T}}(k)\check{E}(k),$$

*then the required $H_\infty$ performance of the closed-loop system* (13.18) *is achieved, that is, $J_\infty < 0$.*

*Proof.* Define

$$J(k) \triangleq \xi^{\mathrm{T}}(k+1)P(k+1)\xi(k+1) - \xi^{\mathrm{T}}(k)P(k)\xi(k). \tag{13.22}$$

Substituting (13.18) into (13.22) leads to

$$\begin{aligned}
\mathbb{E}\{J(k)\} &= \mathbb{E}\{\xi^{\mathrm{T}}(k+1)P(k+1)\xi(k+1) - \xi^{\mathrm{T}}(k)P(k)\xi(k)\} \\
&= \mathbb{E}\{\xi^{\mathrm{T}}(k)(A_f^{\mathrm{T}}(k)P(k+1)A_f(k) - P(k))\xi(k) \\
&\quad + h^{\mathrm{T}}(k)G_f^{\mathrm{T}}(k)P(k+1)G_f(k)h(k) \\
&\quad + 2\xi^{\mathrm{T}}(k)A_f^{\mathrm{T}}(k)P(k+1)D_f(k)\hat{\omega}(k) \\
&\quad + \hat{\omega}^{\mathrm{T}}(k)D_f^{\mathrm{T}}(k)P(k+1)D_f(k)\hat{\omega}(k)\}.
\end{aligned} \tag{13.23}$$

Taking (13.4) into consideration, we can obtain

$$\begin{aligned}
&\mathbb{E}\{h^{\mathrm{T}}(k)G_f^{\mathrm{T}}(k)P(k+1)G_f(k)h(k)\} \\
&= \mathbb{E}\{\mathrm{tr}[G_f^{\mathrm{T}}(k)P(k+1)G_f(k)h(k)h^{\mathrm{T}}(k)]\} \\
&= \mathbb{E}\left\{\mathrm{tr}\left[G_f^{\mathrm{T}}(k)P(k+1)G_f(k)\sum_{i=1}^{q}\Pi^i(k)x^{\mathrm{T}}(k)\Gamma_i(k)x(k)\right]\right\} \\
&= \mathbb{E}\left\{\xi^{\mathrm{T}}(k)\left(\sum_{i=1}^{q}Z^{\mathrm{T}}\Gamma_i(k)Z\mathrm{tr}\left[G_f^{\mathrm{T}}(k)P(k+1)G_f(k)\Pi^i(k)\right]\right)\xi(k)\right\}
\end{aligned} \tag{13.24}$$

and therefore

$$
\begin{aligned}
\mathbb{E}\{J(k)\} = \mathbb{E}\Big\{ & \xi^{\mathrm{T}}(k)\Big(A_f^{\mathrm{T}}(k)P(k+1)A_f(k) - P(k) \\
& + \sum_{i=1}^{q} Z^{\mathrm{T}}\Gamma_i(k)Z\mathrm{tr}\Big[G_f^{\mathrm{T}}(k)P(k+1)G_f(k)\Pi^i(k)\Big]\Big)\xi(k) \\
& + 2\xi^{\mathrm{T}}(k)A_f^{\mathrm{T}}(k)P(k+1)D_f(k)\hat{\omega}(k) \\
& + \hat{\omega}^{\mathrm{T}}(k)D_f^{\mathrm{T}}(k)P(k+1)D_f(k)\hat{\omega}(k)\Big\}.
\end{aligned}
\tag{13.25}
$$

Adding the following zero term

$$
\begin{aligned}
& \|y(k)\|^2 + \|\varepsilon(k)L(k)C(k)x(k)\|^2 - \gamma^2\|\hat{\omega}(k)\|^2 \\
& - (\|y(k)\|^2 + \|\varepsilon(k)L(k)C(k)x(k)\|^2 - \gamma^2\|\hat{\omega}(k)\|^2)
\end{aligned}
\tag{13.26}
$$

to both sides of (13.25) results in

$$
\begin{aligned}
\mathbb{E}\{J(k)\} = \mathbb{E}\Big\{ & \xi^{\mathrm{T}}(k)\Big(A_f^{\mathrm{T}}(k)P(k+1)A_f(k) - P(k) \\
& + \sum_{i=1}^{q} Z^{\mathrm{T}}\Gamma_i(k)Z(\mathrm{tr}[G_f^{\mathrm{T}}(k)P(k+1)G_f(k)\Pi^i(k)] + \mathrm{tr}[\Pi^i_{22}(k)]) \\
& + C_f^{\mathrm{T}}(k)C_f(k) + \varepsilon^2(k)Z^{\mathrm{T}}C^{\mathrm{T}}(k)L^{\mathrm{T}}(k)L(k)C(k)Z\Big)\xi(k) \\
& + \hat{\omega}^{\mathrm{T}}(k)\Big(D_f^{\mathrm{T}}(k)P(k+1)D_f(k) + \hat{E}^{\mathrm{T}}(k)\hat{E}(k) - \gamma^2 I\Big)\hat{\omega}(k) \\
& + 2\xi^{\mathrm{T}}(k)\Big(A_f^{\mathrm{T}}(k)P(k+1)D_f(k) + C_f^{\mathrm{T}}(k)\hat{E}(k)\Big)\hat{\omega}(k) \\
& - \Big(\|y(k)\|^2 + \|\varepsilon(k)L(k)C(k)x(k)\|^2 - \gamma^2\|\hat{\omega}(k)\|^2\Big)\Big\}.
\end{aligned}
\tag{13.27}
$$

Noting that

$$
\begin{aligned}
\mathbb{E}\{A_f^{\mathrm{T}}(k)P(k+1)A_f(k)\} &= \bar{A}_f^{\mathrm{T}}(k)P(k+1)\bar{A}_f(k) + \breve{A}_f^{\mathrm{T}}P(k+1)\breve{A}_f(k), \\
\mathbb{E}\{A_f^{\mathrm{T}}(k)P(k+1)D_f(k)\} &= \bar{A}_f^{\mathrm{T}}(k)P(k+1)\bar{D}_f(k) + \breve{A}_f^{\mathrm{T}}P(k+1)\breve{D}_f(k), \\
\mathbb{E}\{D_f^{\mathrm{T}}(k)P(k+1)D_f(k)\} &= \bar{D}_f^{\mathrm{T}}(k)P(k+1)\bar{D}_f(k) + \breve{D}_f^{\mathrm{T}}(k)P(k+1)\breve{D}_f(k), \\
\mathbb{E}\{C_f^{\mathrm{T}}(k)C_f(k)\} &= \bar{C}_f^{\mathrm{T}}(k)\bar{C}_f(k) + \breve{C}_f^{\mathrm{T}}(k)\breve{C}_f(k), \\
\mathbb{E}\{C_f^{\mathrm{T}}(k)\hat{E}(k)\} &= \bar{C}_f^{\mathrm{T}}(k)\bar{E}(k) + \breve{C}_f^{\mathrm{T}}(k)\breve{E}(k), \\
\mathbb{E}\{\hat{E}^{\mathrm{T}}(k)\hat{E}(k)\} &= \bar{E}^{\mathrm{T}}(k)\bar{E}(k) + \breve{E}^{\mathrm{T}}(k)\breve{E}(k),
\end{aligned}
\tag{13.28}
$$

we can see that

$$\mathbb{E}\{J(k)\} = \xi^{\mathrm{T}}(k) \left( \bar{A}_f^{\mathrm{T}}(k)P(k+1)\bar{A}_f(k) + \check{A}_f^{\mathrm{T}}P(k+1)\check{A}_f(k) - P(k) \right.$$

$$+ \sum_{i=1}^q Z^{\mathrm{T}}\Gamma_i(k) \times Z(\mathrm{tr}[G_f^{\mathrm{T}}(k)P(k+1)G_f(k)\Pi^i(k)]$$

$$+ \bar{C}_f^{\mathrm{T}}(k)\bar{C}_f(k) + \check{C}_f^{\mathrm{T}}(k)\check{C}_f(k) + tr[\Pi_{22}^i(k)]$$

$$+ \varepsilon^2(k)Z^{\mathrm{T}}C^{\mathrm{T}}(k)L^{\mathrm{T}}(k)L(k)C(k)Z \Big) \xi(k)$$

$$+ \hat{\omega}^{\mathrm{T}}(k) \left( \bar{D}_f^{\mathrm{T}}(k)P(k+1)\bar{D}_f(k) + \check{D}_f^{\mathrm{T}}(k)P(k+1)\check{D}_f(k) \right.$$

$$+ \bar{E}^{\mathrm{T}}(k)\bar{E}(k) + \check{E}^{\mathrm{T}}(k)\check{E}(k) - \gamma^2 I \Big) \hat{\omega}(k) \qquad (13.29)$$

$$+ 2\xi^{\mathrm{T}}(k) \left( \check{A}_f^{\mathrm{T}}P(k+1)\check{D}_f(k) + \bar{C}_f^{\mathrm{T}}(k)\bar{E}(k) \right.$$

$$+ \bar{A}_f^{\mathrm{T}}(k)P(k+1)\bar{D}_f(k) + \check{C}_f^{\mathrm{T}}(k)\check{E}(k) \Big) \hat{\omega}(k)$$

$$- \left( \|y(k)\|^2 + \|\varepsilon(k)L(k)C(k)x(k)\|^2 - \gamma^2\|\hat{\omega}(k)\|^2 \right)$$

$$= \begin{bmatrix} \xi(k) \\ \hat{\omega}(k) \end{bmatrix}^{\mathrm{T}} \Lambda(k) \begin{bmatrix} \xi(k) \\ \hat{\omega}(k) \end{bmatrix}$$

$$- \mathbb{E}\{\|y(k)\|^2 + \|\varepsilon(k)L(k)C(k)x(k)\|^2 - \gamma^2\|\hat{\omega}(k)\|^2\}.$$

Summing (13.29) on both sides from 0 to $N-1$ with respect to $k$ results in

$$\mathbb{E}\left\{ \sum_{k=0}^{N-1} J(k) \right\} = \mathbb{E}\{\xi^{\mathrm{T}}(N)P(N)\xi(N)\} - \xi^{\mathrm{T}}(0)P(0)\xi(0)$$

$$= \sum_{k=0}^{N-1} \left( \begin{bmatrix} \xi(k) \\ \hat{\omega}(k) \end{bmatrix}^{\mathrm{T}} \Lambda(k) \begin{bmatrix} \xi(k) \\ \hat{\omega}(k) \end{bmatrix} \right. \qquad (13.30)$$

$$- \mathbb{E}\left\{ \|y(k)\|^2 + \|\varepsilon(k)L(k)C(k)x(k)\|^2 - \gamma^2\|\hat{\omega}(k)\|^2 \right\} \bigg).$$

Hence, the $H_\infty$ performance index defined in (13.16) is given by

$$\bar{J}_\infty = \mathbb{E}\left\{ \sum_{k=0}^{N-1} \left( \|y(k)\|^2 + \|\varepsilon(k)L(k)C(k)x(k)\|^2 - \gamma^2\|\hat{\omega}(k)\|^2 \right) \right\} - \gamma^2\|x(0)\|_\Omega^2$$

$$= \xi^{\mathrm{T}}(0)(P(0) - \gamma^2\bar{\Omega})\xi(0) - \mathbb{E}\{\xi^{\mathrm{T}}(N)P(N)\xi(N)\} \qquad (13.31)$$

$$+ \sum_{k=0}^{N-1} \left( \begin{bmatrix} \xi(k) \\ \hat{\omega}(k) \end{bmatrix}^{\mathrm{T}} \Lambda(k) \begin{bmatrix} \xi(k) \\ \hat{\omega}(k) \end{bmatrix} \right).$$

Noting that $\Lambda(k) < 0$, $P(0) \leq \gamma^2 \bar{\Omega}$, and $P(N) > 0$, we immediately arrive at $\bar{J}_\infty < 0$, which indicates $J_\infty < 0$. The proof is complete. □

### 13.2.2　Covariance Analysis

This subsection discusses the performance index of state covariance for the closed-loop system (13.18). Firstly, define the state covariance matrix of system (13.18) by

$$\Xi(k) \triangleq \mathbb{E}\{\xi(k)\xi^{\mathrm{T}}(k)\}. \tag{13.32}$$

It is easy to see that

$$X(k) = [I \quad 0]\Xi(k)\begin{bmatrix} I \\ 0 \end{bmatrix} = Z\Xi(k)Z^{\mathrm{T}}. \tag{13.33}$$

Next, define a matrix-valued function as

$$\mathcal{F}(Y(k)) \triangleq A_f(k)Y(k)A_f^{\mathrm{T}}(k) + D_f(k)\tilde{W}(k)D_f^{\mathrm{T}}(k) \\ + \sum_{i=1}^{q} G_f(k)\Pi^i(k)G_f^{\mathrm{T}}(k)\mathrm{tr}[\Gamma_i(k)Y(k)], \tag{13.34}$$

where

$$\tilde{W}(k) = \begin{bmatrix} W(k) & 0 \\ 0 & \gamma^{-2}\varepsilon^2(k)L(k)C(k)ZY(k)Z^{\mathrm{T}}C^{\mathrm{T}}(k)L^{\mathrm{T}}(k) \end{bmatrix}.$$

Then the upper bound of the covariance matrix $\Xi(k)$ can be obtained as given in the following Theorem 13.2.2.

**Theorem 13.2.2** *Consider system* (13.18). *Given the controller feedback gains* $A_g(k)$, $B_g(k)$, *and* $C_g(k)$. *If there exist a sequence of positive definite matrices* $\{Q(k)\}_{1 \leq k \leq N}$ *satisfying the following matrix inequality*

$$Q(k+1) \geq \mathcal{F}(Q(k)), \tag{13.35}$$

*with the initial condition* $Q(0) = \Xi(0)$, *then* $Q(k) \geq \Xi(k)$, $\forall k \in \{1, 2, \ldots, N\}$.

*Proof.* The Lyapunov-type equation that governs the evolution of the state covariance $\Xi(k)$ of closed-loop system (13.18) is given by

$$\begin{aligned} \Xi(k+1) &= A_f(k)\Xi(k)A_f^{\mathrm{T}}(k) + \mathbb{E}\{G_f(k)h(k)h^{\mathrm{T}}(k)G_f^{\mathrm{T}}(k)\} + D_f(k)\bar{W}(k)D_f^{\mathrm{T}}(k) \\ &= A_f(k)\Xi(k)A_f^{\mathrm{T}}(k) + D_f(k)\bar{W}(k)D_f^{\mathrm{T}}(k) \\ &\quad + \sum_{i=1}^{q} G_f(k)\Pi^i(k)G_f^{\mathrm{T}}(k)\mathrm{tr}[\Gamma_i(k)\Xi(k)], \end{aligned} \tag{13.36}$$

where

$$\bar{W}(k) = \begin{bmatrix} W(k) & 0 \\ 0 & \gamma^{-2}\varepsilon^2(k)N(k)C(k)Z\Xi(k)Z^{\mathrm{T}}C^{\mathrm{T}}(k)N^{\mathrm{T}}(k) \end{bmatrix}.$$

The following proof is done by induction. Obviously, $Q(0) \geq \Xi(0)$. Suppose $Q(k) \geq \Xi(k)$; then

$$\mathrm{tr}[\Gamma_i(k)Q(k)] \geq \mathrm{tr}[\Gamma_i(k)\Xi(k)]. \tag{13.37}$$

Taking (13.34) into consideration, we can easily see that

$$Q(k+1) \geq \mathcal{F}(Q(k)) \geq \Xi(k+1), \tag{13.38}$$

which completes the proof.                                                                 □

So far, we have analyzed the $H_\infty$ performance and covariance constraint separately for the closed-loop system. In the following stage, a theorem is presented to conclude the above analysis, which intends to handle both $H_\infty$ and covariance performance simultaneously. A unified framework is constructed by means of the recursive matrix inequalities approach. First of all, without loss of any generality, we denote

$$\Pi^i = \pi_i\pi_i^{\mathrm{T}} = \begin{bmatrix} \pi_{1i} \\ \pi_{2i} \end{bmatrix}\begin{bmatrix} \pi_{1i} \\ \pi_{2i} \end{bmatrix}^{\mathrm{T}},$$

where $\pi_i = [\pi_{1i}^{\mathrm{T}} \quad \pi_{2i}^{\mathrm{T}}]^{\mathrm{T}}$ $(i = 1, 2, \ldots, q)$ are column vectors of appropriate dimensions.

**Theorem 13.2.3** *Consider the system* (13.18). *Given the controller feedback gains* $A_g(k)$, $B_g(k)$, *and* $C_g(k)$, *a positive scalar* $\gamma > 0$, *and a matrix* $\Omega > 0$. *If there exist sequences of positive definite matrices* $\{Q(k)\}_{1\leq k\leq N}$, $\{P(k)\}_{1\leq k\leq N}$, *and a sequence of positive scalars* $\{\eta_i(k)\}_{0\leq k\leq N}$ $(i = 1, 2, \ldots, q)$ *satisfying the following recursive matrix inequalities*

$$\begin{bmatrix} -\eta_i(k) & \pi_i^{\mathrm{T}}G_f^{\mathrm{T}}(k) \\ * & -P^{-1}(k+1) \end{bmatrix} < 0, \tag{13.39}$$

$$\begin{bmatrix} \hat{Y}(k) & C_f^{\mathrm{T}}(K)\hat{E}(k) & A_f^{\mathrm{T}}(k) \\ * & -\gamma^2 I + \hat{E}^{\mathrm{T}}(k)\hat{E}(k) & D_f^{\mathrm{T}}(k) \\ * & * & -P^{-1}(k+1) \end{bmatrix} < 0, \tag{13.40}$$

$$\begin{bmatrix} -Q(k+1) & A_f(k)Q(k) & \hat{\Phi}_{13}(k) & D_f(k) \\ * & -Q(k) & 0 & 0 \\ * & * & \hat{\Phi}_{33}(k) & 0 \\ * & * & * & -\hat{W}^{-1}(k) \end{bmatrix} < 0, \tag{13.41}$$

*with the initial condition*

$$\begin{cases} P(0) \leq \gamma^2\bar{\Omega}, \\ Q(0) = \Xi(0), \end{cases} \tag{13.42}$$

*where*

$$\hat{\Upsilon}(k) = - P(k) + C_f^{\mathrm{T}}(k)C_f(k) + \sum_{i=1}^{q} \Gamma_i(k) \times (\eta_i(k) + \mathrm{tr}[\Pi_{22}^i](k))$$

$$+ \varepsilon^2(k)Z^{\mathrm{T}}C^{\mathrm{T}}(k)L^{\mathrm{T}}(k)L(k)C(k)Z,$$

$$\hat{W}(k) = \begin{bmatrix} W(k) & 0 \\ 0 & \gamma^{-2}\varepsilon^2(k)L(k)C(k)ZQ(k)Z^{\mathrm{T}}C^{\mathrm{T}}(k)L^{\mathrm{T}}(k) \end{bmatrix},$$

$$\hat{\Phi}_{13}(k) = [G_f(k)\pi_1 \quad G_f(k)\pi_2 \quad \cdots \quad G_f(k)\pi_q],$$

$$\hat{\Phi}_{33}(k) = \mathrm{diag}\{ - \rho_1 I, -\rho_2 I, \ldots, -\rho_q I\},$$

$$\rho_i = (\mathrm{tr}[\Gamma_i(k)P(k)])^{-1}, \quad i = 1, 2, \cdots, q,$$

*then, for the closed-loop system* (13.18), $J(k) < 0$ *and* $\Xi(k) \leq Q(k)$ *are satisfied simultaneously.*

*Proof.* Based on the analysis of $H_\infty$ performance and system state covariance in Subsections 13.2.1 and 13.2.2, it only needs to show that, under the initial condition (13.42), inequalities (13.39) and (13.40) imply inequality (13.21), while inequality (13.41) is equivalent to inequality (13.35).

Actually, by the Schur Complement Lemma, inequality (13.39) is equivalent to

$$-\eta_i(k) + \pi_i^{\mathrm{T}} G_f^{\mathrm{T}}(k)P(k + 1)G_f(k) < 0, \tag{13.43}$$

which means that

$$\pi_i^{\mathrm{T}} G_f^{\mathrm{T}}(k)P(k + 1)G_f(k) < \eta_i(k). \tag{13.44}$$

By the property of matrix trace, it can be rewritten as

$$tr[G_f^{\mathrm{T}}(k)P(k + 1)G_f(k)\Pi^i(k)] < \eta_i(k). \tag{13.45}$$

Similarly, using the Schur Complement Lemma again, inequality (13.40) is equivalent to

$$\hat{\Lambda}(k) \triangleq \begin{bmatrix} \hat{\Lambda}_{11}(k) & \hat{\Lambda}_{12}(k) \\ \hat{\Lambda}_{12}^{\mathrm{T}}(k) & \hat{\Lambda}_{22}(k) \end{bmatrix} < 0, \tag{13.46}$$

where

$$\hat{\Lambda}_{11}(k) = A_f^{\mathrm{T}}(k)P(k + 1)A_f(k) - P(k) + C_f^{\mathrm{T}}(k)C_f(k)$$

$$+ \sum_{i=1}^{q} \Gamma_i(k) \times (\eta_i(k) + \mathrm{tr}[\Pi_{22}^i]) + \varepsilon^2(k)Z^{\mathrm{T}}C^{\mathrm{T}}(k)L^{\mathrm{T}}(k)L(k)C(k)Z,$$

$$\hat{\Lambda}_{12}(k) = A_f^{\mathrm{T}}(k)P(k + 1)D_f(k) + C_f^{\mathrm{T}}(k)\hat{E}(k),$$

$$\hat{\Lambda}_{22}(k) = - \gamma^2 I + D_f^{\mathrm{T}}(k)P(k + 1)D_f(k) + \hat{E}^{\mathrm{T}}(k)\hat{E}(k).$$

Taking (13.28) into consideration, we can see that inequality (13.21) can be implied by inequalities (13.39) and (13.40) under the same initial condition.

Likewise, employing the Schur Complement Lemma to inequality (13.41), we have

$$Q(k + 1) \geq \mathcal{F}(Q(k)). \tag{13.47}$$

Then it can be easily verified that inequality (13.41) is equivalent to (13.35). Thus, according to Theorem 13.2.1 and Theorem 13.2.2, the $H_\infty$ index defined in (13.16) satisfies $\bar{J}_\infty < 0$ and, at the same time, the covariance of closed-loop system (13.18) achieves $\Xi(k) \leq Q(k), \forall k \in \{0, 1, \cdots, N\}$. The proof is now complete. $\qquad\square$

## 13.3    Robust Finite-Horizon Controller Design

### 13.3.1    Controller Design

In this subsection, an algorithm is presented based on Theorem 13.2.3 to solve the output feedback control problem for the nonlinear stochastic system (13.1). The controller gains can be obtained by solving certain set of RLMIs recursively, which means that a set of LMIs will be solved recursively to obtain the desired controller parameters at each time point.

The following theorem provides the sufficient condition of the solvability of the addressed output feedback controller design problem.

**Theorem 13.3.1** *For a given disturbance attenuation level $\gamma > 0$, a positive definite matrix $\Omega > 0$ and a sequence of pre-specified covariance upper bounds $\{\Theta(k)\}_{0 \leq k \leq N}$, if there exist sequences of positive definite matrices $\{M(k)\}_{1 \leq k \leq N}$, $\{N(k)\}_{1 \leq k \leq N}$, $\{Q_1(k)\}_{1 \leq k \leq N}$, $\{Q_2(k)\}_{1 \leq k \leq N}$, a sequence of positive scalars $\{\eta_i(k)\}_{1 \leq k \leq N}$ $(i = 1, 2, \ldots, q)$, and sequences of real-valued matrices $\{Q_3(k)\}_{1 \leq k \leq N}$, $\{V(k)\}_{1 \leq k \leq N}$, $\{A_g(k)\}_{0 \leq k \leq N}$, $\{B_g(k)\}_{0 \leq k \leq N}$, and $\{C_g(k)\}_{0 \leq k \leq N}$, under the initial condition*

$$\begin{cases} P_1(0) \leq \gamma^2 \Omega, \\ P_2(0) = P_3(0) = 0, \\ X(0) = Q_1(0) \leq \Theta(0), \\ Q_2(0) = Q_3(0) = 0, \end{cases} \tag{13.48}$$

*such that the following recursive LMIs*

$$\begin{bmatrix} -\eta_i(k) & \pi_{1i}^{\mathrm{T}} & \pi_{2i}^{\mathrm{T}} B_g^{\mathrm{T}}(k) \\ * & -M(k+1) & -V(k+1) \\ * & * & -N(k+1) \end{bmatrix} < 0, \tag{13.49}$$

$$\Upsilon(k) \triangleq \begin{bmatrix} \Upsilon_{11}(k) & \Upsilon_{21}^{\mathrm{T}}(k) \\ * & \Upsilon_{22}(k) \end{bmatrix} < 0, \tag{13.50}$$

$$\Phi(k) \triangleq \begin{bmatrix} \Phi_{11}(k) & \Phi_{12}(k) \\ * & \Phi_{22}(k) \end{bmatrix} < 0, \tag{13.51}$$

$$Q_1(k+1) - \Theta(k+1) \leq 0, \tag{13.52}$$

*are satisfied with the parameters updated by*

$$
\begin{cases}
P_1(k+1) = \left(M(k+1) - V(k+1)N^{-1}(k+1)V^{\mathrm{T}}(k+1)\right)^{-1}, \\
P_2(k+1) = (N(k+1) - V^{\mathrm{T}}(k+1)M^{-1}(k+1)V(k+1))^{-1}, \\
P_3(k+1) = -M^{-1}(k+1)V(k+1)(N(k+1) - V^{\mathrm{T}}(k+1)M^{-1}(k+1)V(k+1))^{-1},
\end{cases}
\tag{13.53}
$$

*where*

$$
\Upsilon_{11}(k) = \begin{bmatrix} \bar{\Upsilon}_{11} & -P_3(k) & \hat{C}^{\mathrm{T}}(k)\hat{E}(k) \\ * & -P_2(k) & 0 \\ * & * & -\gamma^2 I + \hat{E}^{\mathrm{T}}(k)\hat{E}(k) \end{bmatrix},
$$

$$
\bar{\Upsilon}_{11} = -P_1(k) + \hat{C}^{\mathrm{T}}(k)\hat{C}(k) + \Sigma_{i=1}^q \Gamma_i(k) \times (\eta_i(k) + \mathrm{tr}[\Pi_{22}^i]),
$$

$$
\Upsilon_{21}(k) = \begin{bmatrix} A(k) & B(k)C_g(k) & \hat{D}(k) \\ B_g(k)\hat{C}(k) & A_g(k) & B_g(k)\hat{E}(k) \end{bmatrix},
$$

$$
\Upsilon_{22}(k) = \begin{bmatrix} -M(k+1) & -V(k+1) \\ * & -N(k+1) \end{bmatrix},
$$

$$
\Phi_{11}(k) = \begin{bmatrix} -Q_1(k+1) & -Q_3(k+1) & \bar{A}(k) & \bar{B}(k) \\ * & -Q_2(k+1) & \bar{C}(k) & \bar{D}(k) \\ * & * & -Q_1(k) & -Q_3(k) \\ * & * & * & -Q_2(k) \end{bmatrix},
$$

$$
\bar{A}(k) = A(k)Q_1(k) + B(k)C_g(k)Q_3^{\mathrm{T}}(k),
$$

$$
\bar{B}(k) = A(k)Q_3(k) + B(k)C_g(k)Q_2(k),
$$

$$
\bar{C}(k) = A_g(k)Q_3^{\mathrm{T}}(k) + B_g(k)\hat{C}(k)Q_1(k),
$$

$$
\bar{D}(k) = A_g(k)Q_2(k) + B_g(k)\hat{C}(k)Q_3(k),
$$

$$
\Phi_{12}(k) = \begin{bmatrix} \pi_{11} & \pi_{12} & \cdots & \pi_{1q} & \hat{D}(k) & 0 \\ B_g(k)\pi_{21} & B_g(k)\pi_{22} & \cdots & B_g(k)\pi_{2q} & B_g(k)\hat{E}(k) & 0 \\ 0 & 0 & \cdots & 0 & 0 & 0 \\ 0 & 0 & \cdots & 0 & 0 & 0 \end{bmatrix},
$$

$$
\Phi_{22}(k) = \mathrm{diag}\{-\rho_1 I, -\rho_2 I, \cdots, -\rho_q I, -\hat{W}^{-1}(k), 0\},
$$

$$
\rho_i = (\mathrm{tr}[\Gamma_i(k)P_1(k)])^{-1}, \quad i = 1, 2, \cdots, q,
$$

*then the addressed variance-constrained $H_\infty$ controller design problem is solved for the nonlinear stochastic system (13.1), and the controller gains $A_g(k)$, $B_g(k)$, and $C_g(k)$ at the sampling instant $k(0 \leq k \leq N)$ can be obtained by solving the corresponding set of RLMIs.*

*Proof.* The proof is based on Theorem 13.2.3. Firstly, suppose that the variables $P(k)$ and $Q(k)$ can be decomposed as follows:

$$P(k) = \begin{bmatrix} P_1(k) & P_3(k) \\ P_3^T(k) & P_2(k) \end{bmatrix},$$

$$P^{-1}(k) = \begin{bmatrix} M(k) & V(k) \\ V^T(k) & N(k) \end{bmatrix}, \tag{13.54}$$

$$Q(k) = \begin{bmatrix} Q_1(k) & Q_3(k) \\ Q_3^T(k) & Q_2(k) \end{bmatrix}.$$

Then it can be easily verified that if (13.53) holds, inequality (13.39) and inequality (13.49) are equivalent to each other. Furthermore, by some tedious but straightforward manipulations, we can also find that inequality (13.40) is equivalent to inequality (13.50), and inequality (13.41) holds if and only if inequality (13.51) holds. Therefore, according to Theorem 13.2.3, $\bar{J}_\infty < 0$ and $\Xi(k) \le Q(k)$ can be guaranteed at the same time under the initial condition (13.48). Moreover, from inequality (13.52), it is easy to see that $X(k) \le Q_1(k) \le \Theta(k)$ is achieved $\forall k \in \{0, 1, \ldots, N\}$. It can now be concluded that the design requirements (R1) and (R2) are simultaneously satisfied and the proof is now complete.                                            □

## 13.3.2  Computational Algorithm

Based on Theorem 13.3.1, the robust controller design (RCD) algorithm can be summarized as follows.

*Step 1.* Given the $H_\infty$ performance index $\gamma$, the positive definite matrix $\Omega$ and the state initial condition $x(0)$. Select the initial values for matrices $\{Q_1(0), Q_2(0), Q_2(0), P_1(0), P_2(0), P_3(0)\}$ that satisfy the condition (13.48) and set $k = 0$.

*Step 2.* Obtain the values of matrices $\{M(k+1), N(k+1), V(k+1), Q_1(k+1), Q_2(k+1), Q_3(k+1)\}$ and the desired controller parameters $\{A_g(k), B_g(k), C_g(k)\}$ for the sampling instant $k$ by solving the LMIs (13.49) to (13.52).

*Step 3.* Set $k = k + 1$ and obtain $\{P_1(k+1), P_2(k+1), P_3(k+1)\}$ by the parameter update formula (13.53).

*Step 4.* If $k < N$, then go to Step 2; else stop the iteration.

**Remark 13.4** *The algorithm above gives a recursive way to get certain values of the desired filter parameters at each time point k. It is obvious that the output feedback control problem can be solved by means of the above-mentioned RCD algorithm with appropriate modifications. On the other hand, it is worth pointing out that the problem might remain unsolved if the LMI is infeasible at a certain time point k, and the calculation burden might be too heavy due to the iterative algorithm. However, with* *the development of fast-speed calculating devices, the problems mentioned above can be well solved by making a tradeoff between the performance of the algorithm and the cost of the computing. A possible research topic in the future is to combine more performance indices and design the closed-loop feedback controller satisfying multiple requirements simultaneously.*

## 13.4   Numerical Example

This section presents a numerical example to demonstrate the effectiveness of the proposed algorithm. Consider the following time-varying nonlinear stochastic system

$$
\begin{cases}
x(k+1) = \left( \begin{bmatrix} 0 & -0.3 \\ 0.1 + 0.1\sin(10k) - 0.1e^{0.01k} & -0.25 \end{bmatrix} \right) x(k) \\
\qquad\qquad + \begin{bmatrix} -1.62 + 0.1\cos(20k) \\ -0.85 + 0.5\cos(k) \end{bmatrix} u(k) + f(x(k), k) \\
\qquad\qquad + \begin{bmatrix} 0.1 + 0.1\sin(k) \\ -0.2 \end{bmatrix} w(k), \\
y(k) = \begin{bmatrix} 0.5 & 0.5 \end{bmatrix} x(k) + g(x(k), k) + 0.15w(k).
\end{cases}
$$

The nonlinear functions $f(x(k), k)$ and $g(x(k), k)$ are taken as follows:

$$
f(x(k), k) = \begin{bmatrix} 0.3 \\ 0.2 \end{bmatrix} (0.2x_1(k)\xi_1(k) + 0.5x_2(k)\xi_2(k)),
$$

$$
g(x(k), k) = 0.1(0.2x_1(k)\xi_1(k) + 0.5x_2(k)\xi_2(k)),
$$

where $x_1(k)$ and $x_2(k)$ are the first, second entry of $x(k)$, and $\xi_1(k)$ and $\xi_2(k)$ are zero mean uncorrelated Gaussian white noise sequences with unity covariances. Assume that $\xi_1(k)$, $\xi_2(k)$ are uncorrelated with $w(k)$. Thus, the above stochastic nonlinearities satisfy

$$
\mathbb{E}\left\{ \begin{bmatrix} f(x(k), k) \\ g(x(k), k) \end{bmatrix} \Big| x(k) \right\} = 0,
$$

$$
\mathbb{E}\left\{ \begin{bmatrix} f(x(k), k) \\ g(x(k), k) \end{bmatrix} \begin{bmatrix} f^{\mathrm{T}}(x(k), k) & g^{\mathrm{T}}(x(k), k) \end{bmatrix} \Big| x(k) \right\}
$$

$$
= \begin{bmatrix} 0.3 \\ 0.2 \\ 0.1 \end{bmatrix} \begin{bmatrix} 0.3 \\ 0.2 \\ 0.1 \end{bmatrix}^{\mathrm{T}} \mathbb{E}\left\{ x^{\mathrm{T}}(k) \begin{bmatrix} 0.04 & 0 \\ 0 & 0.25 \end{bmatrix} x(k) \right\},
$$

which means that

$$
\Gamma_i(k) = \begin{bmatrix} 0.04 & 0 \\ 0 & 0.25 \end{bmatrix}, \qquad \pi_{1i}(k) = \begin{bmatrix} 0.3 \\ 0.2 \end{bmatrix}, \qquad \pi_{2i}(k) = 0.1.
$$

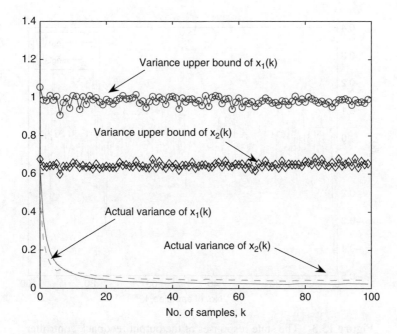

**Figure 13.1**    The variance upper bound and actual variance

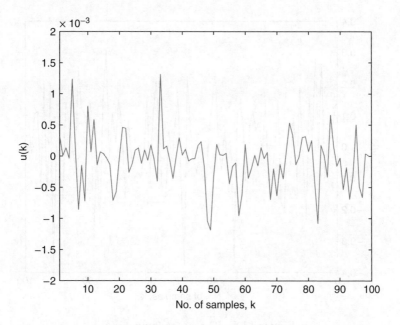

**Figure 13.2**    The input signal $u(k)$

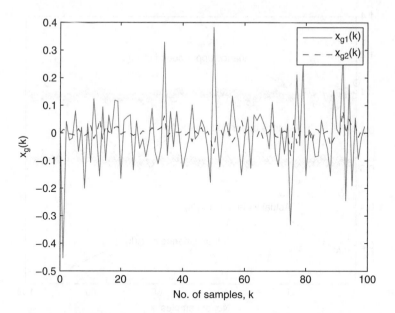

**Figure 13.3**    The state responses of the output feedback controller

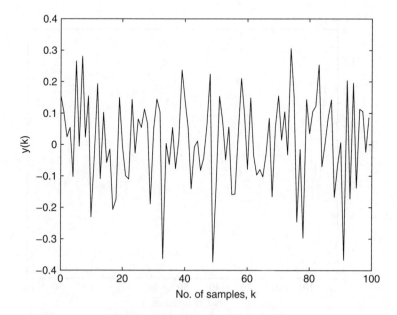

**Figure 13.4**    The system output $y(k)$

Assume that $\text{Prob}\{\theta(k) = 1\} = \mathbb{E}\{\theta(k)\} = \bar{\theta} = 0.5$; then $\kappa_\theta^2 = 0.25$. Let the sensor failure matrix $\Psi(k) = 0.2\sin^2(1.5k)$. It can easily be obtained that $0 \le \Psi(k) \le 0.2$ and $H(k) = 0.1$, $L(k) = 1$.

Set $N = 100$, $\gamma = 1.2$, $\Omega = 0.5I$, $\{\Theta(k)_{1 \le k \le N}\} = \text{diag}\{1.2, 0.9\}$, and choose the parameters initial values so as to satisfy equation (13.48); then set $k = 0$. The desired controller parameters can be obtained via the developed computational algorithm and the solvability of the addressed problem can be checked using the MATLAB LMI Toolbox. The results are shown in Figures 13.1 to 13.4, which confirms that the desired output-feedback controller is well achieved and the proposed RCD algorithm is indeed effective.

## 13.5  Summary

A multi-objective controller design problem for a class of nonlinear stochastic systems has been discussed in this chapter. The proposed system model is widely seen in real-world engineering applications. An algorithm based on recursive linear matrix inequalities is proposed to solve the multi-objective control problem taking the prescribed $H_\infty$ and covariance performance indices into consideration simultaneously. The numerical values of the desired controller parameters can be obtained by the proposed computing algorithm, and an illustrative example is presented to show the effectiveness of the proposed method.

### 9.2.5  Summary

# 14

# Mixed $H_2/H_\infty$ Control with Actuator Failures: the Finite-Horizon Case

This chapter deals with the fault-tolerant control problem for a class of nonlinear stochastic time-varying systems with actuator failures. Both $H_2$ and $H_\infty$ performance requirements are taken into consideration. The proposed actuator failure model is quite general and could cover several frequently seen actuator failure phenomena as special cases. The stochastic nonlinearities are quite general and could represent several types of nonlinear systems. It is the purpose of this chapter to find equilibrium strategies of a two-player Nash game, and meanwhile both $H_2$ and $H_\infty$ performances are achieved via a proposed state feedback control scheme, which is characterized by the solution to a set of coupled matrix equations. The feedback gains can be solved recursively backward in $k$. A numerical computing algorithm is presented and then a simulation example is given to illustrate the effectiveness and applicability of the proposed algorithm.

The rest of the chapter is arranged as follows: Section 14.1 formulates the mixed $H_2/H_\infty$ control problem for the nonlinear stochastic time-varying systems with actuator failures. In Section 14.2, the $H_\infty$ performance is analyzed in terms of the Riccati equation method. Section 14.3 gives the methodology to solve the addressed multi-objective problem and outlines the computational algorithm to recursively obtain the required parameters. A numerical example is presented in Section 14.4 to show the effectiveness and applicability of the proposed algorithm. Section 14.5 draws the summary.

*Variance-Constrained Multi-Objective Stochastic Control and Filtering*, First Edition.
Lifeng Ma, Zidong Wang and Yuming Bo.
© 2015 John Wiley & Sons, Ltd. Published 2015 by John Wiley & Sons, Ltd.

## 14.1   Problem Formulation

Consider the following nonlinear stochastic time-varying system defined on $k \in [0, N-1]$:

$$
\begin{cases}
x(k+1) = A(k)x(k) + \kappa(B(k)u(k)) + D(k)\omega(k) \\
\qquad\qquad + f(x(k), u(k), k), \\
y(k) = C(k)x(k) + E(k)u(k) + g(x(k), u(k), k)
\end{cases}
\tag{14.1}
$$

where $x(k)$, $y(k)$, and $u(k)$ stand for the system state, control output, and control input respectively, $\omega(k)$ represents the external disturbance belonging to $l_2$, and $A(k)$, $B(k)$, $C(k)$, $D(k)$, and $E(k)$ are known time-varying matrices with appropriate dimensions.

For notation simplicity, we denote

$$
f(k) \triangleq f(x(k), u(k), k),
$$

$$
g(k) \triangleq g(x(k), u(k), k),
$$

$$
\varphi(k) \triangleq [f^{\mathrm{T}}(k) \ \ g^{\mathrm{T}}(k)]^{\mathrm{T}}.
$$

For all $x(k)$ and $u(k)$, $\varphi(k)$ is assumed to satisfy

$$
\mathbb{E}\{\varphi(k)|x(k)\} = 0,
\tag{14.2}
$$

$$
\mathbb{E}\{\varphi(k)\varphi^{\mathrm{T}}(j)|x(k)\} = 0, k \neq j,
\tag{14.3}
$$

$$
\begin{aligned}
\mathbb{E}\{\varphi(k)\varphi^{\mathrm{T}}(k)|x(k)\} = \sum_{i=0}^{q} &\begin{bmatrix} \Pi_i^{11}(k) & \Pi_i^{12}(k) \\ \Pi_i^{12}(k) & \Pi_i^{22}(k) \end{bmatrix} \\
&\times \mathbb{E}\left\{ x^{\mathrm{T}}(k)\Gamma_i(k)x(k) \right. \\
&\left. + u^{\mathrm{T}}(k)\Xi_i(k)u(k) \right\},
\end{aligned}
\tag{14.4}
$$

where $\Pi_i^{11}(k) \geq 0$, $\Pi_i^{22}(k) \geq 0$, $\Gamma_i(k) \geq 0$, $\Xi_i(k) \geq 0$, and $\Pi_i^{12}(k)$ $(i = 1, 2, \ldots, q)$ are known matrices with compatible dimensions.

The nonlinear function $\kappa(\cdot) : \mathbb{R}^m \mapsto \mathbb{R}^m$ is defined as follows:

$$
\kappa(B(k)u(k)) = R(k)B(k)u(k),
\tag{14.5}
$$

where

$$
\begin{aligned}
&R(k) = \mathrm{diag}\{r_1(k), r_2(k), \ldots, r_m(k)\}, \\
&0 \leq r_{il}(k) \leq r_i(k) \leq r_{iu}(k) < \infty,
\end{aligned}
\tag{14.6}
$$

with $r_{il}(k) \leq 1$ and $r_{iu}(k) \geq 1$ being lower and upper bounds on $r_i(k)$.

**Remark 14.1** *The time-varying equation (14.5) is used to interpret the occurrence of incomplete information phenomenon due to the possible actuator failures. In such a model, when $r_i(k) = 1$, it means that the actuator is in good condition; otherwise* *there might be partial or complete actuator failure, which leads to incomplete information. The model proposed above could explain several types of failures occurring in engineering practice, including output quantization studied in Ref. [89] and sensor saturation studied in Ref. [132].*

Let $R(k) = H(k) + H(k)N(k)$, where

$$H(k) = \text{diag}\{h_1(k), h_2(k), \ldots, h_m(k)\},$$

$$h_i(k) = \frac{r_{il}(k) + r_{iu}(k)}{2},$$

$$N(k) = \text{diag}\{n_1(k), n_2(k), \ldots, n_m(k)\}, \tag{14.7}$$

$$n_i(k) = \frac{r_i(k) - h_i(k)}{h_i(k)}.$$

Denoting $L(k) = \text{diag}\{l_1(k), l_2(k), \ldots, l_m(k)\}$ with $l_i(k) = (r_{iu}(k) - r_{il}(k))/(r_{iu}(k) + r_{il}(k))$, we can easily obtain

$$N^T(k)N(k) \le L^T(k)L(k) \le I. \tag{14.8}$$

So far, we have successfully converted the influence caused by the actuator failures on the system output sampling into certain norm-bounded parameter uncertainties. Then the original system (14.1) can be reformulated as follows:

$$\begin{cases} x(k+1) = A(k)x(k) + \bar{B}(k)u(k) + D(k)\omega(k) \\ \qquad\qquad + H(k)N(k)B(k)u(k) + f(k), \\ y(k) = C(k)x(k) + E(k)u(k) + g(k), \end{cases} \tag{14.9}$$

where $\bar{B}(k) = H(k)B(k)$.

In the following, we give the design objective of this chapter. Firstly, we consider the following $H_\infty$ performance index for system (14.9):

$$J_\infty = \mathbb{E}\left\{ \sum_{k=0}^{N-1} (\|y(k)\|^2 - \gamma^2\|\omega(k)\|^2) \right\} - \gamma^2\|x(0)\|_W^2, \tag{14.10}$$

where $W > 0$ is a known weighting matrix.

In this chapter, it is our objective to design a state feedback controller of the form $u(k) = K(k)x(k)$ such that:

(R1) For a pre-specified $H_\infty$ disturbance attenuation level $\gamma > 0$, $J_\infty \le 0$ holds for all nonzero $\omega(k)$ and $x(0)$.

(R2) When there exists a worst-case disturbance such that the $H_\infty$ disturbance attenuation level is maximized, the controlled output energy defined by $J_2 = \sum_{k=0}^{N-1} \mathbb{E}\{\|y(k)\|^2\}$ is minimized.

Noticing that there exist parameter uncertainties in (14.9), we shall further reconstruct the system so as to obtain a system without $N(k)$ as follows:

$$\begin{cases} x(k+1) = A(k)x(k) + \bar{B}(k)u(k) + \bar{D}(k)v(k) + f(k), \\ y(k) = C(k)x(k) + E(k)u(k) + g(k), \end{cases} \tag{14.11}$$

where

$$v(k) = \begin{bmatrix} \omega(k) \\ \gamma^{-1}\epsilon(k)N(k)B(k)u(k) \end{bmatrix},$$

$$\bar{D}(k) = \begin{bmatrix} D(k) & \gamma\epsilon^{-1}(k)H(k) \end{bmatrix},$$

with $\epsilon(k) \neq 0$ being any known real-valued constant.

We now consider the following performance index:

$$\bar{J}_\infty = \mathbb{E}\left\{ \sum_{k=0}^{N-1} \left( \|y(k)\|^2 + \|\epsilon(k)L(k)B(k)u(k)\|^2 \right. \right.$$
$$\left. \left. -\gamma^2 \|v(k)\|^2 \right) \right\} - \gamma^2 \|x(0)\|_W^2. \tag{14.12}$$

Next, a lemma is given to reveal the relationship between $J_\infty$ and $\bar{J}_\infty$.

**Lemma 14.1.1** *Given a positive scalar $\gamma$. The performance indices $J_\infty$ and $\bar{J}_\infty$ are defined by (14.10) and (14.12) respectively; then $J_\infty \leq 0$ is implied by $\bar{J}_\infty \leq 0$.*

*Proof.* If $\bar{J}_\infty \leq 0$ holds, it is found that

$$\sum_{k=0}^{N-1} \mathbb{E}\{\|y(k)\|^2 + \|\epsilon(k)L(k)B(k)u(k)\|^2\}$$
$$\leq \gamma^2 \left( \sum_{k=0}^{N-1} \mathbb{E}\{\|v(k)\|^2\} + \|x(0)\|_W^2 \right). \tag{14.13}$$

Expanding both sides of the inequality shown above, we have

$$\sum_{k=0}^{N-1} \mathbb{E}\{\|y(k)\|^2 + \|\epsilon(k)\bar{L}(k)B(k)u(k)\|^2\}$$
$$\leq \gamma^2 \left( \sum_{k=0}^{N-1} \mathbb{E}\{\|\omega(k)\|^2\} + \|x(0)\|_W^2 \right), \tag{14.14}$$

where $\bar{L}(k) = \sqrt{L^T(k)L(k) - N^T(k)N(k)}$.

This indicates that

$$\sum_{k=0}^{N-1} \mathbb{E}\{\|y(k)\|^2\} \leq \left(\sum_{k=0}^{N-1} \mathbb{E}\{\|\omega(k)\|^2\} + \|x(0)\|_W^2\right). \tag{14.15}$$

It follows directly that $J_\infty \leq 0$. Moreover, it is worth pointing out that $N(k)$ can be any arbitrary matrix corresponding to any arbitrary actuator failure. When there is no actuator failure occurring, that is, $r_{il}(k) = r_i(k) = r_{iu}(k) = 1$, then we have $N(k) = L(k) = 0$, and therefore $J_\infty = \bar{J}_\infty$. As a result, we can see that $\bar{J}_\infty$ is a tight upper bound of $J_\infty$. The proof is complete. □

Based on the above discussions, the purpose of this chapter could be restated as follows. In this chapter, it is our objective to design a state feedback controller of the form $u(k) = K(k)x(k)$ such that:

($G1$)  For a pre-specified $H_\infty$ disturbance attenuation level $\gamma > 0$, $\bar{J}_\infty \leq 0$ holds for all nonzero $v(k)$ and $x(0)$.

($G2$)  When there exists a worst-case disturbance $v^*(k)$ such that the $H_\infty$ disturbance attenuation level is maximized, the output energy defined by $J_2 = \sum_{k=0}^{N-1} \mathbb{E}\{\|y(k)\|^2\}$ is minimized.

In other words, we aim at finding equilibrium strategies $u^*(k) = K(k)x(k)$ that satisfy the Nash equilibria defined by

$$\bar{J}_\infty(u^*(k), v(k)) \leq \bar{J}_\infty(u^*(k), v^*(k)) \leq 0, \tag{14.16}$$

$$J_2(u(k), v^*(k)) \geq J_2(u^*(k), v^*(k)). \tag{14.17}$$

Now we are in a situation to deal with the mixed $H_2/H_\infty$ control problem for the nonlinear stochastic system with possible actuator failures.

## 14.2  $H_\infty$ Performance

In this section, we shall firstly analyze the system disturbance attenuation level. A theorem that plays a vital role in the controller design stage will be presented to give a necessary and sufficient condition guaranteeing the pre-specified $H_\infty$ specification. To this end, we obtain the unforced system by setting $u(k) = 0$ in system (14.11) as follows:

$$\begin{cases} x(k+1) = A(k)x(k) + \bar{D}(k)v(k) + f(k), \\ y(k) = C(k)x(k) + g(k). \end{cases} \tag{14.18}$$

We now consider the following backward recursion:

$$\begin{aligned} P(k) = {}& A^T(k)P(k+1)A(k) + C^T(k)C(k) \\ & + A^T(k)P(k+1)\bar{D}(k)\Phi^{-1}(k)\bar{D}^T(k)P(k+1)A(k) \\ & + \sum_{i=0}^{q} \Gamma_i(k)(\text{tr}[P(k+1)\Pi_i^{11}(k)] + \text{tr}[\Pi_i^{22}(k)]), \end{aligned} \tag{14.19}$$

where

$$\Phi(k) = \gamma^2 I - \bar{D}^{\mathrm{T}}(k)P(k+1)\bar{D}(k). \tag{14.20}$$

The following theorem gives a necessary and sufficient condition for system (14.18) capable of satisfying the pre-specified $H_\infty$ requirement.

**Theorem 14.2.1** *Given a positive scalar $\gamma > 0$ and a positive definite matrix W. The pre-specified $H_\infty$ disturbance attenuation level defined in (G1) can be achieved for all nonzero $x(0)$ and $v(k)$ if and only if, with the final condition $P(N) = 0$, there exist solutions $P(k)$ $(0 \leq k < N)$ to (14.19) such that $\Phi(k) > 0$ and $P(0) \leq \gamma^2 W$.*

*Proof* **Sufficiency**
By defining

$$J_k = x^{\mathrm{T}}(k+1)P(k+1)x(k+1) - x^{\mathrm{T}}(k)P(k)x(k), \tag{14.21}$$

we have

$$\begin{aligned}
\mathbb{E}\{J_k\} = \mathbb{E}\Big\{ &\big(A(k)x(k) + \bar{D}(k)v(k) + f(k)\big)^{\mathrm{T}} P(k+1) \\
&\times \big(A(k)x(k) + \bar{D}(k)v(k) + f(k)\big) \Big\} - x^{\mathrm{T}}(k)P(k)x(k) \\
= x^{\mathrm{T}}(k) &\Bigg( A^{\mathrm{T}}(k)P(k+1)A(k) - P(k) \\
&+ \sum_{i=0}^{q} \Gamma_i(k)\big(\mathrm{tr}[P(k+1)\Pi_i^{11}(k)]\big) \Bigg) x(k) \\
&+ 2x^{\mathrm{T}}(k)A^{\mathrm{T}}(k)P(k+1)\bar{D}(k)v(k) \\
&+ v^{\mathrm{T}}(k)\bar{D}^{\mathrm{T}}(k)P(k+1)\bar{D}(k)v(k).
\end{aligned} \tag{14.22}$$

Adding the following zero term

$$\|y(k)\|^2 - \gamma^2\|v(k)\|^2 - (\|y(k)\|^2 - \gamma^2\|v(k)\|^2) \tag{14.23}$$

to both sides of (14.22) and then taking the mathematical expectation results in

$$\begin{aligned}
\mathbb{E}\{J_k\} \\
= x^{\mathrm{T}}(k)&\big(A^{\mathrm{T}}(k)P(k+1)A(k) - P(k) + C^{\mathrm{T}}(k)C(k) \\
&+ \sum_{i=0}^{q} \Gamma_i(k)\big(\mathrm{tr}[P(k+1)\Pi_i^{11}(k)] + \mathrm{tr}[\Pi_i^{22}(k)]\big)\, x(k) \\
&+ 2x^{\mathrm{T}}(k)A^{\mathrm{T}}(k)P(k+1)\bar{D}(k)v(k) \\
&+ v^{\mathrm{T}}(k)\big(-\gamma^2 I + \bar{D}^{\mathrm{T}}(k)P(k+1)\bar{D}(k)\big)v(k) \\
&- \mathbb{E}\big\{\big(\|y(k)\|^2 - \gamma^2\|v(k)\|^2\big)\big\}.
\end{aligned} \tag{14.24}$$

Completing the squares of $v(k)$, we have

$$
\begin{aligned}
\mathbb{E}\{J_k\} = &- (v(k) - v^*(k))^{\mathrm{T}}\Phi(k)(v(k) - v^*(k)) \\
&- \mathbb{E}\{\|y(k)\|^2 - \gamma^2\|v(k)\|^2\},
\end{aligned}
\tag{14.25}
$$

where

$$
v^*(k) = \Phi^{-1}(k)\bar{D}^{\mathrm{T}}(k)P(k+1)A(k)x(k).
\tag{14.26}
$$

Now, summing (14.25) from 0 to $N - 1$ with respect to $k$ leads to

$$
\begin{aligned}
\mathbb{E}\left\{\sum_{k=0}^{N-1} J_k\right\} = &\, \mathbb{E}\{x^{\mathrm{T}}(N)P(N)x(N)\} - x^{\mathrm{T}}(0)P(0)x(0) \\
&- \sum_{k=0}^{N-1}\left((v(k) - v^*(k))^{\mathrm{T}}\Phi(k)(v(k) - v^*(k))\right. \\
&\left. + \mathbb{E}\{\|y(k)\|^2 - \gamma^2\|v(k)\|^2\}\right).
\end{aligned}
\tag{14.27}
$$

Noticing $P(N) = 0$, we can find from (14.27) that

$$
\begin{aligned}
\sum_{k=0}^{N-1}&\left(\mathbb{E}\{\|y(k)\|^2 - \gamma^2\|v(k)\|^2\}\right) - \gamma^2\|x(0)\|_W^2 \\
&= x^{\mathrm{T}}(0)(P(0) - \gamma^2 W)x(0) \\
&\quad - \sum_{k=0}^{N-1}(v(k) - v^*(k))^{\mathrm{T}}\Phi(k)(v(k) - v^*(k)).
\end{aligned}
\tag{14.28}
$$

Therefore, since $\Phi(k) > 0$ and $P(0) \leq \gamma^2 W$, we arrive at

$$
\bar{J}_\infty \leq 0,
\tag{14.29}
$$

which implies $J_\infty \leq 0$. Moreover,

$$
\sup_{\{x(0),\omega(k)\}} J_\infty = \sup_{\{x(0),v(k)\}} \bar{J}_\infty = 0.
\tag{14.30}
$$

**Necessity** See the proof of Ref. [66]. The proof is now complete. □

In this section, we have analyzed the $H_\infty$ performance of the unforced system and obtained an important necessary and sufficient condition that will play a key role in the controller design procedure. This will be discussed in the following section.

## 14.3 Multi-Objective Controller Design

In this section, we aim to find the desired Nash equilibrium strategies with the prescribed $H_\infty$ constraint. We shall present a theorem to give a necessary and sufficient condition of the existence of such a control scheme. Then a computational algorithm will be proposed to obtain the numerical values of the feedback gain at each sampling instant $k$.

## 14.3.1   Controller Design

Firstly, for notational convenience, we denote

$$\tilde{A}(k) = A(k) + \bar{B}(k)K(k), \quad \bar{C}(k) = C(k) + E(k)K(k). \tag{14.31}$$

Implementing $u(k) = K(k)x(k)$ to system (14.11), we obtain the following closed-loop nonlinear stochastic system:

$$\begin{cases} x(k+1) = \tilde{A}(k)x(k) + \bar{D}(k)\upsilon(k) + f(k), \\ \quad y(k) = \bar{C}(k)x(k) + g(k). \end{cases} \tag{14.32}$$

The following theorem gives a necessary and sufficient condition for the existence of the desired multi-objective controller in terms of certain coupled matrix equations.

**Theorem 14.3.1** *Given an $H_\infty$ performance index $\gamma > 0$. With the final condition $P(N) = 0$ and $Q(N) = 0$, there exists a state feedback controller $u(k) = K(k)x(k)$ for system (14.1) such that both the design requirements (G1) and (G2) can be satisfied simultaneously, if and only if there exist solutions $P(k)$ and $Q(k)$ to the following coupled matrix equations:*

$$P(k) = \tilde{A}^{\mathrm{T}}(k)P(k+1)\tilde{A}(k) + \bar{C}^{\mathrm{T}}(k)\bar{C}(k) + \sum_{i=0}^{q} \left( (\Gamma_i(k) + K^{\mathrm{T}}(k)\Xi_i(k)K(k)) \right.$$

$$\times \left( \mathrm{tr}[P(k+1)\Pi_i^{11}(k)] + \mathrm{tr}[\Pi_i^{22}(k)] \right) + \epsilon^2(k)K^{\mathrm{T}}(k)B^{\mathrm{T}}(k)L^{\mathrm{T}}(k)L(k)B(k)K(k)$$

$$+ \left( \bar{D}^{\mathrm{T}}(k)P(k+1)\tilde{A}(k) \right)^{\mathrm{T}} \Phi^{-1}(k) \left( \bar{D}^{\mathrm{T}}(k)P(k+1)\tilde{A}(k) \right), \tag{14.33}$$

$$P(0) \leq \gamma^2 W, \tag{14.34}$$

$$Q(k) = \tilde{A}^{\mathrm{T}}(k)Q(k+1)\tilde{A}(k) + C^{\mathrm{T}}(k)C(k) \sum_{i=0}^{q} \Gamma_i(k) \left( \mathrm{tr}[P(k+1)\Pi_i^{11}(k)] + \mathrm{tr}[\Pi_i^{22}(k)] \right)$$

$$- \left( \bar{B}^{\mathrm{T}}(k)Q(k+1)\tilde{A}(k) + E^{\mathrm{T}}(k)C(k) \right)^{\mathrm{T}} \Psi^{-1}(k) \left( \bar{B}^{\mathrm{T}}(k)Q(k+1)\tilde{A}(k) \right.$$

$$+ E^{\mathrm{T}}(k)C(k) \right), \tag{14.35}$$

$$\Phi(k) = \gamma^2 I - \bar{D}^{\mathrm{T}}(k)P(k+1)\bar{D}(k) > 0, \tag{14.36}$$

$$\Psi(k) = \bar{B}^{\mathrm{T}}(k)Q(k+1)\bar{B}(k) + E^{\mathrm{T}}(k)E(k)$$

$$+ \sum_{i=0}^{q} \Xi_i(k) \left( \mathrm{tr}[P(k+1)\Pi_i^{11}(k)] + \mathrm{tr}[\Pi_i^{22}(k)] \right) > 0, \tag{14.37}$$

$$K(k) = -\Psi^{-1}(k) \left( \bar{B}^{\mathrm{T}}(k)Q(k+1)\tilde{A}(k) + E^{\mathrm{T}}(k)C(k) \right), \tag{14.38}$$

$$T(k) = \Phi^{-1}(k)\bar{D}^{\mathrm{T}}(k)P(k+1)\tilde{A}(k), \tag{14.39}$$

where

$$\bar{A}(k) = A(k) + \bar{D}(k)T(k). \tag{14.40}$$

*Proof* **Sufficiency**

Firstly, according to Theorem 14.2.1, if there exist solutions $P(k)$ to (14.33) such that $\Phi(k) > 0$, system (14.11) will meet the pre-specified $H_\infty$ requirement and the worst-case disturbance can be obtained as $v^*(k) = T(k)x(k)$.

Now we are in a situation to find the feedback controller gain $K(k)$ such that the output energy can be minimized with the worst-case disturbance. To this end, when the worst-case disturbance happens, by applying $v^*(k) = T(k)x(k)$, system (14.11) can be rewritten as

$$\begin{cases} x(k+1) = (A(k) + \bar{D}(k)T(k))x(k) + \bar{B}(k)u(k) + f(k), \\ y(k) = C(k)x(k) + E(k)u(k) + g(k). \end{cases} \tag{14.41}$$

By defining the following quadratic index

$$J_{2k} = x^{\mathrm{T}}(k+1)Q(k+1)x(k+1) - x^{\mathrm{T}}(k)Q(k)x(k), \tag{14.42}$$

we could have

$$\begin{aligned} \mathbb{E}\{J_{2k}\} = \mathbb{E}\Big\{ &\left(\bar{A}(k)x(k) + \bar{B}(k)u(k) + f(k)\right)^{\mathrm{T}} Q(k+1) \\ &\times \left(\bar{A}(k)x(k) + \bar{B}(k)u(k) + f(k)\right)\Big\} \\ &- x^{\mathrm{T}}(k)Q(k)x(k) \\ = x^{\mathrm{T}}(k) &\bigg(\bar{A}^{\mathrm{T}}(k)Q(k+1)\bar{A}(k) - Q(k) \\ &+ \sum_{i=0}^{q} \Gamma_i(k)\mathrm{tr}[P(k+1)\Pi_i^{11}(k)]\bigg) x(k) \\ + u^{\mathrm{T}}(k) &\bigg(\bar{B}^{\mathrm{T}}(k)Q(k+1)\bar{B}(k) \\ &+ \sum_{i=0}^{q} \Xi_i(k)\left(\mathrm{tr}[P(k+1)\Pi_i^{11}(k)]\right)\bigg) u(k) \\ + 2x^{\mathrm{T}}(k) &\bar{A}^{\mathrm{T}}(k)Q(k+1)\bar{B}(k)u(k). \end{aligned} \tag{14.43}$$

It then follows that

$$\begin{aligned} \mathbb{E}\{J_{2k} &+ \|y(k)\|^2 - \|y(k)\|^2\} \\ = \mathbb{E}\Big\{ x^{\mathrm{T}}(k) &\bigg(A^{\mathrm{T}}(k)Q(k+1)A(k) - Q(k) + C^{\mathrm{T}}(k)C(k) \\ &+ \sum_{i=0}^{q} \Gamma_i(k)\left(\mathrm{tr}[P(k+1)\Pi_i^{11}(k)] + \mathrm{tr}[\Pi_i^{22}(k)]\right)\bigg) x(k) \end{aligned}$$

$$+ 2x^T(k)(A^T(k)Q(k+1)\bar{B}(k) + C^T(k)E(k))u(k)$$

$$+ u^T(k) \Bigg( \bar{B}^T(k)Q(k+1)\bar{B}(k) + E^T(k)E(k)$$

$$+ \sum_{i=0}^{q} \Xi_i(k) \left( \text{tr}[P(k+1)\Pi_i^{11}(k)] + \text{tr}[\Pi_i^{22}(k)] \right) \Bigg) u(k)$$

$$- \|y(k)\|^2 \Bigg\}. \tag{14.44}$$

Completing the squares of $u(k)$, we obtain

$$\mathbb{E}\{J_{2k}\} = \mathbb{E}\{(u(k) - u^*(k))^T \Psi(k)(u(k) - u^*(k)) - \|y(k)\|^2\}, \tag{14.45}$$

where $\Psi(k)$ is defined in (14.37) and $u^*(k) = K(k)x(k)$ with $K(k)$ being defined in (14.38). Moreover, we have

$$J_2 = \mathbb{E} \left\{ \sum_{k=0}^{N-1} \|y(k)\|^2 \right\}$$

$$= \mathbb{E} \left\{ \sum_{k=0}^{N-1} (u(k) - u^*(k))^T \Psi(k)(u(k) - u^*(k)) \right\} \tag{14.46}$$

$$+ x^T(0)Q(0)x(0) - x^T(N)Q(N)x(N).$$

Noting $Q(N) = 0$, we can see that, when $u(k) = u^*(k)$, the output energy $J_2$ is minimized at $J_{2\min} = x^T(0)Q(0)x(0)$, which means that the desired Nash game strategies expressed by (14.16) and (14.17) are achieved, and the design requirements $(G1)$ and $(G2)$ are satisfied simultaneously.

**Necessity** If $u^*(k) = K(k)x(k)$ and $v^*(k) = T(k)x(k)$ are the desired equilibrium strategies satisfying

$$\bar{J}_\infty(u^*(k), v(k)) \leq \bar{J}_\infty(u^*(k), v^*(k)),$$
$$J_2(u(k), v^*(k)) \geq J_2(u^*(k), v^*(k)), \tag{14.47}$$

and by implementing $u^*(k) = K(k)x(k)$ to system (14.11), the closed-loop system (14.32) can be obtained. Therefore, according to Theorem 14.2.1, the matrix equation (14.33) admits solutions $P(k) > 0$ $(0 \leq k < N)$. Furthermore, the worst-case disturbance, if it exists, can be represented by $v^*(k) = T(k)x(k)$. Substituting $v^*(k) = T(k)x(k)$ to system (14.11), we have system (14.41). Then the existence of the strategy $u^*(k) = K(k)x(k)$ indicates the existence of $K(k)$ over the finite horizon

[0, N − 1]. Therefore, we can see from (14.38) that $\Psi(k)$ should be nonsingular for all $0 \le k < N$. Observing equation (14.35), given the final condition $Q(N) = 0$, we could learn that $Q(k)$ can always be obtained by solving (14.35) with nonsingular $\Psi(k)$ and backward in time. The proof is now complete.                    □

**Remark 14.2** *The proposed theorem gives a necessary and sufficient condition for the existence of the required multi-objective controller. By means of the Nash game approach, we have successfully converted the controller existence problem into the feasibility of certain coupled matrix equations. We should point out that the actuator failure model considered in this chapter is quite general and therefore can be applied in many branches of reliable control and signal processing problems, such as control with actuator saturation, filtering with missing measurements, and so on.*

### 14.3.2   Computational Algorithm

As mentioned above, the state feedback controller gains $K(k)$ and $T(k)$ can be obtained by solving the presented coupled matrix equations. In the next stage, it is our aim to propose an algorithm to get the numerical values of these desired parameters at each time point $k$ recursively.

#### Mixed $H_2/H_\infty$ Reliable Controller Design Algorithm

*Step 1.* Set the finite time $N$ and $k = N − 1$. Set the pre-specified $H_\infty$ index $\gamma > 0$. Set $P(N) = Q(N) = 0$. Select the positive definite matrices $W > 0$ and $\epsilon(k) > 0$ properly.

*Step 2.* With all the pre-set parameters and the available $P(k + 1)$ and $Q(k + 1)$, solve the set of coupled matrix equations (14.36) to (14.38) to get $\{\Phi(k), \Psi(k), K(k), T(k)\}$.

*Step 3.* With the obtained $\{\Phi(k), \Psi(k), K(k), T(k)\}$, solve (14.33) and (14.35) for $P(k)$ and $Q(k)$ respectively.

*Step 4.* If $k = 0$, go to Step 5. Else, set $k = k − 1$ and go to Step 2.

*Step 5.* Stop.

**Remark 14.3** *The* Mixed $H_2/H_\infty$ Reliable Controller Design Algorithm *gives a recursive way to obtain the numerical values of the feedback gain at each time point $k$. It should be noticed that the existence of the controller is expressed by the feasibility of coupled Riccati equations; therefore, the presented algorithm has a much lower computing complexity than the traditional LMI method. However, since the Riccati equations are backward in time, this algorithm is only suitable for the offline design. The possible research topic in the future is to develop the multi-objective controller design method, which could be expressed forward in time.*

## 14.4 Numerical Example

In this section, numerical examples are presented to demonstrate the effectiveness of the Mixed $H_2/H_\infty$ Reliable Controller Design Algorithm proposed in this chapter.

Set $N = 5$, $\gamma = 1.3$, $W = 1.4$, $\epsilon(k) = 1$ and the final condition $P(5) = Q(5) = 0$.

Consider a one-dimensional system with the following time-varying parameters over the finite horizon $[0, 5]$:

$$A(k) = 0.3 + 0.1\cos(5k), \quad B(k) = 0.6 - 0.2\sin(8k),$$

$$C(k) = 0.4 + 0.1\sin(10k), \quad D(k) = 0.3 - 0.1e^{0.2k},$$

$$E(k) = 0.2, \quad \Pi_i^{11}(k) = 0.25, \quad \Pi_i^{12}(k) = 0.4,$$

$$\Pi_i^{22}(k) = 0.64, \quad \Gamma_i(k) = 0.25, \quad \Xi_i(k) = 0.36.$$

We now consider the following two different actuator failure cases.

***Case 1***: Let the actuator failure matrix be $R(k) = 1.4\sin^2(x(k))$. Therefore, it can be found that $0 \leq R(k) \leq 1.4$ and $H(k) = 0.7$, $L(k) = 1$.

Using the developed computational algorithm, we can check the feasibility of the coupled Riccati-like equations and then calculate the desired controller parameters, which are shown below.

- Set $k = 4$.
- Then we obtain

$$\Phi(4) = \begin{bmatrix} 1.6900 & 0 \\ 0 & 1.6900 \end{bmatrix} > 0, \quad T(4) = \begin{bmatrix} 0 \\ 0 \end{bmatrix},$$

$$\Psi(4) = 0.2704 > 0, \quad K(4) = -0.3510.$$

- With the obtained $\{\Phi(4), \Psi(4), K(4), T(4)\}$, we have

$$P(4) = 0.3519, \quad Q(4) = 0.3519.$$

By similar recursions, it can be found that:
- $k = 3$:

$$\Phi(3) = \begin{bmatrix} 1.6851 & -0.0377 \\ -0.0377 & 1.3986 \end{bmatrix} > 0, \quad T(3) = \begin{bmatrix} 0.0024 \\ 0.0186 \end{bmatrix},$$

$$\Psi(3) = 0.4073 > 0, \quad K(3) = -0.2619,$$

$$P(3) = 0.2647, \quad Q(3) = 0.2653.$$

- $k = 2$:

$$\Phi(2) = \begin{bmatrix} 1.6840 & -0.0363 \\ -0.0363 & 1.4708 \end{bmatrix} > 0, \quad T(2) = \begin{bmatrix} 0.0014 \\ 0.0084 \end{bmatrix},$$

$$\Psi(2) = 0.3504 > 0, \quad K(2) = -0.3584,$$

$$P(2) = 0.3861, \quad Q(2) = 0.3862.$$

- $k = 1$:

$$\Phi(1) = \begin{bmatrix} 1.6778 & -0.0625 \\ -0.0625 & 1.3703 \end{bmatrix} > 0, \quad T(1) = \begin{bmatrix} 0.0119 \\ 0.0609 \end{bmatrix},$$

$$\Psi(1) = 0.3358 > 0, \quad K(1) = -3.3308,$$

$$P(1) = 0.3178, \quad Q(1) = 0.3243.$$

- $k = 0$:

$$\Phi(0) = \begin{bmatrix} 1.6773 & -0.0578 \\ -0.0578 & 1.4268 \end{bmatrix} > 0, \quad T(0) = \begin{bmatrix} 0.0105 \\ 0.0479 \end{bmatrix},$$

$$\Psi(0) = 0.3562 > 0, \quad K(0) = -0.3950,$$

$$P(0) = 0.3441, \quad Q(0) = 0.3487.$$

It can be verified that $P(0) \leq \gamma^2 W$, which means that $\{\Phi(k), \Psi(k), K(k), T(k), P(k), Q(k)\}$ are the solutions of the set of coupled matrix equations (14.33) to (14.38). Therefore, the Nash equilibrium strategies $u^*(k) = K(k)x(k)$ and $v^*(k) = T(k)x(k)$ have been found.

*Case 2*: Let the actuator failure matrix be $R(k) = \text{sign}((x(k)) + 1.2$. Therefore, it can be found that $0.2 \leq R(k) \leq 2.2$ and $H(k) = 1.2, L(k) = 0.8333$.

By a similar simulation, it can be found that:

- $k = 4$:

$$\Phi(4) = \begin{bmatrix} 1.6900 & 0 \\ 0 & 1.6900 \end{bmatrix} > 0, \quad T(4) = \begin{bmatrix} 0 \\ 0 \end{bmatrix},$$

$$\Psi(4) = 0.2704 > 0, \quad K(4) = -0.3510,$$

$$P(4) = 0.3519, \quad Q(4) = 0.3519.$$

- $k = 3$:

$$\Phi(3) = \begin{bmatrix} 1.6851 & -0.0647 \\ -0.0647 & 0.8336 \end{bmatrix} > 0, \quad T(3) = \begin{bmatrix} 0.0006 \\ 0.0080 \end{bmatrix},$$

$$\Psi(3) = 0.6113 > 0, \quad K(3) = -0.2262,$$

$$P(3) = 0.2610, \quad Q(3) = 0.2611.$$

- $k = 2$:

$$\Phi(2) = \begin{bmatrix} 1.6841 & -0.0614 \\ -0.0614 & 1.0548 \end{bmatrix} > 0, \quad T(2) = \begin{bmatrix} -0.0009 \\ -0.0098 \end{bmatrix},$$

$$\Psi(2) = 0.3504 > 0, \quad K(2) = -0.3584,$$

$$P(2) = 0.3861, \quad Q(2) = 0.3862.$$

- $k = 1$:

$$\Phi(1) = \begin{bmatrix} 1.6778 & -0.1069 \\ -0.1069 & 0.7523 \end{bmatrix} > 0, \quad T(1) = \begin{bmatrix} 0.0121 \\ 0.1064 \end{bmatrix},$$

$$\Psi(1) = 0.3948 > 0, \quad K(1) = -0.4090,$$

$$P(1) = 0.3131, \quad Q(1) = 0.3325.$$

- $k = 0$:

$$\Phi(0) = \begin{bmatrix} 1.6775 & -0.0977 \\ -0.0977 & 0.9280 \end{bmatrix} > 0, \quad T(0) = \begin{bmatrix} 0.0069 \\ 0.0535 \end{bmatrix},$$

$$\Psi(0) = 0.4709 > 0, \quad K(0) = -0.4163,$$

$$P(0) = 0.3305, \quad Q(0) = 0.3361.$$

It can be verified that $P(0) \leq \gamma^2 W$, which means that $\{\Phi(k), \Psi(k), K(k), T(k), P(k), Q(k)\}$ are the solutions of the set of coupled matrix equations (14.33) to (14.38). Therefore, the Nash equilibrium strategies $u^*(k) = K(k)x(k)$ and $v^*(k) = T(k)x(k)$ have been found.

## 14.5  Summary

In this chapter, the mixed $H_2/H_\infty$ control problem has been studied for a type of nonlinear stochastic systems against actuator failures. The actuator failure model is quite general and could stand for several common failures as special cases. A state feedback control scheme has been proposed for the stochastic time-varying systems satisfying both $H_2$ and $H_\infty$ performance indices simultaneously. The solvability of the addressed multi-objective control problem is expressed by the feasibility of certain coupled matrix equations. The numerical values of the controller gains can be obtained by the developed computing algorithm. An illustrative example is given to show the effectiveness and applicability of the proposed design strategy.

# 15

# Conclusions and Future Topics

## 15.1  Concluding Remarks

In this book, the variance-constrained multi-objective control and filtering problems have been investigated for nonlinear stochastic systems. Latest results on analysis and synthesis problems for nonlinear stochastic systems with multiple performance constraints have been firstly surveyed. Then, in each chapter, for the different kinds of nonlinear stochastic systems, the stability, robustness, and the pre-specified system performance indices, such as system covariance (or variance) constraints, $H_\infty$ noise attenuation level, and $H_2$ performance index, are analyzed. Sufficient conditions of robust asymptotically stochastic stability or robust exponential stochastic stability are derived for the closed-loop systems in the presence of the parameter uncertainty, external and internal nonlinear disturbances, sensor and actuator failures, missing measurements, output degradation, etc. Subsequently, the controller/filter synthesis problems are investigated, and sufficient conditions for the existence of the desired controller/filter with which all the pre-specified performance requirements can be simultaneously satisfied are given in terms of feasibility of a set of linear matrix inequalities (LMIs) or Riccati equations.

## 15.2  Future Research

This book has established a unified theoretical framework for analysis and synthesis of variance-constrained multi-objective control/filtering problems for nonlinear stochastic systems. However, the obtained results are still quite limited. Some of the possible future research topics are listed as follows.

In practical engineering, there are still some more complicated yet important kinds of nonlinearities that have not been studied. Therefore, the variance-constrained multi-objective control and filtering problems for more general nonlinear systems still remain open and challenging.

*Variance-Constrained Multi-Objective Stochastic Control and Filtering*, First Edition.
Lifeng Ma, Zidong Wang and Yuming Bo.
© 2015 John Wiley & Sons, Ltd. Published 2015 by John Wiley & Sons, Ltd.

Another future research direction is to further investigate new performance indices (e.g., system energy constraints) that can be simultaneously considered with other existing ones. Also, variance-constrained multi-objective modeling, estimation, filtering, and control problems could be considered for more complex systems [151–155].

It would be interesting to study the problems of variance-constrained multi-objective analysis and design for large-scale nonlinear interconnected systems that are frequently seen in modern industries.

A practical engineering application of the existing theories and methodologies would be the target tracking problem.

# References

[1] Z. Wang, Y. Liu, and X. Liu, "$H_\infty$ filtering for uncertain stochastic time-delay systems with sector-bounded nonlinearities", *Automatica*, Vol. **44**, No. 5, pp. 1268–1277, 2008.

[2] Y. Liu, Z. Wang, and X. Liu, "Robust $H_\infty$ control for a class of nonlinear stochastic systems with mixed time-delay", *International Journal of Robust and Nonlinear Control*, Vol. **17**, No. 16, pp. 1525–1551, 2007.

[3] D. Yue and Q. Han, "Delay-dependent exponential stability of stochastic systems with time-varying delay, nonlinearity and Markovian switching", *IEEE Transactions on Automatic Control*, Vol. **50**, No. 2, pp. 217–222, 2005.

[4] H. K. Khalil, *Nonlinear systems*, Third Edition, Publishing House of Electronics Industry, Beijing, 2007.

[5] J. Guckenheimer and P. Holmes, *Nonlinear oscillations, dynamical systems and bifurcations of vector fields*, Springer-Verlag, New York, 1983.

[6] J. Hale and H. Kocak, *Dynamics and bifurcations*, Springer-Verlag, New York, 1991.

[7] S. Strogatz, *Nonlinear dynamics and chaos*, Addison Wesley, Reading, MA, 1994.

[8] J. Wait, L. Huelsman, and G. Korn, *Introduction to operational amplifiers*, McGraw-Hill, New York, 1975.

[9] L. Guo and H. Wang, "Fault detection and diagnosis for general stochastic systems using B-spline expansions and nonlinear filters", *IEEE Transactions on Circuits and Systems–I: Regular Papers*, Vol. **52**, No. 8, pp. 1644–1652, 2005.

[10] Z. Wang, Y. Liu, M. Li, and X. Liu, "Stability analysis for stochastic Conhen–Grossberg neural networks with mixed time delays", *IEEE Transactions on Neural Networks*, Vol. **17**, No. 3, pp. 814–820, 2006.

[11] Y. Wang, L. Xie, and C. De Souza, "Robust control of a class of uncertain nonlinear systems", *Systems and Control Letters*, Vol. **19**, pp. 139–149, 1992.

[12] R. Pearson, "Gray-box modeling of nonideal sensors", Proceedings of American Control Conference, 2001, pp. 4404–4409.

[13] Y. Cao, Z. Lin, and B. Chen, "An output feedback $H_\infty$ controller design for linear systems subject to sensor nonlinearities", *IEEE Transactions on Circuits and Systems–I*, Vol. **50**, pp. 914–921, 2003.

[14] Q. Han, "Absolute stability of time delay systems with sector-bounded nonlinearity", *Automatica*, Vol. **41**, pp. 2171–2176, 2005.

[15] G. Kreisselmeier, "Stabilization of linear systems in the presence of output measurement saturation", *Systems and Control Letters*, Vol. **29**, pp. 27–30, 1996.

[16] Z. Lin and T. Hu, "Semi-global stabilization of linear systems subject to output saturation", *Systems and Control Letters*, Vol. **43**, pp. 211–217, 2001.

[17] M. Fu and L. Xie, "The sector bound approach to quantized feedback control", *IEEE Transactions on Automatic Control*, Vol. **50**, No. 11, pp. 1698–1711, 2005.

[18] Z. Wang, D. W. C. Ho, H. Dong, and H. Gao, "Robust $H_\infty$ finite-horizon control for a class of stochastic nonlinear time-varying systems subject to sensor and actuator saturations", *IEEE Transactions on Automatic Control*, Vol. **55**, No. 7, pp. 1716–1722, 2010.

[19] B. Shen, Z. Wang, H. Shu, and G. Wei, "Robust $H_\infty$ finite-horizon filtering with randomly occurred nonlinearities and quantization effects", *Automatica*, Vol. **46**, No. 11, pp. 1743–1751, 2010.

[20] B. Shen, Z. Wang, H. Shu, and G. Wei, "$H_\infty$ filtering for uncertain time-varying systems with multiple randomly occurred nonlinearities and successive packet dropouts", *International Journal of Robust and Nonlinear Control*, Vol. **21**, No. 14, pp. 1693–1709, 2011.

[21] H. Dong, Z. Wang, D. W. C. Ho, and H. Gao, "Robust $H_\infty$ filtering for Markovian jump systems with randomly occurring nonlinearities and sensor saturation: the finite-horizon case", *IEEE Transactions on Signal Processing*, Vol. **59**, No. 7, pp. 3048–3057, 2011.

[22] H. Dong, Z. Wang, and H. Gao, "Fault detection for Markovian jump systems with sensor saturations and randomly varying nonlinearities", *IEEE Transactions on Circuits and Systems - Part I*, Vol. **59**, No. 10, pp. 2354–2362, 2012.

[23] J. Hu, Z. Wang, H. Gao, and L. K. Stergioulas, "Robust sliding mode control for discrete stochastic systems with mixed time delays, randomly occurring uncertainties, and randomly occurring nonlinearities", *IEEE Transactions on Industrial Electronics*, Vol. **59**, No. 7, pp. 3008–3015, 2012.

[24] D. Ding, Z. Wang, J. Hu, and H. Shu, "$H_\infty$ state estimation for discrete-time complex networks with randomly occurring sensor saturations and randomly varying sensor delays", *IEEE Transactions on Neural Networks and Learning Systems*, Vol. **23**, No. 5, pp. 725–736, 2012.

[25] J. Liang, Z. Wang, B. Shen, and X. Liu, "Distributed state estimation in sensor networks with randomly occurring nonlinearities subject to time-delays", *ACM Transactions on Sensor Networks*, Vol. **9**, No. 1, doi: 10.1145/2379799.2379803, 2012.

[26] G. Wei, Z. Wang, and B. Shen, "Probability-dependent gain-scheduled control for discrete stochastic delayed systems with randomly occurring nonlinearities", *International Journal of Robust and Nonlinear Control*, Vol. **23**, No. 7, pp. 815–826, 2013.

[27] L. Ma, Z. Wang, Y. Bo, and Z. Guo, "A game theory approach to mixed $H_2/H_\infty$ control for a class of stochastic time-varying systems with randomly occurring nonlinearities", *Systems and Control Letters*, Vol. **60**, No. 12, pp. 1009–1015, 2011.

[28] Z. Wang, B. Shen, and X. Liu, "$H_\infty$ filtering with randomly occurring sensor saturations and missing measurements", *Automatica*, Vol. **48**, No. 3, pp. 556–562, 2012.

[29] D. Ding, Z. Wang, J. Hu, and H. Shu, "Dissipative control for state-saturated discrete time-varying systems with randomly occurring nonlinearities and missing measurements", *International Journal of Control*, Vol. **86**, No. 4, pp. 674–688, 2013.

[30] F. Yang, Z. Wang, D. W. C. Ho, and X. Liu, "Robust $H_2$ filtering for a class of systems with stochastic nonlinearities", *IEEE Transactions on Circuits Systems–II: Express Briefs*, Vol. **53**, No. 3, pp. 235–239, 2006.

[31] Y. Yaz and E. Yaz, "State estimation of uncertain nonlinear stochastic systems with general criteria", *Applied Mathematics Letters*, Vol. **14**, pp. 605–610, 2001.

[32] H. Dong, Z. Wang, D. W. C. Ho, and H. Gao, "Variance-constrained $H_\infty$ filtering for nonlinear time-varying stochastic systems with multiple missing measurements: the finite-horizon case", *IEEE Transactions on Signal Processing*, Vol. **58**, No. 5, pp. 2534–2543, 2010.

[33] L. Ma, Y. Bo, Y. Zhou, and Z. Guo, "Error variance-constrained $H_\infty$ filtering for a class of nonlinear stochastic systems with degraded measurements: the finite horizon case", *International Journal of Systems Science*, Vol. **43**, No. 12, pp. 2361–2372, 2012.

[34] L. Ma, Z. Wang, Y. Bo, and Z. Guo, "Robust $H_2$ sliding mode control for non-linear discrete-time stochastic systems", *IET Control Theory and Applications*, Vol. **3**, No. 11, pp. 1537–1546, 2009.

[35] K. Åström, *Introduction to stochastic control theory*, Acadamic Press, New York and London, 1970.

[36] B. Chen and W. Zhang, "Stochastic $H_2/H_\infty$ control with state-dependent noise", *IEEE Transactions on Automatic Control*, Vol. **49**, No. 1, pp. 45–57, 2004.

[37] X. Chen and K. Zhou, "Multi-objective $H_2$ and $H_\infty$ control design", *SIMA Journal of Control Optimization*, Vol. **40**, No. 2, pp. 628–660, 2001.

[38] E. Collins and R. Skelton, "A theory of state covariance assignment for discrete systems", *IEEE Transactions on Automatic Control*, Vol. **32**, No. 1, pp. 35–41, 1987.

[39] B. Huang, "Minimum variance control and performance assessment of time-variant processes", *Journal of Process Control*, Vol. **12**, No. 6, pp. 707–719, 2002.

[40] C. Scherer, "Multi-objective $H_2/H_\infty$ control", *IEEE Transactions on Automatic Control*, Vol. **40**, No. 6, pp. 1054–1062, 1995.

[41] N. Wiener, *The extrapolation, interpolation and smoothing of stationary time series with engineering applications*, John Wiley & Sons, Inc., New York, 1949.

[42] R. Kalman and R. Bucy, "A new approach to linear filtering and prediction problems", *ASME Journal of Basic Engineering*, Vol. **82**, pp. 34–45, 1960.

[43] R. Kalman and R. Bucy, "New results in linear filtering and prediction theory", *ASME Journal of Basic Engineering*, Vol. **83**, pp. 95–107, 1961.

[44] D. Bernstein and W. Haddad, "Steady-state Kalman filtering with an $H_\infty$ error bound", *Systems and Control Letters*, Vol. **12**, No. 1, pp. 9–16, 1989.

[45] Y. Hung and F. Yang, "Robust $H_\infty$ filtering with error variance constraints for uncertain discrete time-varying systems with uncertainty", *Automatica*, Vol. **39**, No. 7, pp. 1185–1194, 2003.

[46] X. Lu, L. Xie, H. Zhang, and W. Wang, "Robust Kalman filtering for discrete-time systems with measurement delay", *IEEE Transactions on Circuits and Systems-II: Express Briefs*, Vol. **54**, No. 6, pp. 522–526, 2007.

[47] H. Rotstein, M. Sznaier, and M. Idan, "Mixed $H_2/H_\infty$ filtering theory and an aerospace application", *International Journal of Robust and Nonlinear Control*, Vol. **6**, pp. 347–366, 1996.

[48] K. Boukas, "Stabilization of stochastic nonlinear hybrid systems", *International Journal of Innovative Computing, Information and Control*, Vol. **1**, No. 1, pp. 131–141, 2005.

[49] B. Huang, S. Shah, and E. Kwok, "Good, bad of optimal? Performance assessment of multivariable processes", *Automatica*, Vol. **33**, No. 6, pp. 1175–1183, 1997.

[50] A. Hotz and R. Skelton, "A covariance control theory", *International Journal of Control*, Vol. **46**, No. 1, pp. 13–32, 1987.

[51] R. Skelton, T. Iwasaki, and K. Grigoriadis, "A unified approach to linear control design", Taylor & Francis, London, 1997.

[52] H. Ruan, E. Yaz, and T. Zhai, "Current output filter for state estimation of nonlinear stochastic systems with linear measurement", *International Journal of Innovative Computing, Information and Control*, Vol. **1**, No. 2, pp. 277–287, 2005.

[53] U. Shaked and C. E. de Souza, "Robust minimum variance filter design", *IEEE Transactions on Signal Processing*, Vol. **43**, No. 11, pp. 2474–2483, 1995.

[54] Z. Wang, D. W. C. Ho, and X. Liu, "Variance-constrained filtering for uncertain stochastic systems with missing measurements", *IEEE Transactions on Automatic Control*, Vol. **48**, No. 7, pp. 1254–1258, 2003.

[55] C. Hsieh, "Extension of unbiased minimum-variance input and state estimation for systems with unknown inputs", *Automatica*, Vol. **45**, pp. 2149–2153, 2009.

[56] U. Shaked, L. Xie, and Y. Soh, "New approach to robust minimum variance filter design", *IEEE Transactions on Signal Processing*, Vol. **49**, No. 11, pp. 2620–2629, 2001.

[57] Z. Wang, D. W. C. Ho, and X. Liu, "Robust filtering under randomly varying sensor delay with variance constraints", *IEEE Transactions on Circuits and Systems - II: Express Briefs*, Vol. **51**, No. 6, pp. 320–326, 2004.

[58] F. Yang, Z. Wang, Y. Hung, and H. Shu, "Mixed $H_2/H_\infty$ filtering for uncertain systems with regional pole assignment", *IEEE Transactions on Aerospace and Electronic Systems*, Vol. **41**, No. 2, pp. 438–448, 2005.

[59] Z. Wang, F. Yang, D. W. C. Ho, and X. Liu, "Robust variance constrained $H_\infty$ control for stochastic systems with multiplicative noises", *Journal of Mathematical Analysis and Applications*, Vol. **328**, pp. 487–502, 2007.

[60] K. Chang and W. Wang, "Robust covariance control control for perturbed stochastic multivariable system via variable structure control", *Systems and Control Letters*, Vol. **37**, pp. 323–328, 1999.

[61] K. Chang and W. Wang, "$H_\infty$ norm constraints and variance control for stochastic uncertain large-scale systems via the sliding mode concept", *IEEE Transactions Circuits Systms I: Fundamental Theory and Applications*, Vol. **46**, pp. 1275–1280, 1999.

[62] W. Wang and K. Chang, "Variable structure based covariance assignment for stochastic multivariable model reference systems", *Automatica*, Vol. **36**, pp. 141–146, 2000.

[63] A. Subramanian and A. Sayed, "Multiobjective filter design for uncertain stochastic time-delay systems", *IEEE Transactions on Automatic Control*, Vol. **49**, No. 1, pp. 149–154, 2004.

[64] L. Ma, Z. Wang, and Z. Guo, "Robust fault-tolerant control for a class of nonlinear stochastic systems with variance constraints", *Journal of Dynamic Systems, Measurement, and Control*, Vol. **132**, 044501-1–, 2010.

[65] L. Ma, Z. Wang, and Z. Guo, "Robust variance-constrained filtering for a class of nonlinear stochastic systems with missing measurements", *Signal Processing*, Vol. **90**, No. 6, pp. 2060–2071, 2010.

[66] L. Ma, Z. Wang, Y. Bo, and Z. Guo, "Finite-horizon $H_2/H_\infty$ control for a class of nonlinear Markovian jump systems with probabilistic sensor failures", *International Journal of Control*, Vol. **84**, No. 11, pp. 1847–1857, 2011.

[67] L. Guo, F. Yang, and J. Fang, "Multi-objective filtering for nonlinear time-delay systems with nonzero initial conditions based on convex optimizations", *Circuits, Systems and Signal Processing*, Vol. **25**, No. 5, pp. 591–607, 2006.

[68] P. Shi, M. Mahmoud, J. Yi, and A. Ismail, "Worst case control of uncertain jumping systems with multi-state and input delay information", *Information Sciences*, Vol. **176**, No. 2, pp. 186–200, 2005.

[69] E. Tian and C. Peng, "Delay-dependent stability and synthesis of uncertain T-S fuzzy systems with time-delay", *Fuzzy Sets & Systems*, Vol. **157**, pp. 544–559, 2005.

[70] D. Yue and J. Lam, "Non-fragile guaranteed cost control for uncertain descriptor systems with time-varying state and input delays", *Optimal Contorl Applications and Methods*, Vol. **26**, No. 2, pp. 85–105, 2005.

[71] G. Zames, "Feedback and optimal sensitivity: model reference transformations, multiplicative seminorms, and approximate inverses", *IEEE Transactions on Automatic Control*, Vol. **26**, pp. 301–320, 1981.

[72] J. Doyle, K. Glover, P. Khargonekar, and B. Francis, "State-space solutions to the standard $H_2$ and $H_\infty$ control problems". *IEEE Transaction on Automatic Control*, Vol. **34**, No. 8, pp. 31–47, 1989.

[73] A. Isidori and W. Kang, "$H_\infty$ control via measurement feedback for general nonlinear systems", *IEEE Transactions on Automatic Control*, Vol. **40**, pp. 466–472, 1995.

[74] A. Van der Schaft, "$L2$-gain analysis of nonlinear systems and nonlinear control", *IEEE Transactions on Automatic Control*, Vol. **37**, No. 6, pp. 770–784, 1992.

[75] Z. Wang, D. Ding, H. Dong, and H. Shu, "$H_\infty$ consensus control for multi-agent systems with missing measurements: the finite-horizon case", *Systems and Control Letters*, Vol. **62**, No. 10, pp. 827–836, 2013.

[76] L. Ma, Z. Wang, and Z. Guo, "Robust $H_\infty$ control of time-varying systems with stochastic nonlinearities: the finite-horizon case", *Proceedings of the Institution of Mechanical Engineers, Part I, Journal of Systems and Control Engineering*, Vol. **224**, pp. 575–585, 2010.

[77] L. Ma, Z. Wang, Y. Bo, and Z. Guo, "Robust $H_\infty$ sliding mode control for nonlinear stochastic systems with multiple data packet losses", *International Journal of Robust and Nonlinear Control*, Vol. **22**, No. 5, pp. 473–491, 2012.

[78] D. Ding, Z. Wang, B. Shen, and H. Shu, "$H_\infty$ state estimation for discrete-time complex networks with randomly occurring sensor saturations and randomly varying sensor delays", *IEEE Transactions on Neural Networks and Learning Systems*, Vol. **23**, No. 5, pp. 725–736, 2012.

[79] D. Ding, Z. Wang, H. Dong, and H. Shu, "Distributed $H_\infty$ state estimation with stochastic parameters and nonlinearities through sensor networks: the finite-horizon case", *Automatica*, Vol. **48**, No. 8, pp. 1575–1585, 2012.

[80] J. Hu, Z. Wang, Y. Niu, and L. K. Stergioulas, "$H_\infty$ sliding mode observer design for a class of nonlinear discrete time-delay systems: a delay-fractioning approach", *International Journal of Robust and Nonlinear Control*, Vol. **22**, No. 16, pp. 1806–1826, 2012.

[81] Z. Wang, H. Dong, B. Shen, and H. Gao, "Finite-horizon $H_\infty$ filtering with missing measurements and quantization effects", *IEEE Transactions on Automatic Control*, Vol. **58**, No. 7, pp. 1707–1718, 2013.

[82] H. Dong, Z. Wang, and H. Gao, "Distributed $H_\infty$ filtering for a class of Markovian jump nonlinear time-delay systems over lossy sensor networks", *IEEE Transactions on Industrial Electronics*, Vol. **60**, No. 10, pp. 4665–4672, 2013.

[83] Z. Wang, D. W. C. Ho, Y. Liu, and X. Liu, "Robust $H_\infty$ control for a class of nonlinear discrete time-delay stochastic systems with missing measurements", *Automatica*, Vol. **45**, No. 3, pp. 684–691, 2009.

[84] Z. Shu, J. Lam, and J. Xiong, "Non-fragile exponential stability assignment of discrete-time linear systems with missing data in actuators", *IEEE Transactions on Automatic Control*, Vol. **54**, No. 3, pp. 625–630, 2009.

[85] G. Wei, Z. Wang, and H. Shu, "Robust filtering with stochastic nonlinearities and multiple missing measurements", *Automatica*, Vol. **45**, No. 3, pp. 836–841, 2009.

[86] F. Yang and Y. Li, "Set-membership filtering for systems with sensor saturation", *Automatica*, Vol. **4**, pp. 1896–1902, 2009.

[87] G. Yang and D. Ye, "Adaptive reliable $H_\infty$ filtering against sensor failures", *IEEE Transactions on Signal Processing*, Vol. **55**, No. 7, pp. 3161–3171, 2007.

[88] M. Fu and L. Xie, "The sector bound approach to quantized feedback control", *IEEE Transactions on Automatic Control*, Vol. **50**, No. 11, pp. 1698–1711, 2005.

[89] M. Fu and C. E. de Souza, "State estimation for linear discrete-time systems using quantized measurements", *Automatica*, Vol. **45**, No. 12, pp. 2937–2945, 2012.

[90] N. Xiao, L. Xie, and M. Fu, "Stabilization of Markov jump linear systems using quantized state feedback", *Automatica*, Vol. **46**, No. 10, pp. 1696–1702, 2010.

[91] X. Dong, "Robust strictly dissipative control for discrete singular systems", *IET Control Theory Applications*, Vol. **1**, No. 4, pp. 1060–1067, 2007.

[92] D. Hill and P. Moylan, "The stability of nonlinear dissipative systems", *IEEE Transaction on Automatic Control*, Vol. **21**, No. 5, pp. 708–711, 1976.

[93] Z. Li, J. Wang, and H. Shao, "Delay-dependent dissipative control for linear time-delay systems", *Journal of Franklin Institution*, Vol. **339**, pp. 529–542, 2002.

[94] Z. Tan, Y. Soh, and L. Xie, "Dissipative control for linear discrete-time systems", *Automatica*, Vol. **35**, pp. 1557–1564, 1999.

[95] J. Willems, "Dissipative dynamical systems, part 1: general theory; part 2: linear systems with quadratic supply rate", *Archive for Rational Mechanics and Analysis*, Vol. **45**, No. 5, pp. 321–393, 1972.

[96] S. Xie, L. Xie, and C. E. de Souza, "Robust dissipative control for linear systems with dissipative uncertainty", *International Journal of Control*, Vol. **70**, pp. 169–191, 1998.

[97] Z. Wang, J. Lam, L. Ma, Y. Bo, and Z. Guo, "Variance-constrained dissipative observer-based control for a class of nonlinear stochastic systems with degraded measurements", *Journal of Mathematical Analysis and Applications*, Vol. **377**, pp. 645–658, 2011.

[98] H. Dong, Z. Wang, and H. Gao, "$H_\infty$ fuzzy control for systems with repeated scalar nonlinearities and random packet losses", *IEEE Transactions on Fuzzy Systems*, Vol. **17**, No. 2, pp. 440–450, 2009.

[99] J. Liang, Z. Wang, and X. Liu, "On passivity and passification of stochastic fuzzy systems with delays: the discrete-time case", *IEEE Transactions on Systems, Man, and Cybernetics - Part B*, Vol. **40**, No. 3, pp. 964–969, 2010.

[100] H. Dong, Z. Wang, D. W. C. Ho, and H. Gao, "Robust $H_\infty$ fuzzy output-feedback control with multiple probabilistic delays and multiple missing measurements", *IEEE Transactions on Fuzzy Systems*, Vol. **18**, No. 4, pp. 712–725, 2010.

[101] H. Dong, Z. Wang, J. Lam, and H. Gao, "Fuzzy-model-based robust fault detection with stochastic mixed time-delays and successive packet dropouts", *IEEE Transactions on Systems, Man, and Cybernetics - Part B*, Vol. **42**, No. 2, pp. 365–376, 2012.

[102] B. K. Lee, C. H. Chiu, and B. S. Chen, "Adaptive minimum variance control for stochastic fuzzy T-S ARMAX model", *International Conference on Machine Learning and Cybernetics*, pp. 3811–3816, 2008.

[103] W. J. Chang and S. M. Wu, "State variance-constrained fuzzy controller design for nonlinear TORA systems with minimizing control input energy", *Proceedings of IEEE International Conference on Robotics and Automation*, 2003, pp. 2616–2621.

[104] W. J. Chang and S. M. Wu, "Continuous fuzzy controller design subject to minimizing input energy with output variance constraints", *European Journal of Control*, Vol. **11**, pp. 269–277, 2005.

[105] W. J. Chang and B. J. Huang, "Passive fuzzy controller design with variance constraint for nonlinear synchronous generator systems", *IEEE 10th International Conference on Power Electronics and Drive Systems*, 2013, pp. 1251–1256.

[106] V. Utkin, "Variable structure control systems with sliding mode", *IEEE Transactions on Automatic Control*, Vol. **22**, pp. 212–222, 1977.

[107] Y. Niu, D. W. C. Ho, and J. Lam, "Robust integral sliding mode control for uncertain stochastic systems with time-varying delay", *Automatica*, Vol. **41**, pp. 873–880, 2005.

[108] Y. Niu and D. W. C. Ho, "Robust ovserver design for Itô stochastic time-delay systems via sliding mode control", *Systems and Control Letters*, Vol. **55**, pp. 781–793, 2006.

[109] K. Abidi, J. Xu, and X. Yu, "On the discrete-time integral sliding-mode control", *IEEE Transactions on Automatic Control*, Vol. **52**, No. 4, pp. 709–715, 2007.

[110] C. Bonivento, M. Sandri, and R. Zanasi, "Discrete variable structure integral controllers", *Automatica*, Vol. **34**, pp. 355–361, 2007.

[111] C. Chan, "Discrete adaptive sliding-mode control of a class of stochastic systems", *Automatica*, Vol. **35**, pp. 1491–1498, 1999.

[112] X. Chen and T. Fukuda, "Robust adaptive quasi-sliding mode controller for discrete-time systems", *Systems and Control Letters*, Vol. **35**, pp. 165–173, 1998.

[113] C. Cheng, M. Lin, and J. Hsiao, "Sliding mode controllers design for linear discrete-time systems with matching pertubations", *Automatica*, Vol. **36**, pp. 1205–1211, 2000.

[114] C. Hwang, "Robust discrete variable structure control with finite-time approach to switching surface", *Automatica*, Vol. **38**, pp. 167–175, 2002.

[115] N. Lai, C. Edwards, and S. Spurgeon, "Discrete output feedback sliding mode control with integral action", *International Journal of Robust and Nonlinear Control*, Vol. **16**, pp. 21–43, 2006.

[116] N. Lai, C. Edwards, and S. Spurgeon, "On output tracking using dynamic output feedback discrete-time sliding-mode controllers", *IEEE Transactions on Automatic Control*, Vol. **52**, No. 10, pp. 1975–1981, 2007.

[117] W. Gao, Y. Wang, and A. Homaifa, "Discrete-time variable structure control systems", *IEEE Transactions on Industrial Electron*, Vol. **42**, No. 2, pp. 117–122, 1995.

[118] Y. Xia, G. Liu, P. Shi, J. Chen, D. Rees, and J. Liang, "Sliding mode control of uncertain linear discrete time systems with input delay", *IET Control Theory and Applications*, Vol. **1**, No. 4, pp. 1169–1175, 2007.

[119] M. Yan, and Y. Shi, "Robust discrete-time sliding mode control for uncertain systems with time-varying state delay", *IET Control Theory and Applications*, Vol. **2**, No. 8, pp. 662–674, 2008.

[120] D. Bernstein, and W. Haddad. "LQG control with an $H_\infty$ performance bound: A Riccati equation approach", *IEEE Transactions on Automatic Control*, Vol. **34**, No. 3, pp. 293–305, 1989.

[121] D. Limebeer, B. Anderson, and B. Hendel, "A Nash game approach to mixed $H_2/H_\infty$ control", *IEEE Transactions on Automatic Control*, Vol. **39**, No. 1, pp. 69–82, 1994.

[122] P. Khargonekar, and M. Rotea, "Mixed $H_2/H_\infty$ control: a convex optimization approach", *IEEE Transactions on Automatic Control*, Vol. **36**, No. 7, pp. 824–837, 1994.

[123] L. Zhang, B. Huang, and J. Lam, "LMI synthesis of $H_2$ and mixed $H_2/H_\infty$ controllers for singular systems", *IEEE Transactions on Circuits and Systems, Part II: Analog and Digital Signal Processing*, Vol. **50**, No. 9, pp. 615–626, 2003.

[124] C. Scherer, P. Gahient, and M. Chilali, "Multi-objective output-feedback control via LMI optimization", *IEEE Transactions on Automatic Control*, Vol. **42**, pp. 896–911, 1997.

[125] M. Khosrowjerdi, R. Nikoukhah, and N. Safari-Shad, "Fault detection in a mixed $H_2/H_\infty$ setting", *Proceedings of IEEE Conference on Decision and Control*, 2003, pp. 1461–1466.

[126] C. Yang and Y. Sun, "Mixed $H_2/H_\infty$ state-feedback design for microsatellite attitude control", *Control Engineering Practice*, Vol. **10**, pp. 951–970, 2002.

[127] X. Chen and K. Zhou, "Multi-objective filtering design", *Proceedings of IEEE Canadian Conference on Electrical and Computer Engineering*, 1999, pp. 708–713.

[128] Y. Theodor and U. Shaked, "A dynamic game approach to mixed $H_2/H_\infty$ estimation", *International Journal of Robust and Nonlinear Control*, Vol. **6**, No. 4, pp. 331–345, 1996.

[129] Z. Wang and B. Huang, "Robust mixed $H_2/H_\infty$ filtering for linear systems with error variance constraints", *IEEE Transactions on Signal Processing*, Vol. **48**, pp. 2463–2467, 2000.

[130] W. Lin, "Mixed $H_2/H_\infty$ control of nonlinear systems", *Proceedings of IEEE Conference on Decision and Control*, 1995, pp. 333–338.

[131] B. Chen, C. Tseng, and H. Uang, "$H_2/H_\infty$ fuzzy output feedback control design for nonlinear dynamic systems: an LMI approach", *IEEE Transactions on Fuzzy Systems*, Vol. **8**, No. 6, pp. 249–265, 2008.

[132] F. Yang, Z. Wang, and D. W. C. Ho, "Robust mixed $H_2/H_\infty$ control for a class of nonlinear stochastic systems", *IEE Proceedings on Control Theory Applications,* Vol. **153**, No. 2, pp. 175–184, 2006.

[133] Y. Yaz and E. Yaz, "On LMI formulations of some problems arising in nonlinear stochastic system analysis", *IEEE Transactions on Automatic Control*, Vol. **44**, No. 4, pp. 813–816, 1999.

[134] L. Xie, Y. Soh, and C. De Souza, "Robust Kalman filtering for uncertain discrete time systems", *IEEE Transactions on Automatic Control*, Vol. **39**, pp. 1310–1314, 1994.

[135] E. K. Boukas and A. Benzaouia, "Stability of discrete-time linear systems with Markovian jumping parameters and constrained control", *IEEE Transactions on Automatic Control*, Vol. **47**, pp. 516–521, 2002.

[136] E. K. Boukas and Z. K. Liu, "Robust $H_\infty$ control of discrete-time Markovian jump linear systems with mode-dependent time-delays", *IEEE Transactions on Automatic Control*, Vol. **46**, pp. 1918–1924, 2001.

[137] Y. Wang, L. Xie, and C. E. de Souza, "Robust control of a class of uncertain nonlinear systems", *Systems and Control Letters*, Vol. **19**, pp. 139–149, 1992.

[138] S. Boyd, L. Ghaoui, E. Feron, and V. Balakrishnan, *Linear matrix inequalities in system and control theory*, Philadelphia: SIAM Studies in Applied Mathematics, 1994.

[139] R. Field and L. Lawrence, "Reliability-based covariance control design", *Proceedings of American Control Conference*, 1997, pp. 11–15.

[140] G. Yang, J. Wang, and Y. Soh, "Reliable LQG control with sensor failures", *IEE Proceedings-Control Theory Applications*, Vol. **147**, No. 4, pp. 433–439, 2000.

[141] M. Oliveira and J. Geromel, "Numerical comparison output feedback design methods", *Proceedings of American Control Conference*, Albuquerque, New Mexico, 1997, pp. 72–76.

[142] H. Gao, J. Lam, and Z. Wang, "Discrete bilinear stochastic systems with time-varying delay: stability analysis and control synthesis", *Chaos, Solitons and Fractals*, Vol. **34**, No. 2, pp. 394–404, 2007.

[143] Z. Wang, F. Yang, D. W. C. Ho, and X. Liu, "Robust $H_\infty$ filtering for stochastic time-delay systems with missing measurements", *IEEE Transactions on Signal Processing*, Vol. **54**, No. 7, pp. 2579–2587, 2006.

[144] F. Yang, Z. Wang, D. W. C. Ho, and M. Gani, "Robust $H_\infty$ control with missing measurements and time delays", *IEEE Transactions on Automatic Control*, Vol. **52**, No. 9, pp. 1666–1672, 2007.

[145] T. Tarn and Y. Rasis, "Observers for nonlinear stochastic systems", *IEEE Transactions on Automatic Control*, Vol. **21**, No. 6, pp. 441–447, 1976.

[146] J. Rawson and C. Hsu, "Reduced order $H_\infty$ filters for discrete linear systems", *Proceedings of 36th Conference on Decision and Control*, San Diego, California, USA, 1997, pp. 3311–3316.

[147] E. Gershon, U. Shaked, and I. Yaesh, "$H_\infty$ control and filtering of discrete-time stochastic systems with multiplicative noise", *Automatica*, Vol. **37**, pp. 409–417, 2001.

[148] Z. Wang, F. Yang, and X. Liu, "Robust filtering for systems with stochastic non-linearities and deterministic uncertainties", *Proceedings of the Institute of Mechanical Engineers, Part I: Journal of Systems and Control Engineering*, Vol. **220**, pp. 171–182, 2006.

[149] W. Zhang, Y. Huang, and H. Zhang, "Stochastic $H_\infty/H_2$ control for discrete-time systems with state and disturbance dependent noise", *Automatica*, Vol. **43**, pp. 513–521, 2007.

[150] S. H. Hong, "Scheduling algorithm of data sampling times in the integrated communication and control systems", *IEEE Transactions on Control System Technology*, Vol. **3**, No. 2, pp. 225–231, 1995.

[151] S. Elmadssia, K. Saadaoui, and M. Benrejeb, "New delay-dependent stability conditions for linear systems with delay", *Systems Science and Control Engineering: An Open Access Journal*, Vol. **1**, No. 1, pp. 2–11, 2013.

[152] Y. Chen and K. A. Hoo, "Stability analysis for closed-loop management of a reservoir based on identification of reduced-order nonlinear model", *Systems Science and Control Engineering: An Open Access Journal*, Vol. **1**, No. 1, pp. 12–19, 2013.

[153] S. R. Desai and R. Prasad, "A new approach to order reduction using stability equation and big bang big crunch optimization", *Systems Science and Control Engineering: An Open Access Journal*, Vol. **1**, No. 1, pp. 20–27, 2013.

[154] A. Mehrsai, H. R. Karimi, and K.-D. Thoben, "Integration of supply networks for customization with modularity in cloud and make-to-upgrade strategy", *Systems Science and Control Engineering: An Open Access Journal*, Vol. **1**, No. 1, pp. 28–42, 2013.

[155] M. Darouach and M. Chadli, "Admissibility and control of switched discrete-time singular systems", *Systems Science and Control Engineering: An Open Access Journal*, Vol. **1**, No. 1, pp. 43–51, 2013.

# Index